A First Course in

DIGITAL SYSTEMS DESIGN

An Integrated Approach

John P. Uyemura

Georgia Institute of Technology

Brooks/Cole Publishing Company

I(T)P® *An International Thomson Publishing Company*

Pacific Grove • Albany • Belmont • Boston • Cincinnati • Johannesburg • London • Madrid
Melbourne • Mexico City • New York • Scottsdale • Singapore • Tokyo • Toronto

Publisher: *Bill Stenquist*
Sponsoring Editor: *Eric Frank*
Product Development Editor: *Suzanne Jeans*
Marketing Representative: *Nathan Wilbur*
Editorial Assistant: *Shelley Gesicki*
Production Editor: *Kelsey McGee*
Manuscript Editor: *Connie Day*

Cover Design: *Roy R. Neuhaus*
Cover Image: *Steven Hunt/Image Bank*
Photo Editor: *Terry Powell*
Composition: *Robin Gold/Forbes Mill Press*
Manufacturing Buyer: *Kris Waller*
Cover Printing: *Phoenix Color Corporation*
Printing and Binding: *R.R. Donnelley & Sons/Willard*

For more information, contact:

BROOKS/COLE PUBLISHING COMPANY
511 Forest Lodge Road
Pacific Grove, CA 93950
USA

International Thomson Editores
Seneca 53
Col. Polanco
11560 México, D. F., México

International Thomson Publishing Europe
Berkshire House 168-173
High Holborn
London WC1V 7AA
England

International Thomson Publishing GmbH
Königswinterer Strasse 418
53227 Bonn
Germany

Thomas Nelson Australia
102 Dodds Street
South Melbourne, 3205
Victoria, Australia

International Thomson Publishing Asia
60 Albert Street
#15-01 Albert Complex
Singapore 189969

Nelson Canada
1120 Birchmount Road
Scarborough, Ontario
Canada M1K 5G4

International Thomson Publishing Japan
Palaceside Bulding, 5F
1-1-1 Hitotsubashi
Chiyoda-ku, Tokyo 100-0003
Japan

Printed in the United States of America

10   9   8   7   6   5   4   3   2

Photo credits: p. 4—IDT; 107, 220, 467—Intel®/C & I Photography, Inc.

Library of Congress Cataloging-in-Publication Data

Uyemura, John P. (John Paul)
        A first course in digital systems design : an integrated approach
/ John P. Uyemura
        p.   cm.
  Includes index.
  ISBN: 0-534-93412-9
  1.   Integrated circuits—Very large scale integration—Design and
construction—Data processing. 2. Computer-aided design.
1. Title.
TK7874.75.U94   1999
621.39'5—dc21   98-50230

Contents

Chapter 3 **Combinational Logic Design 57**

Preface

This book is designed as a text for a first course in digital systems design as taken by students in electrical engineering, computer engineering, and computer sciences. It has been written to provide a new paradigm for the teaching of the subject, one that covers classical topics but also integrates modern technology into the discussion to provide the student with a real-world viewpoint of modern digital design.

Topical Coverage

A quick reading of the table of contents will verify that the "usual" topics have been included in the text. However, several non-standard chapters have been integrated into the discussion with equal emphasis. The general flow of topics follows.

Chapter 1, titled *Concepts in Digital Systems*, is a general introduction to the use of binary numbers. It provides sections on encoding and binary represention of base-10 numbers and quantities. It also introduces the concept of hierarchical design, which is propagated through the entire book. Fundamentals of *Boolean Algebra and Logic Gates* are covered in **Chapter 2**, and the principles are extended to *Combinational Logic Design* in **Chapter 3**. These three chapters provide the theoretical foundations of the subject.

Switching characteristics and delay times are critical in modern system design. These are introduced in **Chapter 4,** *Digital Hardware*. The coverage is quite broad, and the main purpose of the chapter is to illustrate the behavior of real-world electronic switching networks. This sets the stage for **Chapter 5**, *First Concepts in VHDL*, which presents the structure and main ideas of an important tool in modern design. Placing VHDL here allows the student to see how theory and physical implementations are related and also provides a mechanism for additional practice in the ideas of digital logic.

Concepts in modern VLSI design that are unique to this text are presented in Chapters 6 and 7. **Chapter 6,** *CMOS Logic Circuits,* introduces MOSFETs as basic switching elements and then teaches how CMOS logic gates are designed. Although this subject may seem a bit out of place, this is simply switch logic as realized in a modern technology. **Chapter 7,** *Silicon Chips and VLSI*, provides an entry into the field of VLSI engineering and silicon concepts that is tailored for the student of digital design. It is introduced as the bottom of the design hierarchy and emphasizes the main ideas in physical design and VLSI that are important to large systems engineering. Both chapters have been written to be accessible to all students in electrical engineering, computer engineering, and computer science. They provide a real-world perspective that is often missing from more standard treatments.

Chapter 8, *Logic Components,* is directed toward examining useful network functions that are created using primitive gates. The list of components includes, among others, decoders, multiplexors, and adders. These are introduced as motivation for moving upward in the system hierarchy. **Chapter 9,** *Memory Elements and Arrays,* introduces the concepts of storage via latches, flip-flops, RAM, and ROM. The material in Section 9.9 is an introduction to the operation of CD-ROMs as a real-world example that illustrates many advanced concepts. **Chapter 10** introduces *Sequential Logic Networks.*

Computer fundamentals are covered in the last two chapters of the book in a manner that teaches the basics while simultaneously reinforcing the ideas of system hierarchies. **Chapter 11,** *Computer Basics,* covers the operations that define a computer, then continues on to discuss the primary components and simple architectural concepts. This treatment is quite general, but is designed to provide a solid understanding of the principles. **Chapter 12***, Advanced Computer Concepts,* covers topics that are important in modern microcomputers, such as pipelining, cache memory, and superscalar designs. It ends with an introduction to the primary concepts involved in parallel processing and ideas for the evolution of computing.

Level of the Treatment

Even though the book introduces some relatively unique aspects of the field to the beginner, every effort was made to keep the discussion at a uniform level. It should be accessible to any student in electrical engineering, computer engineering, and computer science who has completed the standard freshman curriculum.

The book was specifically written for a first course in the subject, and the material has been class-tested to ensure that the treatment is coherent. The goal was to provide enough details so that the student would have an understanding of both the qualitative and quantitative aspects of every topic. It was not, however, designed to provide advanced discussions and detailed analyses; complex issues should be covered in higher-level courses. This book can be used to establish the background for taking advanced courses in the subjects of digital design, computer architecture, VHDL, electronic circuits, and VLSI. The exception to this might be VHDL, as we find it useful to incorporate the subject into the same course using a separate text and computer tools. Alternatively, the book is sufficient by itself as an introduction to digital design for the non-specialist.

Use of the Text

It is possible to use the text in several ways that vary with the intended emphasis. The outline provides three distinct topical groups: logic design, integration and VLSI, and basic computer architecture.

The fundamentals of logic design are covered in Chapters 1, 2, 3, 8, 9, and 10. These can be used as the basis for such a course by omitting some sections in the later chapters that deal with CMOS and VLSI. Including Chapter 5 on VHDL is highly recommended; in fact, if time permits, it would be worthwhile to expand the treatment in labs employing a specialized text. Chapter 4 deals with the real-world aspects of hardware delays, and the instructor should consider Sections 4.1 through 4.3 as a minimum introduction.

The details of hardware are contained in Chapters 4, 6, and 7. Each chapter has been written to present the most important concepts in the early sections. This allows the instructor to present some of the material without having to cover entire chapters. In Chapter 6, Sections 6.1 through 6.5 provide the basics, while the remaining sections take the reader to a more advanced level. The material in Chapter 7 through Section 7.4.1 gives the reader an overview of silicon integrated circuits. Section 7.6 (*Cells, Libraries, and Hierarchical Design*) is worth consideration as it emphasizes some important points about the design process. The other sections in the chapter are optional but enhance the overall theme. These chapters have been used independently as a VLSI primer for senior and graduate students and working engineers.

Chapters 11 and 12 are an introduction to computers and computer architectures. They discuss the fundamentals of what a computer is and how it is built and set the foundation for more advanced studies. However, they are sufficiently detailed to ensure that the students have seen the important concepts even if they do not take another course in the subject.

This book was written with the student in mind. In less than 500 pages it presents many fields that are traditionally taught only in advanced courses, where the student often is left with the task of linking diverse material and concepts. There is an emphasis on design hierarchies throughout the text. Using this view as a central theme, the book introduces design process at all levels, from the bottom-up and from the top-down.

It is important to restate that this book is intended for use in a **first course** on digital systems. Students who are specializing in any (or all) of the areas need to take more advanced courses to achieve the level expected of a graduating engineer or computer scientist. Every topic could have been taken to a much higher level but that would have defeated the main purpose. There are many fine textbooks that are more advanced and can be used in a follow-up course for specialists. The lower level material in an advanced book can be be treated as review or ignored completely. This allows the instructor to place emphasis on the more difficult and time-consuming presentations. Topics such as advanced sequential circuits, VHDL, CMOS VLSI systems, and computer architecture fall into this category. A solutions manual for the exercises is available to the instructor.

Philosophy

This book covers the material that I feel every student in electrical engineering, computer engineering, and computer science should know before graduating.

As a teacher, I often worry that our program delegates many important topics to elective courses. And for an engineer facing the challenges of the new century, understanding basic concepts in digital networks and computers ranks among the most critical. Not that every one needs to know how to design a high-performance computer, just how the system is put together and how it works. I tell my "non-digital" students (in fields such as circuit design, electromagnetics, and fiber optics) that I think they should be able to read and understand technical articles at the "popular" level—such as those found in *Byte* magazine. If they can accomplish a task at this level, then I am more at ease with the job we are doing as educators.

The decade of the 1990s has proved to be one of remarkable technological advances. The "computer revolution" has affected every aspect of society. Problems that were once thought intractable can now be solved and our imaginations run rampant with new possibilities. Few would disagree with these observations. Yet our approach to teaching the fundamentals of digital logic remain almost the same as that which was used 25 years ago when I first learned the material. By "approach" I mean to say that digital logic has generally been taught as a stand-alone fundamental course with only a small amount of hardware and system design. Students who want to go deeper into related subjects must take additional courses. While this works fine in principle, it ignores the fact that the subject is no longer practiced as a stand-alone art. Modern digital design relies on engineering groups made up of individuals that have an understanding of all aspects of the problem, from the top to the bottom in the hierarchical chain, with expertise in one or two areas.

A recent example illustrates this point. Several interviewers from a major integrated circuit company contacted me to discuss their "ideal" engineering candidate. They said that the great majority of the students they interviewed wanted to be involved in microprocessor design, but only about 25% of them had any background in VLSI. They were not necessarily looking for circuit designers, but felt that it should be obvious to every potential employee that microprocessors are made in silicon, and that even the system architects and tool designers needed to know some of the basics of VLSI. Although this is admittedly a specialized case, the applications of digital VLSI devices such as ASICs and FPGAs have expanded at a furious pace. If we as educators do not provide the background for future growth, then our graduates risk becoming obsolete very quickly.

With this book, I put forth the view that modern digital logic consists of several interacting areas that combine in a cohesive fashion. This includes traditional topics such as Boolean algebra, logic formalisms, Karnaugh maps, and most of the subjects one would expect to find. But it goes well beyond these traditional subject areas by including VHDL, CMOS, VLSI and RISC computer architectures to show what the field looks like to a modern logic designer.

There is a deeper motivation for writing a book of this type. The integration of diverse topics into a single course addresses some of the new curriculum issues that are beginning to tear away at existing programs. At Georgia Tech, we have been immersed in redesigning the electrical engineering/computer engineering curriculum to address the needs of the next generation. Overall, we have concluded that we must change our current educational paradigms to accomplish this goal. However, we are still working on the details. This book is one idea for the change. It provides the fundamentals of *modern* digital systems design while still working within *classical* boundaries. The difference is that many so-called "advanced" topics have migrated to the introductory level. But this is normal evolution. One does not need a good memory to recall when computer architecture was exclusively a graduate level course.

Digital systems design has expanded well beyond what can be covered within the classical paradigms. But it is not alone. Virtually every field that traditionally forms the foundation of electrical and computer engineering has changed markedly during the past decade. It seems obvious that our approach to education in these areas needs to evolve in a similar manner. I thus offer this book as my contribution to the effort.

Acknowledgments

I would like to extend my appreciation to Suzanne Jeans, product development editor at Brooks/Cole, for her help during all phases of the project. Kelsey McGee, the production editor for the book, has been a pleasure to work with. Her diligence and keen eye, coupled with a wonderful personality and sense of humor, made a long and tedious task enjoyable! I cannot imagine having anyone finer for the job! Roy Neuhaus did a spectacular job creating the cover design (wouldn't you agree?) and deserves a hearty "well done!"

Many thanks to the reviewers who read through the various drafts of the book and were able to see the big picture as discussed in the preface: Donald W. Bouldin, University of Tennessee/Knoxville; Tri Caohuu, San Jose State University; Krishnendu Chakrabarty, Duke University; M. U. Farooq, West Virginia University Institute of Technology; Kenneth J. Hintz, George Mason University; Ben M. Huey, Arizona State University; Martin Kaliski, California Polytechnic State University; Ramakrishna Nunna, Stevens Institute of Technology; Samuel H. Russ, Mississippi State University; Kewal J. Saluja, University of Wisconsin/Madison; Richard R. Schultz, University of North Dakota; William G. Sutton, George Mason University; Val G. Tareski, North Dakota State University/Fargo; and Peter J. Warter, University of Delaware. It would have been easy to dismiss the approach as being too radical or far-afield, but they instead provided pages and pages of helpful comments, corrections, and suggestions that have greatly improved the contents and flow. Finding such a large group of highly qualified people with open minds was especially refreshing.

My wife Melba and my daughters, Christine and Valerie, are at the center of my life, and have always supported my writing projects. Were it not for them, a book such as this never would have found its way to the printer. So, I thank them once again for their endless love.

John P. Uyemura
Atlanta, Georgia

This book is dedicated to my wife

Melba Valerie

for her never-ending love and support
during our many years together.

植　村

CHAPTER 1

Concepts in Digital Systems

Digital systems have become a part of everyday life. Endless examples of digital hardware can be found all around us: automatic teller machines, compact-disc players, telephone systems, personal computers, video games—the list seems to go on forever!

In this chapter, we will begin our excursion into the world of digital networks by examining a few concepts that form the basis of the entire field. We will look at the binary number system and learn how a digital system is constructed.

1.1 What Is a Digital System?

Explaining exactly what constitutes a digital system can be either a very easy task or a very difficult chore, depending on whom we are talking to. In the basic sense, we can write that

- A **digital system** is an electronic network that "processes" information using only **digits** (numbers) to implement calculations and operations.

A **binary** number system uses only the digits 0 and 1. These binary digits are called **bits**.

However, the digits used are those from a special numbering scheme where only two values are possible: either the digit has a value of **0** or it has a value of **1**. This characteristic defines the **binary** or **base-2 number system**, and the numbers themselves (0 and 1) are called **bits**, short for "binary digits." Although this seems very simple, digital system design is complicated by the fact that *only* 0s and 1s can be used to perform all operations and calculations. In particular, a digital machine must perform the following tasks:

1. Translate information from our world into a binary "language" that can be understood by the digital network.

2. Perform the required calculations and operations using only the binary digits 0 and 1.

3. Return an answer to our world in a form that can we understand.

How we accomplish these tasks is the subject matter of this book. Although the process may sound quite complicated at this point, we will see that digital systems design can always be broken down into small **units** (or **modules**) that are straightforward to characterize and design. Once the small units are created, they can be used to build larger systems. This idea is shown by the conceptual illustration in Figure 1.1. Primitive units are used as the basic building blocks. They are combined to create larger, more complex units A, B, and C, which are in turn used to create even larger units X and Y. The size of any particular unit is arbitrary; its interior can be simple or complex and may consist of any simpler units needed. This illustrates what we will call a **hierarchical design approach,** wherein simple digital units are used to create more complex ones. The complexity is increased as necessary to obtain a final system that has the desired characteristics.

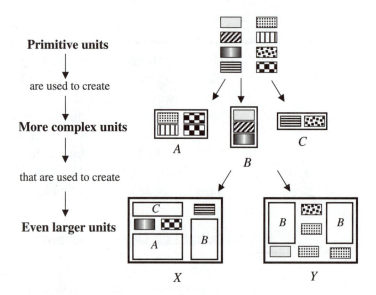

Figure 1.1 Hierarchical design levels

1.2 Views of a Digital System

There are several equivalent descriptions of a digital system. In this book we will continuously stress the idea that a digital network can be viewed in different ways. The level that is appropriate or convenient depends on the problem at hand.

1.2.1 Hierarchies

As we have noted, the concept of hierarchies is quite useful for designing a complex digital network. It allows us to start simply and build up toward levels of great complexity. Digital networks can be described by a fairly well-defined set of **hierarchical levels**, each being an individual area of detailed study. In our approach, however, we will view each level merely as an alternative viewpoint on the problem at hand. This will enable us to understand both the operation and the design of a digital system. The primary hierarchical levels used in this book are discussed in the following paragraphs.

Logic Networks

Digital logic is based on the behavior of binary numbers. It is possible to describe any digital network in terms of the fact that groups of binary variables can be used to represent virtually any set of data. For example, we may use binary numbers to represent an audio music track (such as that on a compact disc), scientific numbers in physics calculations, genetic codes, or information on stars or to catalog the findings from an archeological excavation in ancient Egyptian ruins. There is no limit to the number of different situations we can describe. Although this may sound a bit abstract at first, we will quickly learn how to use binary numbers to achieve these goals by means of a type of algebra that describes logic networks.

Electrical Circuits

The term **hardware** refers to the physical construction of a digital system. The physical realization of a digital network is accomplished by using electronic components that control the flow of electric current in a manner that implements logic operations. This is the most familiar form of a digital network in the everyday world. For example, when a computer system is mentioned, we usually visualize a piece of electronic equipment.

A realistic digital system is designed using digital logic and then implemented in hardware with electronic components. This merging of viewpoints is very important to understand, because it shows how theory is used to build an actual system.

Formal Description

It is possible to describe the behavior of a digital system by using only descriptive phrases that are defined within a context of a "language." This concept has been formalized in the development of **hardware description languages** (HDLs). An HDL allows us to define logic operations using statements that conform to a set of predefined rules and syntax. This concept of a language is the same as that used for a high-level programming language, such as Java or C++, except that it is applied to the problem of digital system design and analysis.

1.2.2 The Personal Computer

The desktop personal computer provides an excellent example of a digital system. When purchased as a complete system, it usually consists of the following units or boxes:

- **The System Unit.** This contains the main electronic "motherboard" with the CPU (central processing unit) chip that is the heart of the computer. In addition, the motherboard has the system memory for storing programs and data, and it has input and output circuitry for communicating with other components. The system unit itself contains a disk storage device, the main power supply, and other auxiliary components such as a modem.

- **The Keyboard.** The keyboard is the main device for providing input to the computer, such as command input and data. A mouse also serves this purpose as long as the appropriate software is installed.

- **The Monitor.** A monitor is just a CRT (cathode ray tube) that provides a visual output from the computer.

Your system may appear similar to that in Figure 1.2. Although other peripheral devices are available, these three provide the main functions of the computer.

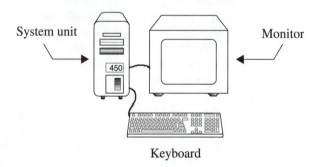

Figure 1.2 Components of a personal computer

Overall, the system may be viewed at several different hierarchical levels. We may treat it as a single unit that runs our programs, or we may want to dig deeper into the workings of the hardware. If you open up the system unit, you will discover many of the building blocks that are used to create the computer. Some of the more important parts are the **central processing unit** (CPU) which is the "heart" of the computer and provides all of the main operations, system memory units, the power supply unit, storage devices such as disk drives, and many other components. We may choose to go even deeper by studying how each component is made. This trek may take us inside of the CPU, where we will find a tiny piece of silicon that acts as an extremely complex set of electronic switches.

The most important comment that we can make at this point is that the computer is built using the same basic ideas as those used for a digital communication system, the guidance system of a space ship, control of the production line in a factory, or maintenance of records in a bank or business. Once we learn about digital logic, an entire new world of complex systems is within our grasp. Let us therefore begin our study by examining the simple numbering system that is at the root of all digital systems.

1.3 Introduction to Binary Numbers

The binary number system provides the basis for digital design. Let us examine a few basic ideas concerning the properties and usage of binary digits.

A **binary variable** represents a binary number that does not have a pre-defined value.

First, we introduce the concept of a **binary variable** A. Since there are only two binary digits, 0 and 1, it is easy to see that, by definition, A can have only values of

$$A = 0 \quad or \quad A = 1 \tag{1.1}$$

No other values of A are permitted. This property enables us to define a useful operation. Consider the value of A. If A is not equal to 0, then A must be equal to 1; conversely, if A is not equal to 1, then A must be equal to 0. This simple argument is used to define the **NOT** operation that is denoted by the overbar symbol

$$NOT(A) = \overline{A} \tag{1.2}$$

such that

$$NOT(0) = 1$$
$$NOT(1) = 0 \tag{1.3}$$

The NOT operation is also called **inversion**, and $\overline{A}$ is called the **complement** of A.

A single bit allows us to describe only two possible values (for example, YES or NO), so a single binary variable by itself is of limited use. However, we can use **groups of bits** to describe more complex situations. To see how this might work, suppose we take four individual bits that we will label a_3, a_2, a_1, a_0. Each bit a_i for $i = 0, 1, 2, 3$ can have a value of 0 or 1. We then construct the 4-bit segment *data*

$$data = a_3 a_2 a_1 a_0, \tag{1.4}$$

where the order of the bits is important, i.e., a_3 must be on the left and a_0 must be on the right, with sequential subscripts as shown. When written in this form, the quantity *data* is treated as a single object that can take on any of 16 distinct values:

$$data = 0000, 0001, 0010, 0011$$
$$0100, 0101, 0110, 0111$$
$$1000, 1001, 1010, 1011$$
$$1100, 1101, 1110, 1111 \tag{1.5}$$

The notation used is easy to understand if we just remember that the order of the bits is important. For example, *data* = 0101 means that

$$a_3 = 0 \qquad a_2 = 1 \qquad a_1 = 0 \qquad a_0 = 1 \tag{1.6}$$

These combinations can be verified by examination. The number of combinations is obtained by noting that there are 4 bits and that each bit has 2 possible values, giving a total of $2^4 = 16$ different permutations. Thus, even though a binary digit can be only 0 or 1, the 4-bit segment *data* can be used to describe a situation where 16 different possibilities are involved.

This line of reasoning can be use to construct groups with arbitrary numbers of bits. For example, the 8-bit segment

$$Info = X_7 X_6 X_5 X_4 X_3 X_2 X_1 X_0 \tag{1.7}$$

is defined to be a single object that has $2^8 = 256$ possible combinations. Similarly, a 16-bit group has $2^{16} = 65,536$ distinct values, and so on. Groups of binary digits can be used to represent any situation, so long as we use a sufficient number of bits. This is the key to using binary numbers for solving real-world problems.

It is useful to introduce some terminology to deal with groups of bits. A group of bits is often called a **word**, regardless of the number of bits involved. In the above examples, *data* is a 4-bit word, and *Info* is an 8-bit word. The number of bits (b) in a word may or may not have significance, depending on the system. In the world of microcomputers, a **byte** (**B**) means an 8-bit word.

When discussing binary systems, we often introduce abbreviations for certain powers of 2 as summarized below.

Word Size	Number of Values	Abbreviation for Value
8b	$2^8 = 256$	
10b	$2^{10} = 1024$	1Kb (kilobit)
16b	$2^{16} = 65,536$	64Kb
20b	$2^{20} = 1,048,576$	1Mb (megabit)
28b	$2^{28} = 268,435,456$	256Mb
30b	$2^{30} = 1,073,741,820$	1Gb (gigabit)

Because these abbreviations are based on the binary numbering system, the prefixes correspond to different values than when they are applied to the usual base-10 numbers. For example, in the binary system, 1Kb means 1024 bits, whereas in conventional usage (say, in physics and electronics), $1K = 1000 = 10^3$. Similarly,

$$1Mb = 1024Kb$$

holds in binary, whereas in conventional usage, 1M is equal to $10^6 = 1,000,000$ and is shorthand for one million. Although this may be confusing now, you will soon get used to it. Another word of caution that should be mentioned at the onset is the distinction between a bit (b) and a byte (B). Since

$$1B = 8b$$

it is important to keep the two separate. For example, the size of the main system memory in a personal computer is usually specified by the number of bytes that it has.

1.4 Data Representations

We can use binary words to **represent** anything that we want by defining them in an appropriate manner. For example, suppose we want to describe the four directions Left, Right, Forward, and Backward using a binary word. Since $2^2 = 4$, we will require a two-bit word to describe the four directions. Let us create the "direction" word $D = D_1 D_0$, where D_1 and D_0 are the individual bits. Once this is established, we can **define** the following associations:

$$D = 00 \Rightarrow \text{Left}$$

$$D = 01 \Rightarrow \text{Right}$$

$$D = 10 \Rightarrow \text{Forward}$$

$$D = 11 \Rightarrow \text{Backward}$$

These assignments are completely arbitrary. However, once they are introduced, we want to keep them intact so that we will know what the different values for D mean. For example, the word $D = 01$ translates to mean "Right." Even though the bits themselves have nothing to do with direction, the definitions provide the proper associations.

The process of giving meaning to a group of bits is called **encoding,** and it can be applied to any situation we want. Numbers and letters are the items most commonly encoded, but we may represent any set of objects by definitions. The reverse process, wherein a binary number is interpreted for our use, is called **decoding**. In this operation, the encoding information is used to extract the meaning. You must use the encoding scheme to decode a binary word. Otherwise, you have a group of meaningless bits!

Encoding and decoding can be likened to having a dictionary that translates between English and binary. To encode an object, you find the binary equivalent in the dictionary. When you want the object back in its original form, you need to look up the binary word and find out what it means in English. This can be a tedious process, so encoding/decoding information is usually contained in a digital logic network that provides automatic translations. This process is illustrated schematically in Figure 1.3. Information is transformed into binary form using the encoder unit. The digital network uses binary to process the data, resulting in a binary output. This output then passes through the decoder, which translates it back into a form that can be recognized by the user.

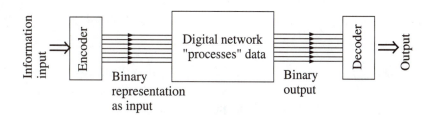

Figure 1.3 The encoding and decoding process

1.5 Binary and Decimal Numbers

Although we live in a world where there are ten digits (0 through 9), we can easily translate these numbers into binary form so that we can use logic gates to perform basic arithmetic operations. After an operation is completed, the final result must be translated back into ordinary digits. These conversions are achieved by using number theory.

Consider first the concept of a **base** or **radix** r for a number system; r is a number that specifies how many digits are available for counting. We say that the ordinary (decimal) number system is **base-10**, since there are ten digits: 0, 1, 2, 3, 4, 5, 6, 7, 8, and 9. There are only two binary numbers, 0 and 1, so this property defines a **base-2** (or binary) system. In order to translate a number from decimal to base-2 and back, we must develop a formal relationship between the two numbering systems. To avoid confusion, we often use subscripts to distinguish a base-10 number N_{10} from a base-2 number b_2. However, when the value of the base is clear from the content, the subscripts are omitted. For example, we immediately recognize that the number 42 cannot be a base-2 value, because both of the digits are out of the binary range.

1.5.1 Binary-to-Decimal Conversion

Now let us examine the details of using binary words to represent the everyday base-10 digits 0 through 9. If we want to represent each of the ten digits by a distinct binary word, then we need to use at least 4 bits; this is because $2^3 < 10 < 2^4$. To this end, let us construct the 4-bit binary word

$$N = N_3 N_2 N_1 N_0$$

to represent base-10 digits. Although the assignment of word definitions in the encoding process is often arbitrary, conversions between two numbering systems must follow a body of specified rules to be consistent with set theory. In particular, for a conversion from base-2 to base-10, we will interpret this word by the following expression:

$$
\begin{aligned}
N &= N_3 N_2 N_1 N_0 \\
&= N_3 \times 2^3 + N_2 \times 2^2 + N_1 \times 2^1 + N_0 \times 2^0 \quad \text{(decimal)} \\
&= N_3 \times 8 + N_2 \times 4 + N_1 \times 2 + N_0 \times 1 \quad \text{(decimal)}
\end{aligned}
\tag{1.8}
$$

This shows that a binary digit in N_j has a base-10 weighting of 2^j.

Example 1-1

Let us convert the 4-bit binary number $N = 0110$ to its base-10 equivalent. Applying the formula gives

$$
\begin{aligned}
N &= 0110 \\
&= (0 \times 8) + (1 \times 4) + (1 \times 2) + (0 \times 1) \quad \text{(decimal)} \\
&= 4 + 2 \quad \text{(decimal)} \\
&= 6 \quad \text{(decimal)}
\end{aligned}
\tag{1.9}
$$

which shows that 0110 is equal to decimal 6. As another example, the binary word $N = 1011$ can be shown, by the same procedure, to have the decimal value

$$
\begin{aligned}
N &= 1011 \\
&= (1 \times 8) + (0 \times 4) + (1 \times 2) + (1 \times 1) \quad \text{(decimal)} \\
&= 8 + 2 + 1 \quad \text{(decimal)} \\
&= 11 \quad \text{(decimal)}
\end{aligned}
\tag{1.10}
$$

The general formula can be used to translate any 4-bit binary word. In particular, we can construct the counting sequence from 0 to 15 shown in Figure 1.4 by directly applying the formula. Note that we *always* start with the digit 0 in a numbering system.

Decimal	Binary
0	0000
1	0001
2	0010
3	0011
4	0100
5	0101
6	0110
7	0111
8	1000
9	1001
10	1010
11	1011
12	1100
13	1101
14	1110
15	1111

Figure 1.4
Some equivalent numbers in the decimal and binary systems

Extending this relationship to a larger-size word is straightforward. An n-bit binary word has 2^n different combinations, so it can be used to represent the decimal numbers from 0 to $(2^n - 1)$. The k-th bit has a decimal weighting of 2^k. For example, an 8-bit word has $2^8 = 256$ different permutations representing base-10 digits from 0 to 255 via the equation

$$
\begin{aligned}
B &= b_7 b_6 b_5 b_4 b_3 b_2 b_1 b_0 \\
&= b_7 \times 2^7 + b_6 \times 2^6 + b_5 \times 2^5 + b_4 \times 2^4 + \\
&\quad b_3 \times 2^3 + b_2 \times 2^2 + b_1 \times 2^1 + b_0 \times 2^0
\end{aligned}
\tag{1.11}
$$

where we note that the weighting factors are given by powers of 2:

$$
2^7 = 128; \ 2^6 = 64; \ 2^5 = 32; \ 2^4 = 16; \ 2^3 = 8; \ 2^2 = 4; \ 2^1 = 2; \ 2^0 = 1 \tag{1.12}
$$

Since the decimal number is obtained by first multiplying each weighting factor by either a 0 or a 1 and then adding the result, we can devise an easy way to convert binary words to decimal numbers using the scheme shown below. To use this chart,

Weighting	128	64	32	16	8	4	2	1
Bit	b_7	b_6	b_5	b_4	b_3	b_2	b_1	b_0

simply note that a 0 in a given position indicates zero contribution to the decimal value, whereas a 1 indicates a full contribution. The decimal number is thus obtained by summing the decimal weights for all "1" locations.

Example 1-2

Consider the 8-bit binary number

$$X = 01011100$$

To find the decimal equivalent value, we use the chart and sum

$$\begin{aligned} X &= 0 + 64 + 0 + 16 + 8 + 4 + 0 + 0 \\ &= 92 \end{aligned} \tag{1.13}$$

This technique can be used for any binary word that has 8 bits. For a longer word size, the higher-order weighting factors 2^9, 2^{10}, and so on must be computed to expand the chart. In general, for an n-bit word

$$A = A_{n-1}A_{n-2} \cdots A_2A_1A_0 \tag{1.14}$$

the decimal equivalent is computed from

$$A = A_{n-1} \times 2^{n-1} + A_{n-2} \times 2^{n-2} + \ldots + A_2 \times 2^2 + A_1 \times 2^1 + A_0 \times 2^0 \tag{1.15}$$

The ability to translate a binary word into a base-10 value allows us to decode the output from a digital logic system. The reverse encoding process provides the opposite capabilities.

1.5.2 Decimal-to-Binary Conversion

Encoding a base-10 (decimal) number into an equivalent binary number requires a bit more work than the reverse operation. In this section, we will introduce a technique called **successive divisions** that allows us to perform this operation.

Consider a decimal number N. In order to find the equivalent binary word b, we will successively divide by 2 and look at the remainder R after each division; the remainder will be a bit in b. This is most easily illustrated by example.

Example 1-3

Let us examine the technique as applied to converting the base-10 number 19_{10} into its binary equivalent. First, we divide 19 by 2.

$$\frac{19}{2} = 9 + \text{Remainder } R_0 = 1 \tag{1.16}$$

The remainder $R_0 = 1$ is the least significant bit (LSB) in the binary representation; by definition, the LSB is the bit with the lowest weighting, which is $2^0 = 1$.

Next, we divide the result of the first step, 9, by 2 and look at the remainder R_1:

$$\frac{9}{2} = 4 + \text{Remainder } R_1 = 1 \tag{1.17}$$

Continuing to divide in the same manner, we obtain the results

$$\frac{4}{2} = 2 + \text{Remainder } R_2 = 0$$

$$\frac{2}{2} = 1 + \text{Remainder } R_3 = 0 \tag{1.18}$$

$$\frac{1}{2} = 0 + \text{Remainder } R_4 = 1$$

The algorithm is completed when the division leaves a value of 0; this defines the most significant bit (MSB) in the word. The MSB is the bit with the highest weighting, which is 2^4 in this case. The equivalent binary representation of 19_{10} is then constructed by using the remainder bits to write

$$\begin{aligned} b &= R_4 R_3 R_2 R_1 R_0 \\ &= 10011 \end{aligned} \tag{1.19}$$

The result can be verified by reverse-translating (decoding) 10011_2 back to base 10. The technique is very general and is quite easy to remember. Two important points are that

- The first remainder is the LSB (least significant bit) of the binary word.
- The last remainder is the MSB (most significant bit).

Example 1-4

Let us convert the decimal number 56 to binary form. The sequence of steps is

$$\frac{56}{2} = 28 + \text{Remainder } R_0 = 0$$

$$\frac{28}{2} = 14 + \text{Remainder } R_1 = 0$$

$$\frac{14}{2} = 7 + \text{Remainder } R_2 = 0$$

$$\frac{7}{2} = 3 + \text{Remainder } R_3 = 1 \tag{1.20}$$

$$\frac{3}{2} = 1 + \text{Remainder } R_4 = 1$$

$$\frac{1}{2} = 0 + \text{Remainder } R_5 = 1$$

which shows that $56_{10} = 111000_2$ as can be verified by the reverse calculation.

1.5.3 Fractions

Although many programs require integer values only, many calculations require fractional quantities. To deal with fractions, we will write a binary fraction in the form

$$b = 0.b_{-1}b_{-2}b_{-3}\cdots \tag{1.21}$$

This can be translated into a base-10 number F_{10} by extending our decoding formula for integers using negative powers of 2 for weighting the bits to the right of the decimal point. This gives the general form

$$
\begin{aligned}
F &= b_{-1} \times 2^{-1} + b_{-2} \times 2^{-2} + b_{-3} \times 2^{-3} + \cdots \\
&= b_{-1} \times (0.5) + b_{-2} \times (0.25) + b_{-3} \times (0.125) + \cdots
\end{aligned}
\tag{1.22}
$$

for the conversion. Since 2^{-x} is less than 1, it is easy to see that this yields a value of $F_{10} < 1$ also.

Example 1-5

Let us convert the binary fraction $b = 0.10110$ into its base-10 equivalent. We have

$$
\begin{aligned}
F &= b_{-1} \times 2^{-1} + b_{-2} \times 2^{-2} + b_{-3} \times 2^{-3} + b_{-4} \times 2^{-4} + b_{-5} \times 2^{-5} \\
&= 1 \times (0.5) + 0 \times (0.25) + 1 \times (0.125) + 1 \times (0.0625) + 0 \times (0.03125) \\
&= 0.5 + 0.125 + 0.0625 \\
&= 0.6875
\end{aligned}
\tag{1.23}
$$

so $0.10110_2 = 0.6875_{10}$.

To convert a decimal fraction into a binary number, we modify the technique used for integer numbers to that of **successive multiplications**. Starting with the base-10 number F, we multiply by 2 and examine the result, which will be in the form $m = b.xxxxx$ with $b = 1$ or $b = 0$ being the binary digit. If $m < 1$, then we multiply by 2 again and interpret the next result. If, on the other hand, $m \geq 1$, then we first subtract 1 and then multiply the difference by 2.

Example 1-6

Consider the decimal number 0.6875_{10}. Multiplying by 2 gives

$$(0.6875) \times 2 = 1.375 \qquad \text{so } b_{-1} = 1 \tag{1.24}$$

Since 1.375 is greater than 1, we subtract 1 to obtain $(1.375 - 1.000) = 0.375$ and then use this number in the next step:

$$(0.375) \times 2 = 0.75 \qquad \text{so } b_{-2} = 0 \tag{1.25}$$

Continuing, we find that

$$(0.75) \times 2 = 1.50 \qquad \text{so } b_{-3} = 1 \qquad \textbf{(1.26)}$$

Thus we must once again subtract $(1.50 - 1.00) = 0.5$ to perform the next step:

$$(0.50) \times 2 = 1.00 \qquad \text{so that } b_{-4} = 1 \qquad \textbf{(1.27)}$$

If we subtract 1, this leaves a value of 0, indicating that our algorithm is completed. Therefore,

$$\begin{aligned} 0.6875_{10} &= 0.b_{-1}b_{-2}b_{-3}b_{-4} \\ &= 0.1011 \end{aligned} \qquad \textbf{(1.28)}$$

is the final answer.

Round-Off Error

Although binary numbers can be used to represent base-10 decimal values, the accuracy of the translation from decimal to binary depends on the number of bits that are used in the base-2 word. This introduces a problem called **round-off error**.

To understand the problem, consider a 4-bit fraction of the form $0.x_{-1}x_{-2}x_{-3}x_{-4}$. In general, this can be used to represent a base-10 number X:

$$\begin{aligned} X &= 0.x_{-1}x_{-2}x_{-3}x_{-4} \\ &= x_{-1} \times (0.5) + x_{-2} \times (0.25) + x_{-3} \times (0.125) + x_{-4} \times (0.0625) \end{aligned} \qquad \textbf{(1.29)}$$

where the actual value depends on the values of the bits x_3 through x_0. Let's use this to calculate the decimal value of two binary words, $X_a = 0.1110$ and $X_b = 0.1111$. Substituting gives

$$\begin{aligned} X_a &= 0.8750 \\ X_b &= 0.9375 \end{aligned} \qquad \textbf{(1.30)}$$

for the two words. The smallest resolution allowed is 0.0625, so it is not possible to represent numbers between these decimal values using the 4-bit binary fraction. For example, the decimal fraction 0.9270 cannot be represented exactly using the form $0.x_{-1}x_{-2}x_{-3}x_{-4}$. The closest value possible would be $X_b = 0.1111 = 0.9375 > 0.9270$, which introduces an error of

$$\frac{0.9375 - 0.9270}{0.9270} \times 100 \approx 1.13\% \qquad \textbf{(1.31)}$$

The only way to overcome this is to add more bits to the binary representation. However, this can result in a binary word that is very long and/or an inconvenient word length. To understand this comment, consider the 5-bit word $X_c = 0.11101$. This has a decimal value of 0.90625, still short of the 0.9270 value. Adding another bit to construct $X_d = 0.111011$ gives a decimal value of 0.921875, and so on. Because round-off error of this type can be a problem in some applications, other techniques of treating fractions have been developed. The most common approach uses binary groups called floating-point representations; these are somewhat complicated and are not discussed until Chapter 11.

1.5.4 Hexadecimal Numbers

The hexadecimal numbering system is a base-16 scheme that uses the following 16 symbols as basic digits:

$$0, 1, 2, 3, 4, 5, 6, 7, 8, 9, A, B, C, D, E, F \tag{1.32}$$

Note that A, B, C, D, E, and F are used as digits, not letters in the usual sense; lowercase digits (a, b, c, d, e, f) are also used in practice. In general usage, these digits are often referred to as "hex" numbers for simplicity. A word in base 16 consists of hex digits arranged in a specified order, with the most significant digit on the left and the least significant digit on the right. For example, $H = 304E$ is a valid hexadecimal word where 3 is the most significant digit and E is the least significant digit. To find the base-10 (decimal) equivalent N of a 4-digit hex word $H = h_3\, h_2\, h_1\, h_0$, we use the formula

$$N = h_3 \times 16^3 + h_2 \times 16^2 + h_1 \times 16^1 + h_0 \times 16^0 \tag{1.33}$$

which may be extended to hex words of arbitrary length. To translate a base-10 number into hexadecimal, we use successive division by 16. The table in Figure 1.5 provides the translation among hex, decimal, and binary numbers.

Hex	Decimal	Binary
0	0	0000
1	1	0001
2	2	0010
3	3	0011
4	4	0100
5	5	0101
6	6	0110
7	7	0111
8	8	1000
9	9	1001
A	10	1010
B	11	1011
C	12	1100
D	13	1101
E	14	1110
F	15	1111

Figure 1.5
The hexadecimal numbering system uses digits 0 through F.

In our treatment of digital logic, we will generally restrict our usage of hexadecimal numbers to representing 4-bit binary words. This will simplify our notation considerably, so it is worth the effort to learn. The equivalence between a hex digit and a 4-bit binary word is easily seen from the table. Moreover, this can be applied to binary words of arbitrary length by grouping the bits into groups of 4 bits each, starting from the right side.

Example 1-7

Let us start with the 16-bit binary word 1001110011100101. When we break this into individual 4-bit groups we get

$$1001\ 1100\ 1110\ 0101_2 = 9CE5_{16}$$
$$= 0x\ 9CE5$$

(1.34)

by directly associating each group with a separate hexadecimal digit. We have introduced the alternative notation "0x *nnnn*" to denote that *nnnn* is a hexadecimal number.[1]

1.6   Cells and Hierarchy

To create a digital system, we will use fundamental building blocks called **cells**. In general, a cell provides some useful operation needed in the system design. The system itself is created by connecting cells in the proper order to provide all of the necessary properties.

Graphical representations of digital networks are quite useful for describing the types of cells used and how they are connected. Drawings of this type are called **logic diagrams**, and they provide a viewpoint that is useful for both analysis and design. The concept of a cell is shown in Figure 1.6. This takes binary input variables A, B, and C and produces an output function $f(A, B, C)$. The input and output points of the cell are called **ports**; they allow the cell to be connected to other cells.

As an example of a function, suppose that we wish to have a cell that has an output of $f = 1$ if and only if a single input is 1, and $f = 0$ otherwise. This can be described by writing

$$f = 1 \quad \text{if } A = 1 \text{ or } B = 1 \text{ or } C = 1;$$
$$f = 0 \quad \text{Otherwise.}$$

As we will see later, Boolean algebra can be used to create equations that implement functions of this type.

In order to create a large digital system, we use small cells to build larger cells with more complex functions. Interconnections between cells are shown in our

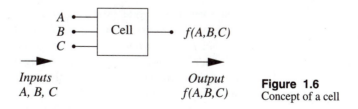

Inputs
A, B, C

Output
f(A,B,C)

Figure 1.6
Concept of a cell

---

[1] The notation 0x for identifying hexadecimal numbers is the convention used in the C programming language.

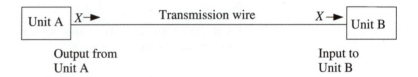

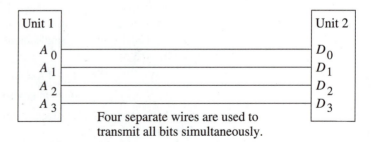

Figure 1.7 Serial data flow path connecting two units

logic diagrams as lines that represent the **signal flow path**. The lines are also called **transmission wires** or simply **interconnects**. When we construct a system using electronic parts, the signal flow lines are equivalent to wires. Figure 1.7 illustrates the use of a signal flow path. The output of Unit A is denoted as X and is transmitted to the input of Unit B by the transmission wire. In this scheme, the bits are sent from Unit A to Unit B in a sequential manner in time. This is called a **serial** transmission arrangement.

In many systems, our work deals primarily with binary words. In this type of design, it is more convenient to transmit all of the bits in a word simultaneously. This is called a **parallel** transmission arrangement. A parallel signal flow path simply consists of several distinct serial paths, with each serial line used to transmit one bit. Figure 1.8 shows the main idea in a 4-bit parallel transmission link. This can be extended to an arbitrary number of bits.

Figure 1.8 Parallel data flow connection

When we draw signal flow diagrams, two signal flow lines often cross one another, and it is necessary to distinguish between the case where the lines are connected and the case where they are completely separate. This is accomplished by using a "dot" as shown in Figure 1.9. In Figure 1.9(a), the two lines are separate, with X on the horizontal line and Y on the vertical line. Figure 1.9(b) shows the case where the lines are connected, and thus have the same variable, X.

In digital system design, small cells are connected to create larger functions. The photograph in Figure 1.10 shows a commercial integrated microprocessor chip.[2] Each section is designed to provide a specific set of functions; the complete system is created by interfacing the sections together.

Another example of this concept is shown in Figure 1.11, where several cells have been used to create a multiple-input, multiple-output system. Note that we

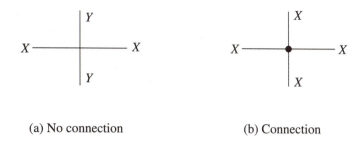

(a) No connection   (b) Connection

Figure 1.9 Wire drawing conventions

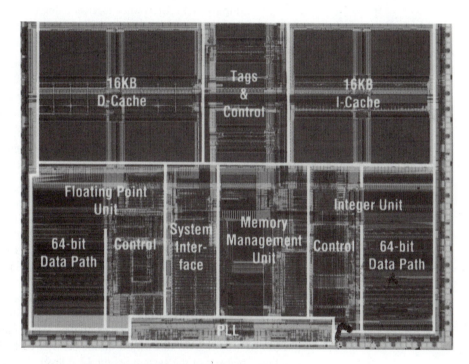

Figure 1.10 Microprocessor chip photograph

have also included a set of **control signals** (at the top of the drawing); control bits provide us the ability to change the cell functions for different situations. The idea of using cells as building blocks is called **hierarchical design**. This concept gives us a structured technique for analyzing and designing complex digital systems.

---

[2] This is a microphotograph of the Orion™ 64-bit microprocessor provided by IDT (Integrated Device Technologies) for use in this book.

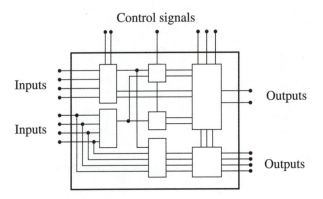

Figure 1.11 Units wired together to form a system

There are two opposite approaches to designing a digital network.

- **Top-down design.** In this approach, we start with the large-scale system specifications and then choose the cells that are needed to synthesize the system. The cells may be very complicated, and they are usually made of smaller, more primitive cells.

- **Bottom-up design.** This is exactly opposite to the top-down approach. In the bottom-up philosophy, basic cells are used to build more complex cells, which in turn provide the foundation for even more complicated functions.

Our treatment will be somewhere in between the top-down and the bottom-up approaches, say "middle of the road." We will not adopt any particular philosophy but will attack the various aspects of the subject in a manner that provides the background to understand the entire field. It is important to learn the details of binary numbers and algebras, because these form the basis for all digital systems. However, modern digital systems can be extremely complex in both structure and operation. When we study large systems, new variables and considerations start to outweigh many of the considerations that were important at the bottom level.

The hierarchies are portrayed in Figure 1.12. At the highest (system) level, the interior of the box is unimportant. Only the overall functions are of concern at the TOP. As we progress downward on the scale, we obtain more information on the internal makeup of the system. At the Unit level, the drawing shows smaller blocks that perform more basic operations than the entire system. At the next step shown, the large units are broken down into more Basic Units so that even more of the detail becomes apparent. At the Component Level, interest is directed toward the fundamental "building blocks" that are used to create the Basic Units. This illustrates how the hierarchical view works. The detail of interest to you at a particular time depends on the level where you are working. Sometimes you will be interested only in the overall function of a complex unit, whereas at other times you may need to understand every element that goes into making a basic unit. The power in this approach derives from the fact that the important aspects vary with the level and that the operation is nested from the bottom to the top of the hierarchy. Note that the drawing shows only representative levels. The viewpoint needed in a particular problem can be defined as necessary.

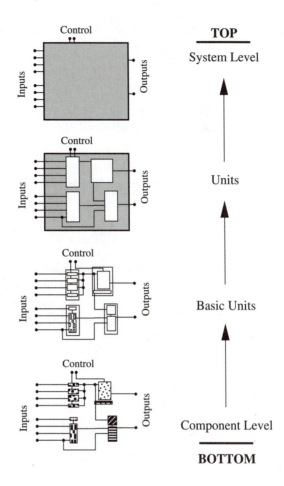

Figure 1.12 Levels in the design hierarchy

1.7 System Primitives

A **system primitive** is a basic function that is used several times to create the entire unit. There are several levels of primitives that are useful, depending on the complexity of the system. A few of these are briefly described below; they are included to illustrate some of the basic concepts. Don't worry about the detailed operation yet. That is the subject of this book!

Clocks

The first concept that we will examine is that of clocks and clocking signals. This is a signal that periodically makes a transition from 0 to 1 and back, as shown in Figure 1.13. Clocks are use to provide a reference for timing operations inside of a digital network. This allows us to synchronize operations in a complex network, making it easier to track the movement of the data. The clock is defined by the

period T, which is the time for one complete cycle; note that T has units of seconds (abbreviated *sec*). It is useful to define the clock **frequency** by

$$f = \frac{1}{T} \tag{1.35}$$

which has strict units of (1/*sec*). The frequency f is usually given in units of hertz (abbreviated *Hz*) such that $1\ Hz = 1\ cycle/sec$.

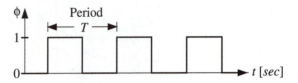

Figure 1.13 Clock signal

The clock frequency is an important parameter. It provides a measure of how fast operations are taking place in the system. In many digital systems, increasing the clock frequency results in a proportional increase in the number of operations performed per second. Even in situations where this is not the case, the clock frequency is a measure of how fast a unit can be expected to perform a particular task.

Example 1-8

A clocking signal has a frequency of 50 MHz, where 1 MHz (1 megahertz) is 10^6 Hz, or one million hertz. Calculate the clock period T.

The clock period is found from

$$T = \frac{1}{f} = \frac{1}{50 \times 10^6} = 2 \times 10^{-8}\ sec \tag{1.36}$$

or $T = 20\ ns$, where "*ns*" is a **nanosecond**, and $1\ ns = 10^{-9}\ sec$. Other useful scaling factors for clocks are the **millisecond** "*ms*" ($1\ ms = 10^{-3}\ sec$), and the **microsecond** "μs" ($1\ \mu s = 10^{-6}\ sec$).

Logic Gates

Logic gates take "input" bits and produce an "output" bit as defined by the logic operation. An example of a logic gate is shown in Figure 1.14, where A and B are the inputs and f is the output. Mathematically, we write that f is a function of A and B, using the notation $f = f(A,B)$. This means that the value of f (which is either a 0 or a 1) is determined by the values of the input variables A and B (each of which is also restricted to have the values of 0 or 1). The "flow" of the data is read from the inputs through the gate to the output.

Memory

A memory cell is capable of capturing and holding the value of a binary variable. A simplified drawing of a memory cell is shown in Figure 1.15. The input is denoted

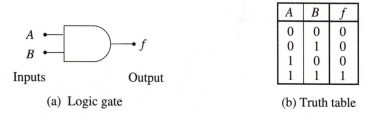

(a) Logic gate

(b) Truth table

A	B	f
0	0	0
0	1	0
1	0	0
1	1	1

Figure 1.14 Example of a digital logic gate

by D, and the output is labeled X. The operation of a memory cell is straightforward. Suppose the input has a value $D = 1$. Then "1" is **stored** in the memory cell; this is also termed **writing** the bit or **loading** the cell. Once a data bit is stored in the cell, it can be accessed (or **read**) at the output point X. The clock input ϕ is used to allow the operation to be "synchronized" with the rest of the system. This means that the read and/or write operations can take place only at certain times defined by the clock signal.

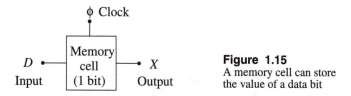

Figure 1.15
A memory cell can store
the value of a data bit

Registers

A register is a block of memory cells that can be used to store words. One type of 8-bit parallel register is shown in Figure 1.16. This allows for parallel loading and reading of an 8-bit word, synchronized by a clocking signal ϕ.

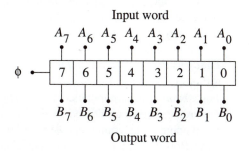

Figure 1.16 A register is designed to store an entire binary word

1.8   Metrics

Digital systems engineering is concerned with creating a digital network that performs a specific task using a defined set of input parameters. One challenging aspect of design is that there are usually several approaches that can be used to solve the problem. However, it is common to find that one approach is better than another because of one or more specific characteristics.

To compare different design solutions, we introduce the concept of a **metric**. In general, a metric is unit of measurement that can be applied in different situations to compare them. An obvious example of a metric is **length** as measured by a meter stick. However, many other metrics are useful in problem solving.

Temporal Metric

In high-speed computing, the time interval needed to complete a calculation is often the critical parameter. Temporal[3] metrics are therefore quite important in the field of digital logic design. The time interval of interest depends on the situation being studied. At the lowest level, it is the time delay for a logic signal through a basic logic gate, which affects the timing of the entire system. A system designer, on the other hand, is more concerned with the overall speed of a logic network when performing a particular set of tasks, such as running a program.

Size

In modern digital systems design, most of the electronics are based on items that are generically referred to as *computer chips*. This is a popularized name for an **integrated circuit**, which is a type of electronic network that is made on a tiny piece of silicon[4] using techniques from the field of **microelectronics**. In general, we attempt to build digital integrated circuits that are very densely packed. An example is the **microprocessor**, which is an entire computer that can be created on a single "chip" of silicon that is typically less than 2 centimeters on a side.

Size is an important metric for two reasons. First, high-density integration leads to physically small electronic circuits, so the large systems that use these chips are easier to build. The second reason is more subtle. The speed at which an electronic network can switch depends on moving electrons around on the chip. Electron velocities are limited by the atoms in the silicon, so we can speed up the signal transfer by making the path lengths small.

Electric Power Consumption

All electrical networks consume electrical energy, which leads to heating. In old, large *mainframe* computers, the heating problem was so bad that the systems required their own air-cooled room. This was perfectly acceptable in the early days of computers; the computers were so large that they required an entire room anyway! However, microelectronics has changed all that. It is now possible to have enormous computing power in a very small case.

---

[3] "Temporal" means having to do with time.

[4] Other materials are also used, but they are less common than silicon.

Digital electronic circuits consume electrical energy anytime they are "powered up," that is, connected to a power supply through an electrical outlet. Furthermore, every time a digital network makes a decision, additional energy is used. Electrical power consumption is important for two reasons. First, we must be concerned about excessive heating of the electronics, which might lead to instabilities and eventual failure. Second, portable systems such as notebook computers are powered by batteries. A low-power circuit allows us much more operating time before the battery must be recharged.

1.9 Hierarchical Plan for the Book

In this chapter we have examined some of the concepts involved in digital logic systems. The most basic idea is that we use the binary numbering system, in which there are only two digits, 0 and 1. In order to make a powerful data processing system, we group several bits together to form words and then use encoding and decoding to give the words meaning both inside and outside of the system.

This book is based on the concept of hierarchical design, whereby we may view a system at different levels depending on the details that are important at the time. Because every complex system is created by starting with individual bits, it makes sense to break the treatment into various levels. The hierarchical approach allows us to "zoom in" to study the details and then "zoom out" to examine the behavior of the entire system. The important point to remember is that each level in the hierarchy is just a different view of the same network. One may start at the bottom level and work upward from simple to complex, or one may start at the top level and work down. Each level can be related to every other level. Sometimes the relationships will be obvious; at other times they will be obscured by details.

The levels that have been chosen as benchmarks for this book are illustrated in Figure 1.17. At the highest level, we have the **system**, which defines the "TOP." This could be, for example, an entire computer. Working downward, we hit a level that has been called **logic units**. These are relatively complex logic blocks that perform a specific function, such as algebraic addition. A logic unit is created by using **logic gates**, which is the next step down in the hierarchy. Logic gates are often taken as the most fundamental building blocks in a digital network, and we will learn an entire algebra based on their behavior. Continuing downward leads us to **CMOS** (Complementary Metal on Oxide on Semiconductor) **circuits**. At this level, we will see how logic gates are created by using electronic switches called transistors. Finally, at the lowest level is the **silicon IC**, where IC is an acronym that stands for *integrated circuit* (it is read by just repeating the letters). An integrated circuit is the device that is often referred to as a "chip" in the popular literature. A silicon IC represents what is known as the physical design of the network, and it is the "BOTTOM" of the hierarchy used here. Two other viewpoints, labeled **VHDL** and **Hardware,** can be applied to any level desired. VHDL is a formal language that enables us to simulate a digital design, and hardware refers to the actual electronic components and parts that are used to build the electronic system.

The treatment in this book actually starts in **Chapter 2** with digital logic gates and binary numbers, a viewpoint located in the middle of the hierarchy. This provides the foundation we need to progress up or down in the hierarchy. More

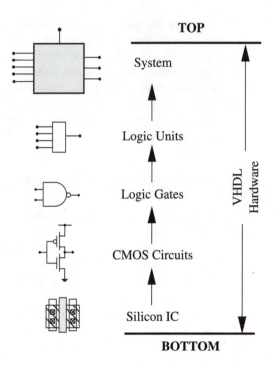

Figure 1.17 Hierarchical levels used in this book

advanced logic design is presented in **Chapter 3**. General hardware concepts are introduced in **Chapter 4**, and **Chapter 5** introduces the formalisms involved in VHDL. The first five chapters thus provide the important concepts.

Chapter 6 is unique in this type of book in that it is dedicated to examing the use of transistors to implement logic gates in what is called a CMOS technology. CMOS circuits are used to design complex networks such as those used in microprocessors, and they are well worth learning about. Moreover, CMOS provides an easy transition to the lowest level where the circuits are created in tiny sections of silicon. The translation of logic circuits into silicon is called **physical design;** it is introduced in **Chapter 7**. The discussion is directed toward studying and understanding high-density logic chips. This takes us to the bottom level of the hierarchy.

Chapter 8 represents a change in gears. It is concerned with the design and construction of several different logic units that are useful in many systems. **Chapter 9** continues at this level and provides a detailed examination of memory circuits, that is, digital units that have the ability to store binary values. Although the emphasis is on digital logic units in both chapters, the connection to VHDL, CMOS circuits, and silicon ICs is an integral part of the discussion. The final area in this group of topics is that of sequential circuits; these are the subject of **Chapter 10**.

The final chapters of the book are concerned with the system level and deal with computers and computer architectures. **Chapter 11** explains how digital logic building blocks can be used to create large components, which are then wired together to form a computer. **Chapter 12** advances these concepts and discusses how simple systems can be extended and developed into the high-performance systems of today.

To help you determine the approximate level in the hierarchy at which the discussion is taking place, icons representing the various levels are provided at certain locations in the book. The discussion starts at the logic level, which is indicated by the logic gate icon in Figure 1.18. Circuits and CMOS transistors are denoted by the drawing in Figure 1.19, which is the circuit for a NOT logic gate. At the lowest level, our discussion will be directed toward transistors on silicon and will be indicated by the odd-looking icon in Figure 1.20. This was chosen because it is a transistor on a silicon chip.

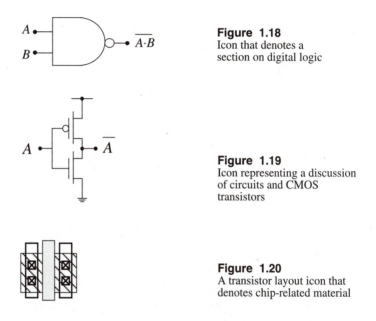

Figure 1.18
Icon that denotes a
section on digital logic

Figure 1.19
Icon representing a discussion
of circuits and CMOS
transistors

Figure 1.20
A transistor layout icon that
denotes chip-related material

As the level progresses upward from the logic level, the networks get more complicated. In this case, the icon that will be used is for a unit called a multiplexor (MUX); see Figure 1.21. Finally, at the highest level, we will use the system icon shown in Figure 1.22; the designation CPU is short for central processing unit, which is the part of the system that provides all of the computing and data processing circuits. It is important to note that the highest level of the hierarchy views the external characteristics of complete systems. These are composed of units, which are in turn made up of gates, which are created using transistors, which are devices made in silicon. In other words, the levels are nested within each other. Keeping this hierarchical feature in mind will enable you to appreciate all of the important considerations that make digital systems design.

Figure 1.21
Multiplexor icon used to
indicate discussions on
moderately complex units

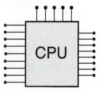

Figure 1.22
CPU icon that indicates a
system-level treatment

1.10 Problems

[1.1] What is the maximum number of combinations for each of the following binary words?

(a) $X = x_1x_0$
(b) $A = a_4a_3a_2a_1a$
(c) $Y = y_5y_4y_3y_2y_1y_0$

[1.2] How many **bits** of memory would be found in a personal computer that has the following system memory sizes?

(a) 16MB
(b) 64MB
(c) 256MB

[1.3] A disk storage device on a desktop computer is advertised to hold 8.1 GB of data.

(a) How many bits is this?
(b) How many 32b words can the disk hold?

[1.4] Suppose that you wanted to establish an encoding scheme that deals with the following set of compass directions: N, S, E, W, NE, NW, SE, SW. Define a binary word called *direction* using the necessary number of bits, and create an encoding/decoding table. What modifications must you make if you want to add the additional directions NNW, NNE, SSW, SSE to your set?

[1.5] A paint store claims to have 43 different shades of white paint. What is the size of the binary word *shade_of_white* that we could use to describe the different shades? How would we deal with the unused binary combinations?

[1.6] Convert the following 4-bit binary numbers into their decimal values.

(a) $W = 0101$
(b) $X = 1100$
(c) $Z = 1001$

[1.7] Convert the following binary numbers into their decimal values.

(a) 101010
(b) 011011
(c) 110001
(d) 011011

[1.8] Convert the following 8-bit binary numbers into their decimal values.

(a) $A = 01010101$
(b) $B = 11001100$
(c) $C = 10100011$

[1.9] Devise a general table for converting 16-bit binary words to decimal values.

[1.10] Convert the following 16-bit words into their decimal values.

(a) 0101 1000 1110 1010
(b) 0100 1101 0001 0100
(c) 1000 1000 1000 1000

[1.11] Convert the following 32-bit words into their decimal equivalents.

(a) 0001 1111 0101 1010 1100 1100 0110 1010
(b) 0100 0110 0101 1111 0101 1010 0000 0100

[1.12] Convert the base-10 decimal numbers below into their binary equivalents.

(a) 8
(b) 14
(c) 23
(d) 36
(e) 18
(f) 9
(g) 16

[1.13] Convert the following decimal numbers into their binary equivalents.

(a) 32
(b) 42
(c) 76
(d) 67
(e) 95

[1.14] Convert the following decimal numbers into their binary equivalents.

(a) 103
(b) 155
(c) 225
(d) 187
(e) 283
(f) 384

[1.15] Convert the following hexadecimal numbers into binary form and then find the equivalent decimal value. (Recall that 0x denotes a hexadecimal value.)

(a) 0x 1F
(b) 0x A8
(c) 0x 7B
(d) 0x 67

[1.16] Convert the following hexadecimal numbers into binary form and then find the equivalent decimal value.

(a) 0x 1F20
(b) 0x 0ABC
(c) 0x 70D2
(d) 0x 86BA

[1.17] Consider the base-10 number 1000. Determine the smallest number of bits n needed to represent a number that is less than or equal to the decimal 1000.

[1.18] A simple calculator can display decimal numbers to a maximum of 6 digits. What is the size of the binary word needed to represent such numbers?

[1.19] Find the decimal value of each of the binary fractions listed below.

  (a) 0.10010
  (b) 0.11010
  (c) 0.01011
  (d) 0.10101

[1.20] Convert the following binary numbers into decimal form.

  (a) 1101.0110
  (b) 1001.0011
  (c) 0101.1001

[1.21] Convert the following base-10 decimals into binary form.

  (a) 0.500
  (b) 0.550
  (c) 0.625
  (d) 0.650

[1.22] Suppose that we want to represent the decimal fraction 0.6 in binary using a 4-bit word.

  (a) Find the binary fraction that gives the closest value.
  (b) What would be the percentage error if we used this representation?
  (c) Repeat the problem for a binary word size of 8 bits.

[1.23] Consider the base-10 number $x = 0.3141$. What is the smallest binary word that can be used to represent this value with a round-off error less than 0.3%?

[1.24] A clock signal has a period of $T = 0.3$ μs. Find the frequency.

[1.25] A clock signal has a period of $T = 150$ ns. Find the frequency.

[1.26] The frequency of a particular clock signal is known to be $f = 400$ MHz. Find the clock period T. Express your answer in convenient scaling units, such as milliseconds.

[1.27] When comparing two desktop computers, you note that one operates with a clock frequency of 500 MHz, while another comparable unit has a central clock of 550 MHz. What is the percent decrease in the length of the clock period when going from the 500 MHz system to the 550 MHz system?

[1.28] Study an advertisement for a personal computer in a newspaper or magazine. Can you use the information provided in the ad and identify the major system components? What about the hierarchical levels in the design?

[1.29] Suppose that we define an automobile as a "system" as discussed in the text. We may apply the hierarchical view to the car and view it as being made up of large primitive units, such as the engine. Each of these is made up of smaller parts; for example, the engine has pistons. Draw a hierarchical block diagram for the car.

CHAPTER 2

Boolean Algebra and Logic Gates

$$A \cdot \overline{A \cdot B}$$

A
B

Digital operations are carried out using the binary number system where a variable x can only have a value of 0 or 1. In this chapter, we will study the behavior of binary numbers using the switching algebra developed by the British mathematician George Boole in the nineteenth century. The branch of mathematics is contained within the concepts of **Boolean algebra**, and provides the basis for modern logic design.

2.1 Data Representation and Processing

In the binary number system, information (or **data**) is represented entirely by using the binary digits (**bits**) 0 and 1. The fundamental purpose of a digital system is to process information. These simple observations lead us to note that there are two basic problems that we must examine before we can proceed to a study of digital systems. How do we represent data using only 0s and 1s. And how can we manipulate (or *process*) the data to provide a useful system?

Consider first the problem of data representation. In the binary number system, data are represented entirely by using only the binary digits (**bits**) 0 and 1. Consider a binary variable x; by definition, x can assume only the value 0 or 1. Thus the single variable x can be useful in describing a situation where there are only two possible outcomes. For example, in classical logic, a quantity such as x is often used to denote the state TRUE or FALSE.

Realistic situations are much more complicated. To represent data when there are more than two possibilities, we form **binary words**, which are groups of bits that are in a defined order, and then provide a meaning for each possible combination. For example, if we let a_3, a_2, a_1, a_0 be individual bits, then we can form the 4-bit word *data* by writing

$$data = a_3 a_2 a_1 a_0 \tag{2.1}$$

Because each bit can be either a 0 or a 1, there are $2^4 = 16$ possible combinations for *data*. In this context, the individual bits a_i that are used to form the word *data* can be viewed as being associated with a particular type of data.

Data processing at the bit level is a very different problem. In this case, the binary variables are used as inputs to a cell that performs a particular mathematical operation, resulting in an output value of 0 or 1. Figure 2.1 illustrates the problem graphically. The inputs are taken to be the bit variables A, B, and C, and each bit can have a value of 0 or 1. The output is a function $f(A,B,C)$ that is itself a dependent binary variable that is allowed to assume a value of only 0 or 1.

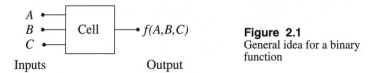

A
B Cell $f(A,B,C)$
C

Inputs   Output

Figure 2.1
General idea for a binary function

The concept of a binary function is straightforward to understand. The value of f depends on the values of the three inputs A, B, C. Given the inputs, f provides a rule that assigns an output of $f = 0$ or $f = 1$. Because a binary function has the property that both the inputs and the outputs are restricted to the values of 0 and 1, the algebra of binary variables is quite different from that used for "ordinary" numbers that range from 0 to 9.

Several approaches can be used to define a binary function. **Truth tables** provide a straightforward way to accomplish this task. A truth table is simply a listing of all possible input bit combinations with the value of the output for each possibility. For n input variables, there are 2^n possible combinations for the n bits, so the truth table will have 2^n entries. A truth table is illustrated in Figure 2.2 for the case of 3 variables A, B, C. Because $2^3 = 8$, there are 8 entries given. The value of the output function f is shown in the right column. Don't worry about the exact meaning of the function at this point; just examine the form of the truth table for now.

Inputs   Output

A	B	C	f
0	0	0	0
0	0	1	1
0	1	0	1
0	1	1	0
1	0	0	1
1	0	1	0
1	1	0	0
1	1	1	1

Figure 2.2
A truth table

2.2 Basic Logic Operations

Boolean algebra is based on a set of logic operations that define basic functions. These are performed using one or more input variables, and they result in a single output bit. The operations presented here are fundamental to all of digital logic. If you take the time to learn the basics thoroughly, then the entire field of digital logic will follow quite naturally. Even the most advanced designer returns to the basics.

2.2.1 The NOT Operation

Consider a binary variable A. The NOT operation transforms A into the output NOT(A) with the following effect:

The NOT operation simply changes the value of a variable from 0 to 1, or vice versa.

$$\text{If } A = 0, \text{ then NOT}(A) = 1;$$

$$\text{If } A = 1, \text{ then NOT}(A) = 0.$$

The truth table in Figure 2.3 defines the NOT operation. With an input A, we usually denote NOT(A) by the simplified notation $\overline{A}$ or A'. Note that, with only one input variable, there are $2^1 = 2$ possible input combinations ($A = 0$ or $A = 1$). Following this observation, we may also call $\overline{A}$ the **complement** of A.

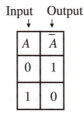

A	$\overline{A}$
0	1
1	0

Input Output

Figure 2.3
The NOT truth table

Another way to express a binary operation is to write it symbolically in algebraic form. For example, let us write the NOT function as $f(A)$ where A is the input. Then

$$f(A) = \overline{A} \tag{2.2}$$

defines f as the NOT operation. By inspection, this gives the same information as the truth table.

Algebraic functions and table listings are useful when we are dealing with the theoretical aspects of a digital system. However, if we want to design the logic networks using electronic gates, it is often more convenient to use pictorial descriptions, called **logic diagrams**, that represent logic functions by various types of graphical symbols. When a digital system is described in this manner, it both shows the logic blocks used to make up the system and specifies how the units are interconnected.

The graphical symbol for a NOT function is shown in Figure 2.4(a); this is also referred to as an **inverter**, and the operation is called **inversion.**[1] By definition of the

---

[1] The term *inverter* is used because an input of 0 gives an output of 1, and vice versa.

NOT symbol

(a) Inverter symbol

(b) Buffer symbol

Figure 2.4 NOT and buffer symbols

symbol, an input variable A gives an output of $\overline{A}$. The triangular shape is common to electronics and represents what we will call a **buffer**; alternatively, one may use a more generalized symbol as described in Section 2.9. In general, a buffer gives an output that is logically the same as the input, as shown in Figure 2.4(b): if A is applied to the input, then the output is also A. Although a buffer does not change the logic state, it is used to represent additional electrical "power drive" needed in certain types of electronic networks. Whenever we want to represent the NOT operation in a logic diagram, we use the "bubble" notation shown in the diagram. The bubble may appear at either the input or the output of a gate, and this shorthand notation is used quite frequently.

2.2.2 The OR Gate

Now let us consider the situation where there are two independent input bits, which we'll call A and B. Each of the 2 bits can assume a value of 0 or 1, so $2^2 = 4$ possible combinations can occur. These can be listed as $AB = 00, 01, 10, 11$.

The first 2-input gate we will examine is that for the **OR operation**; because it has 2 inputs, this particular case will be called an OR2 gate in our discussion. With inputs A and B, this is described as follows:

The OR operation requires two or more inputs.

If $A = 1$, or $B = 1$, or both $A = 1$ and $B = 1$,

then

A OR $B = 1$

else

A OR $B = 0$

This can be restated by saying that if any input has a value of 1, then A OR $B = 1$. Since the function can have a value of only 0 or 1, this implies that if both $A = 0$ and $B = 0$, then A OR $B = 0$. We will usually employ the symbol "+" to denote the OR operation:

$$A \text{ OR } B = A + B \tag{2.3}$$

This is very common when there is no possibility of confusing the binary operation with standard addition. The truth table for the OR function is shown in Figure 2.5; note that each of the four input combinations is described by the separate line listing. Another symbol used for the OR operation is

$$A \text{ OR } B = A \vee B \tag{2.4}$$

but we will not use it in our treatment.

A	B	A + B
0	0	0
0	1	1
1	0	1
1	1	1

Figure 2.5
OR2 truth table

The graphical symbol that is commonly used to describe the OR function is illustrated in Figure 2.6. The inputs A and B are shown on the left, and the output A OR B is on the right, with the values assigned in the truth table. Note that the distinctive shape of the OR "logic gate" allows us to recognize the operation very quickly.

OR gate symbol
for 2 inputs

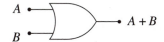

Figure 2.6
OR2 logic gate symbol

In this chapter, we will learn how to manipulate binary variables using a special type of algebra. As our first example, we may construct a 4-input OR operation by writing

$$f = A + B + C + D \tag{2.5}$$

The symbol for the OR4 logic gate illustrated in Figure 2.7.

The OR operation
compares the values
of two or more inputs.

Figure 2.7
OR4 gate symbol

This function may be defined by

　　　If one or more inputs is a 1,

　　then

　　　　$f = 1$

　　else

　　　　$f = 0$.

Alternately, we may write that

　　　$f = 0$ if and only if (iff) all inputs are 0

which gives the same result. Although the number of inputs is not limited by the theory, practical electronic gates rarely have more than 3 or 4 inputs.

2.2.3 The AND Gate

Our next basic operation is called the AND operation. It detects the situation where all of the inputs are equal to 1 and produce an output of 1 for this case. For two inputs A and B, we define the AND2 (a 2-input AND gate) by writing

If both $A = 1$ and $B = 1$,

then

A AND $B = 1$

else

A AND $B = 0$

Alternatively, we may say that

If $A = 0$ or $B = 0$,

then

A AND $B = 0$

else

A AND $B = 1$

The truth table for this case is given in Figure 2.8, where we have used the notation

$$A \ \text{AND} \ B = A \cdot B \qquad \textbf{(2.6)}$$

A	B	$A \cdot B$
0	0	0
0	1	0
1	0	0
1	1	1

Figure 2.8
AND2 truth table

Other common notations are

$$A \ \text{AND} \ B = AB$$
$$= A \wedge B \qquad \textbf{(2.7)}$$

with $A \cdot B$ and AB being the most common choices in this book. The shape-specific logic gate symbol for the AND operation is shown in Figure 2.9.

AND gate symbol
for 2 inputs

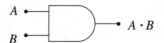

Figure 2.9
AND2 logic gate symbol

The AND operation can be extended to the case where there are more than 2 inputs. For example, a 4-input AND gate (AND4) with inputs A, B, C, D is specified by writing

$$g = A \cdot B \cdot C \cdot D \tag{2.8}$$

and is represented by the graphical symbol in Figure 2.10. By definition, this gives a logic 1 output only if all of the inputs are equal to 1; otherwise, the output is a logic 0. Although theoretically one can create AND gates with any number of inputs, practical electronic AND gates rarely have more than 3 or 4 inputs.

Figure 2.10
AND4 logic gate symbol

2.3   Basic Identities

Boolean algebra describes the behavior of binary variables that are subjected to the multiple NOT, OR, and AND operations. There are some basic identities that are particularly useful to remember, because they help simplify complex logic expressions.

2.3.1   NOT Identity

Performing two NOT operations on a binary variable provides a basic identity of the form

$$\overline{(\overline{A})} = A \tag{2.9}$$

which is referred to as the **involution theorem**. This is equivalent to writing

$$\text{NOT}[\text{NOT}(A)] = A \tag{2.10}$$

and says that since $\text{NOT}(A) = \overline{A}$, $\text{NOT}(\overline{A}) = A$. This result is obvious; A can have a value of only 0 or 1. Although the theorem may seem trivial at first sight, it is very useful to remember when dealing with complex Boolean equations.

2.3.2   OR Identities

The OR function has an output of 1 if any of the inputs are 1s. This leads to several useful identities. Using the truth table in Figure 2.5, we see that

$$\begin{aligned} A + 1 &= 1 \\ A + 0 &= A \end{aligned} \tag{2.11}$$

The first expression says that ORing 1 with anything gives an output of 1, as substantiated by definition: a 1 at any input yields a 1 at the output. The second equation, on the other hand, notes that applying a 0 to the input of an OR gate does not force the output to any specified value; instead, the value of A determines the outcome.

The behavior of an OR gate is further understood by the **idempotent theorem**

$$A + A \ = A \tag{2.12}$$

which says that ORing a variable with itself yields the same variable. This can be verified by noting the two cases

$$\begin{aligned} 0 + 0 &= 0 \\ 1 + 1 &= 1 \end{aligned} \tag{2.13}$$

verifying the theorem. Similarly, we may write

$$A + \overline{A} \ = 1 \tag{2.14}$$

which is known as the **complementary** property and is based on the fact that $1 + 0 = 1$ is always true.

2.3.3 AND Identities

The AND function provides a similar set of identities, all based on the characteristic that a single 0 at the input of any AND gate results in a 0 at the output.

Consider first ANDing a variable with the constants 0 and 1. Using the truth table in Figure 2.8, we can easily verify that

$$\begin{aligned} A \cdot 0 &= 0 \\ A \cdot 1 &= A \end{aligned} \tag{2.15}$$

describes the behavior for any variable A. The idempotent theorem becomes

$$A \cdot A = A \tag{2.16}$$

and the complementary property is now

$$A \cdot \overline{A} = 0 \tag{2.17}$$

The AND identities, like those associated with the NOT and OR operations, are useful enough that they should be committed to memory.

It is important to note that the identities in this section are just generalized expressions that provide the same information as that found in the truth tables. However, writing them as equations provides the basis for a powerful algebra that is used to design digital switching networks.

2.4 Algebraic Laws

The next group of operations that are of interest are those that deal with ordering and precedence. These fall into three main types dealing with commutation, association, and distribution. Verification of the laws can be obtained using the definitions of the AND and OR functions, with specific examples from the truth table listings.

2.4.1 Commutative Laws

The laws of commutation allow us to arrange variables in any order without changing the result. With two variables A and B, these are given by

$$A + B = B + A$$
$$A \cdot B = B \cdot A \tag{2.18}$$

and the extension to multiple variables is also true. For example,

$$X + Y + Z = Y + Z + X \tag{2.19}$$

and

$$X \cdot Y \cdot Z = Y \cdot X \cdot Z = X \cdot Z \cdot Y \tag{2.20}$$

can easily be obtained from the original statements.

2.4.2 Associative Laws

Associative laws define the order in which the operations are performed. In both the OR and the AND operations, the grouping does not affect the result. This gives identities such as

$$A + B + C = (A + B) + C$$
$$= A + (B + C) \tag{2.21}$$

for the OR operation, and identities such as

$$A \cdot B \cdot C = (A \cdot B) + C$$
$$= A \cdot (B \cdot C) \tag{2.22}$$

for the AND expressions. Since the associative law applies to both the OR and the AND operations, we usually omit parentheses when writing logic expressions.

Now suppose that both the AND and the OR operations are contained in a single expression. The question that needs to be answered is "Which operation has precedence?" The answer is quite simple. In the absence of grouping using parentheses, the AND has precedence over the OR when both operations appear in the same equation. For example, the function

$$F = A \cdot (B + C) \tag{2.23}$$

is evaluated by first finding the OR result

$$X = (B + C) \tag{2.24}$$

and then ANDing X with A to arrive at

$$F = A \cdot X \tag{2.25}$$

On the other hand, the expression

$$G = A \cdot B + C \tag{2.26}$$

is written to mean that the AND operation must be computed first using

$$Y = A \cdot B \tag{2.27}$$

such that

$$G = Y + C \tag{2.28}$$

gives the final result. These two examples illustrate the importance of maintaining proper grouping in a logical equation.

2.4.3 Distributive Laws

The distribution of AND and OR operations is governed by the following laws. It is important to remember the rule of precedence that, within a grouping, the AND operation always precedes the OR.

The two important distributive laws are given by

$$\begin{aligned} A \cdot (B + C) &= (A \cdot B) + (A \cdot C) \\ A + (B \cdot C) &= (A + B) \cdot (A + C) \end{aligned} \tag{2.29}$$

The first equation describes how the AND operation is distributed between two ORed variables; it is identical to the distributive law from ordinary algebra using multiplication and addition. Figure 2.11 shows two gate diagrams that, by the distributive theorem, give the same output f. These are created by using the output of a logic gate as the input to the next gate. This type of wiring scheme is called a **logic cascade** and forms the basis of all digital logic systems design. To understand the concept of this type of **logic diagram**, simply follow the paths of the variables through each gate from input to output.

The distribution of the OR between two ANDed variables in the second line of equation (2.29) deserves some additional study. Gate diagrams for both sides of the

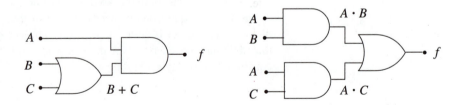

Figure 2.11 Equivalent logic cascades

identity are shown in Figure 2.12, where the output has been labeled as g. Comparing this with Figure 2.11, we see that the *structure* of the logic is similar, i.e., the placement of the gates and the wiring are identical. The difference between the first and second lines in equation (2.29) is that the AND and OR gates are interchanged. The two are said to be **duals** of one another. To generalize this concept, we note that if we start with a logic function k, then the dual of k, which we will call K, is obtained by interchanging the AND and OR operations. For example, if

$$k = A \cdot B + C \tag{2.30}$$

then the dual function is

$$K = A + B \cdot C \tag{2.31}$$

Duals are often useful in simplifying complex logic equations.

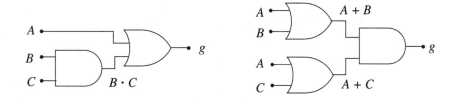

Figure 2.12   Example of the distributive law

2.5   NOR and NAND Gates

Up to this point, we have examined the basic operations of NOT, OR, and AND. Theoretically, we can create any logic function using only these three gates. However, when we attempt to build a digital network using electronic components, we find that these may not be the easiest group to deal with. Most technologies can be used to construct simple inverters, but OR and AND gates are not necessarily the simplest circuits that can be made. Because of this consideration, we define two more "basic" gates that are created by combining the fundamental ones.

Consider first a **NOR2** gate with inputs A and B. The name "NOR" is shortened from NOT-OR and means that the output is defined as the complement of the OR gate. With inputs A and B, the NOR2 operation is denoted by $\overline{A + B}$ and is defined by the truth table in Figure 2.13. This is equivalent to the statement

If either input is 1, then
$$\overline{A + B} = 0$$
else
$$\overline{A + B} = 1$$

A	B	$\overline{A + B}$
0	0	1
0	1	0
1	0	0
1	1	0

Figure 2.13
NOR2 truth table

The NOR gate appears so often that is has a symbol of its own. This is obtained by simplifying the NOT-OR logic cascade shown in Figure 2.14(a) to the form in Figure 2.14(b), which replaces the inverter with a bubble at the output.

The NOR operation has its own symbol.

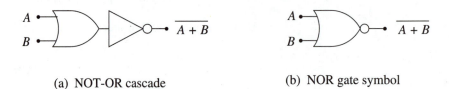

(a) NOT-OR cascade

(b) NOR gate symbol

Figure 2.14 Creation of the symbol for the NOR function

The **NAND** gate is constructed in a similar manner. For two inputs A and B, the NAND operation is denoted by $\overline{A \cdot B}$. Logically, the NAND gate is equivalent to the AND followed by a NOT, as verified by the truth table in Figure 2.15. This shows that the output is a 1 unless all of the inputs are 1s.

A	B	$\overline{A \cdot B}$
0	0	1
0	1	1
1	0	1
1	1	0

Figure 2.15
NAND2 truth table

One way to state this behavior is by writing

If either input is 0,

   then

$$\overline{A \cdot B} = 1$$

   else

$$\overline{A \cdot B} = 0$$

The evolution of the NAND symbol is shown in Figure 2.16(a) and (b). Once again, we see an example where the NOT operation is denoted by just inserting a bubble into the path.

For simplicity, the
NAND operation also
has its own symbol.

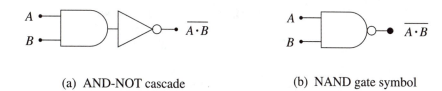

(a) AND-NOT cascade

(b) NAND gate symbol

Figure 2.16   NAND2 logic gate

2.5.1 DeMorgan Theorems

The DeMorgan theorems provide alternative expressions that relate the NOR and
NAND operations to each other. They are very useful in practical logic design
problems and are part of the fundamental set of techniques that are used by logic
designers.

Consider first a NOR2 gate with inputs A and B. The first DeMorgan theorem
that we will study allows us to express the NOR operation as

$$\overline{(A + B)} = \overline{A} \cdot \overline{B} \tag{2.32}$$

In other words, the NOR operation is equivalent to ANDing the complements of the
inputs. This can be proved by directly comparing the truth tables for the two expres-
sions $\overline{(A + B)}$ and $\overline{A} \cdot \overline{B}$. It is left as an exercise that is well worth attempting!
This DeMorgan relation can be extended to an arbitrary number of variables. For
example, applying the theorem to a NOR3 gate with inputs of A, B, and C gives the
relation

$$\overline{(A + B + C)} = \overline{A} \cdot \overline{B} \cdot \overline{C} \tag{2.33}$$

and so on.

It is often useful to visualize theorems graphically by means of logic diagrams. Figure 2.17 illustrates the NOR3 DeMorgan equivalence. Figure 2.17(a)
shows the standard NOR gate symbol, whereas Figure 2.17(b) provides the equivalent network using three inverters (NOT) and one AND gate. Tracing the direction of the logic flow shows the formation of the complements, followed by the
ANDing operation.

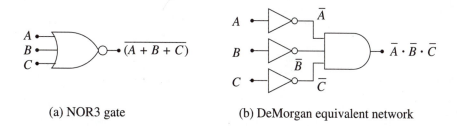

(a) NOR3 gate

(b) DeMorgan equivalent network

Figure 2.17   DeMorgan equivalence

The DeMorgan relations arise so often that it is convenient to introduce simplified symbols to represent them. Recalling that the "bubble" denotes inversion in these diagrams gives rise to the reduction shown in Figure 2.18.

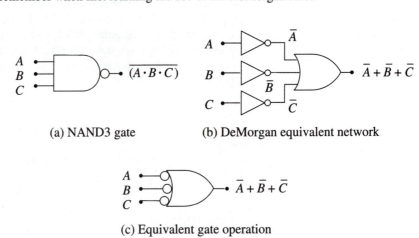

A
B $\overline{(A + B + C)}$
C

A
B $\overline{A} \cdot \overline{B} \cdot \overline{C}$
C

(a) NOR3 gate   (b) Equivalent gate operation

Figure 2.18   DeMorgan NOR equivalent networks

Now, suppose that we have a NAND2 gate with inputs A and B. The second DeMorgan theorem states that

$$\overline{(A \cdot B)} = \overline{A} + \overline{B} \tag{2.34}$$

In this case, the NAND operation is equivalent to ORing the complements of the inputs. This result can be verified by comparing $(A \cdot B)$ with $\overline{A} + \overline{B}$ for every combination of A and B.

For a NAND3 gate with inputs of A, B, and C, the theorem reads as

$$\overline{(A \cdot B \cdot C)} = \overline{A} + \overline{B} + \overline{C}, \tag{2.35}$$

and the theorem can be extended to an arbitrary number of inputs. The logic gate diagrams in Figure 2.19 illustrate this identity in a graphical manner. Figure 2.19(a) and (b) are identical by the DeMorgan theorem, and Figure 2.19(c) gives the simplified equivalent gate operation. Reductions of this type are often easier to remember when first learning the use of the DeMorgan rules.

A
B $\overline{(A \cdot B \cdot C)}$
C

A $\overline{A}$
B $\overline{A} + \overline{B} + \overline{C}$
       $\overline{B}$
C $\overline{C}$

(a) NAND3 gate   (b) DeMorgan equivalent network

A
B $\overline{A} + \overline{B} + \overline{C}$
C

(c) Equivalent gate operation

Figure 2.19   A 3-variable DeMorgan example

Note that we may also perform reductions when input variables are already complemented. For example,

$$\overline{(x \cdot y \cdot z)} = \bar{x} + y + \bar{z} \tag{2.36}$$

where we have used the fact that

$$\overline{(\bar{y})} = y \tag{2.37}$$

in the reduction.

2.6 Useful Boolean Identities

To complete our discussion of basic binary relationships, we note that several useful Boolean identities can be obtained using the relations discussed above.

The first set that we will study is

$$A + A \cdot B = A$$
$$A + \bar{A} \cdot B = A + B \tag{2.38}$$

The first identity is referred to as **redundancy** and shows that the value of B does not affect the result. The second is equally important in performing algebraic manipulations.

Reductions can be verified in different ways, from creating a truth table to taking a purely algebraic approach using the theorems. Application of algebra is the preferred technique. However, it is useful to see how the truth table technique works by proving the relations. Figure 2.20 shows the setup of the table. With two inputs A and B, there are $2^2 = 4$ different combinations for the pair of variables. For each possibility, we find $A \cdot B$ and then compute the value of $A + A \cdot B$. Comparing this result with the input values shows that the function is always equal to the value of A, regardless of the value of B, which confirms the identity.

A	B	$A \cdot B$	$A + A \cdot B$
0	0	0	0
0	1	0	0
1	0	0	1
1	1	1	1

Figure 2.20
The output $A + A \cdot B$ always has the same value as A.

Another approach to proving the relationship is to use substitution of values for one of the variables. Consider the expression

$$A + A \cdot B \tag{2.39}$$

for the two cases $B = 0$ and $B = 1$. If $B = 0$, then $A \cdot B = 0$, since anything ANDed with 0 gives 0. Then we are left with

$$A + 0 = A \tag{2.40}$$

as expected. If $B = 1$, then $A \cdot B = A$, and the expression reduces to

$$A + A \cdot B = A + A = A \tag{2.41}$$

which again confirms the validity of the identity.

The best way to prove the identity is to use Boolean identities. The following steps illustrate the sequence for the first relation.

$$
\begin{aligned}
A + A \cdot B &= A \cdot (1 + B) \\
&= A \cdot (1) \\
&= A
\end{aligned}
\tag{2.42}
$$

The first line uses the distributive law to factor A out of the two terms. Noting that $(1 + B) = 1$ gives the second line, and the identity $A \cdot 1 = A$ completes the proof.

2.7 Algebraic Reductions

We are now in a position to study how Boolean algebra can be applied to the design of digital switching systems.

One common problem in combinational logic design is reduction of a logic expression to the "simplest" possible form; "simplest" usually means that we want to implement the function using the smallest number of gates. The reduction is accomplished by applying the basic identities in a step-by-step manner. The basic rules are summarized in Figure 2.21 to aid in this task until you commit them to memory.

OR Identities	AND Identities
$A + 0 = A$	$A \cdot 0 = 0$
$A + 1 = 1$	$A \cdot 1 = A$
$A + A = A$	$A \cdot A = A$
$A + \overline{A} = 1$	$A \cdot \overline{A} = 0$
$\overline{\overline{A}} = A$	
$A + B = B + A$	$A \cdot B = B \cdot A$
$A + (B + C) = (A + B) + C$	$A \cdot (B \cdot C) = (A \cdot B) \cdot C$
$A \cdot (B + C) = A \cdot B + A \cdot C$	$A + (B \cdot C) = (A + B) \cdot (A + C)$
$\overline{(A + B)} = \overline{A} \cdot \overline{B}$	$\overline{(A \cdot B)} = \overline{A} + \overline{B}$
$A + A \cdot B = A$	$A + \overline{A} \cdot B = A + B$

Figure 2.21 Summary of Boolean relations

The technique for reducing logic expressions is straightforward: just apply the appropriate identity and simplify! However, the easiest way to illustrate the idea is by examples.

Example 2-1

Let us reduce the equation

$$f = A \cdot B + A \cdot \bar{B} \tag{2.43}$$

by first using the distributive law to group the B-terms together such that

$$f = A \cdot (B + \bar{B}) \tag{2.44}$$

Then, because the expression in parentheses is always 1, we have

$$f = A \cdot 1 = A \tag{2.45}$$

as the simplest form. Note that the original expression contained two AND functions, one OR operation, and one NOT, whereas the algebra shows that the entire equation can be replaced just by A.

Example 2-2

The function

$$F = A \cdot B \cdot C + B \cdot C \tag{2.46}$$

requires one AND3 gate, one AND2 gate, and one OR2 gate. However, if we group the factors $B \cdot C$, then we may write

$$\begin{aligned} F &= A \cdot (B \cdot C) + (B \cdot C) \\ &= [A + 1](B \cdot C) \\ &= B \cdot C \end{aligned} \tag{2.47}$$

where the second line follows from noting that A is redundant. With this reduction, we see that F can be expressed as a single AND operation.

Example 2-3

Suppose that we are given the function

$$g = \overline{(a + \bar{b} + c) + (b + \bar{c})} \tag{2.48}$$

Applying DeMorgan's rule for the NOR operation gives

$$g = \overline{(a + \bar{b} + c)} \cdot \overline{(b + \bar{c})} \tag{2.49}$$

which can be rewritten, using the other DeMorgan relation, to

$$\begin{aligned} g &= (\bar{a} \cdot b \cdot \bar{c}) \cdot (\bar{b} \cdot c) \\ &= 0 \end{aligned} \tag{2.50}$$

so that $g = 0$ since the identity $X \cdot \overline{X} = 0$ applies to both the b and c variables. This means that g is always zero regardless of any variable; that is, g is a constant.

Example 2-4

Consider the function

$$h = (A + B + C) \cdot (A + B) \tag{2.51}$$

As written, this uses one OR2 gate, one OR3 gate, and one AND2 gate. To see if this simplifies, we first expand the function using the distributive law to write

$$h = A \cdot A + A \cdot B + B \cdot A + B \cdot B + C \cdot A + B \cdot C \tag{2.52}$$

Since $X \cdot X = X$ and $Y + Y = Y$, we can remove the superfluous terms to arrive at

$$h = A + A \cdot B + B + A \cdot C + B \cdot C \tag{2.53}$$

We can reduce the first three terms by noting that

$$(A + A \cdot B) + B = (A + B) \tag{2.54}$$

Thus

$$
\begin{aligned}
h &= A + B + A \cdot C + B \cdot C \\
&= (A + A \cdot C) + (B + B \cdot C) \\
&= (A + C) + (B + C) \\
&= A + B + C
\end{aligned} \tag{2.55}
$$

This shows that h can be implemented by using a single OR3 gate, which is much simpler than what is required in the original expression.

Example 2-5

Consider the function

$$T = A \cdot B + A \cdot B \cdot C + A \cdot B \cdot \overline{C} \tag{2.56}$$

The last two terms can be grouped together to eliminate C, allowing the following reduction:

$$
\begin{aligned}
T &= A \cdot B + A \cdot B \cdot (C + \overline{C}) \\
&= A \cdot B + A \cdot B \\
&= (A \cdot B)
\end{aligned} \tag{2.57}
$$

which shows the simplification possible.

Example 2-6

Suppose that we have

$$V = 1 + x \cdot (x \cdot \bar{y} + y \cdot \bar{z} + y \cdot z) + x \cdot \bar{y} \cdot z + g(x, y, z) \tag{2.58}$$

This automatically reduces to $V = 1$ because of the first term.

Example 2-7

In the same manner,

$$H = [(x + y \cdot z + \bar{x} \cdot z) \cdot (a + \bar{b} + c)] \cdot 0$$

evaluates to

$$H = 0$$

because ANDing 0 with anything always results in 0.

2.8 Complete Logic Sets

We have seen that complex logic functions can be created by combining basic logic operations such as NOT and AND. An important problem that we face in a realistic design environment arises when we want to choose a basic set of logic operations on which to design a system. A **complete set** of logic operations is one that allows us to create every possible logic function using only those in the set.

Let us examine this in the context of the NOT, AND, and OR operations. We can show that these operations can be used to create three complete logic sets:

- NOT, AND
- NOT, OR
- NOT, AND, and OR

Any group is sufficient to build up arbitrary logic functions, the first two being minimal sets. Note that the group

- AND, OR

is *not* a complete set. In other words, you cannot create arbitrary logic functions using only the AND and OR operations; the NOT must be included.

Example 2-8

An obvious example of a function that cannot be implemented using only the AND and OR operations is

$$f(a, b) = \bar{a} \cdot b \qquad \textbf{(2.59)}$$

This is a valid expression, but we must be able to produce $\bar{a}$ from a, which requires the NOT operation.

With the above discussion, it is easily seen that the following are both complete logic sets:

- NAND
- NOR

Thus, for example, any logic function can be created using only NAND gates. This consideration is important when we examine the electronic circuits that are used to create practical digital networks.

2.8.1 NAND-Based Logic

Let us examine the idea that NAND gates form a complete set. The key to this concept is the DeMorgan relation

$$\overline{A \cdot B} = \bar{A} + \bar{B} \tag{2.60}$$

which shows that the NAND can be expressed using the OR operation with complemented inputs. Note in particular that if $B = A$, then this becomes

$$\overline{A \cdot A} = \bar{A} \tag{2.61}$$

so that we may create the NOT operation by tying the inputs together as in Figure 2.22.

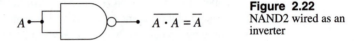

Figure 2.22
NAND2 wired as an inverter

Although there are different approaches to NAND-based logic, one simple approach is to start with the desired function and then apply DeMorgan rules to eliminate any OR operations. For example, suppose that we want to use NAND gates to implement the function

$$f = x + \bar{y} \tag{2.62}$$

Taking the complement twice and applying the DeMorgan rule give

$$\begin{aligned} f &= \overline{\overline{x + \bar{y}}} \\ &= \overline{\bar{x} \cdot y} \end{aligned} \tag{2.63}$$

This can be built using logic diagrams as shown in Figure 2.23.

Figure 2.23 NAND logic example

2.8.2 NOR-Based Logic

NOR-gate logic proceeds in the same manner. It is based on the DeMorgan relation

$$\overline{A + B} = \bar{A} \cdot \bar{B} \tag{2.64}$$

which shows that the NOR can be viewed as the AND operation between the complemented variables. Since

$$\overline{A + A} = \bar{A} \cdot \bar{A} = \bar{A} \tag{2.65}$$

the NOT function can be obtained by tying the NOR inputs together as in Figure 2.24; this is analogous to the approach used in NAND-based logic.

$$A \quad \overline{A + A} = \overline{A}$$

Figure 2.24
NOR2 wired as an inverter

As a simple example, suppose that we want to construct the function

$$g = a \cdot b + c \tag{2.66}$$

using only NOR gates. To cast this into an appropriate form, we apply DeMorgan reductions of

$$
\begin{aligned}
g &= \overline{\overline{a \cdot b + c}} \\
&= \overline{\overline{(\bar{a} + \bar{b}) \cdot \bar{c}}} \tag{2.67} \\
&= \overline{\overline{(\bar{a} + \bar{b})} + c}
\end{aligned}
$$

which expresses g by using only NOT and NOR operations. The equivalent logic diagram is shown in Figure 2.25. Since the reduction has given an OR operation, we first form

$$\bar{g} = \overline{\overline{(\bar{a} + \bar{b})} + c} \tag{2.68}$$

and then complement to arrive at g.

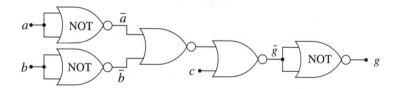

Figure 2.25 NOR-gate logic example

2.9 IEEE Logic Gate Symbols

The logic gate symbols introduced in this chapter and used throughout most of the book are called **shape-specific** symbols, since the operations are distinguished from each other by the different geometrical shapes. Other symbols can be used to construct logic diagrams so long as they are defined. The IEEE standard symbols are very common in practice and are summarized in this section.

In IEEE standard symbols, all gates are represented by rectangular boxes. The operation for each gate is specified by symbols and notation within the box itself. The basic operations are summarized in Figure 2.26. In general, the notation "1" means multiplication by 1, "≥1" defines the OR operation, and the ampersand "&" is used to denote the AND operation. Complements are obtained by placing a bubble (NOT) at the output, changing the first-row Buffer, OR, and AND gates into the second-row NOT, NOR, and NAND gates.

It is worthwhile to learn IEEE notation as this will allow you to read and understand almost every logic diagram you come across. In our treatment, we will maintain our use of shape-specific symbols for the simple gates but will use IEEE notation for more complex functions where it becomes very convenient.[2]

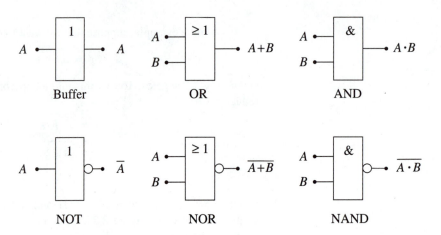

Figure 2.26 IEEE standard logic gate symbols

2.10 Problems

In the problems below, assume that all inputs are the uncomplemented variables.

[2.1] Verify that $\overline{(\overline{x})} = x$ by using a simple cascade of two NOT gates in a logic diagram.

[2.2] Construct the logic symbol for an OR3 gate and give the truth table.

[2.3] Construct the logic symbol for an AND3 gate and give the truth table.

[2.4] Draw the logic network that implements the function

$$f = \overline{a + \overline{b}}$$

with inputs a and b using individual OR and NOT gates. Then construct the function table, listing each possible entry and the output value.

[2.5] Draw the logic network for the function

$$H(x, y, z) = x \cdot y + x \cdot z$$

---

[2] IEEE is an acronym that stands for the Institute of Electrical and Electronics Engineers, a professional organization for electrical engineers.

Can this function be obtained using fewer gates? If so, do the simplification and draw the simpler logic network.

[2.6] Construct the logic network for the function

$$w(a, b, c) = a \cdot b + b \cdot c$$

Can this be simplified to a smaller number of gates? If so, show the steps and the final result.

[2.7] Draw the logic network for the function

$$g(A, B, C) = A \cdot \bar{B} + C \cdot \bar{A}$$

if A, B, and C are the inputs. Use AND, OR, and NOT gates in your circuit.

[2.8] Determine the function f for the logic network below.

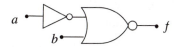

[2.9] Find the expression for g in the logic diagram below.

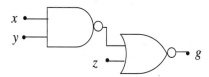

[2.10] Find the expressions for f_1 and f_2 in the logic diagram shown.

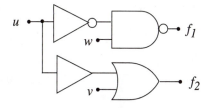

[2.11] Construct the 2-variable truth table for the function

$$F = \bar{A} \cdot B + A \cdot \bar{B} + \bar{A} \cdot \bar{B}$$

Study your results. Is it possible to express F using a simpler expression?

[2.12] Construct the truth table for the following expression.

$$T(a, b, c) = a \cdot b + \bar{b} \cdot (a + c)$$

[2.13] Consider the function

$$h(x, y, x) = (x + y) \cdot (\bar{x} + z) \cdot (y + z)$$

Construct the logic diagram for this function.

[2.14] Consider a 4-input NAND gate with inputs a, b, c, and d. Draw the logic symbol, and then describe the operation of the gate with a statement in the form

> If (Condition)
>
> then
>
>> (Result)
>
> else
>
>> (Other result)

[2.15] Consider a 4-input NOR gate with inputs x, y, z, and w. Draw the logic symbol, and then describe the operation of the gate with a statement in the form

> If (Condition)
>
> then
>
>> (Result)
>
> else
>
>> (Other result)

[2.16] Consider a 3-input AND gate with independent inputs a and b and a clock signal ϕ applied to the third input. The clock signal ϕ is periodic in time as shown. This is an example of a clock-controlled gate that is used to help move data around a complex system in an orderly manner.

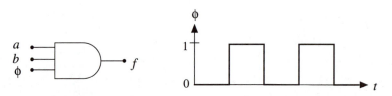

Write a description of the gate operation using the following as a starting point:

> If $\phi = 1$
>
> then
>
>> (Result)
>
> else if
>
>> $\phi = 0$
>
> then
>
>> (Result)

What does this tell you about the "important" values of the independent inputs a and b relative to the clocking signal?

[2.17] Construct the truth table for the function

$$Q = x \cdot y + z$$

[2.18] Construct the truth table for the function

$$h = a \cdot b + \bar{a} \cdot b$$

Can this be reduced to a simpler form? If so, state the property that makes the simplification possible, and find the reduced expression.

[2.19] Construct the truth table for the function

$$F = \bar{A} \cdot B + A \cdot \bar{B} \cdot C$$

Can this be reduced to a simpler form? If so, state the property that makes the simplification possible, and find the reduced expression.

[2.20] Find the simplest form for the function G that is created by the following logic network.

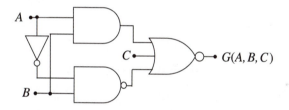

[2.21] Find the simplest form for the function F that is created by the following logic network.

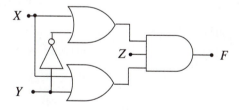

[2.22] Find the simplest form for the function Z defined by the following logic network.

[2.23] Consider the function

$$X = \bar{A} \cdot B + A \cdot \bar{B}$$

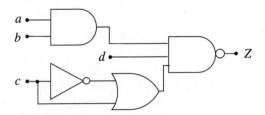

Create the truth table for this expression, and compare it to the OR operation. Do the same with the function

$$Y = A \cdot B + \bar{A} \cdot \bar{B}$$

What is the relationship between X and Y?

[2.24] Reduce the function

$$f = a \cdot b + a \cdot b \cdot \bar{c} + a \cdot b \cdot c$$

to simplest form.

[2.25] Simplify the function

$$g = x + y + \bar{x} \cdot y + \bar{y}$$

using Boolean reductions.

[2.26] Reduce the expression

$$J = \bar{A} \cdot B + A \cdot \bar{B} + A \cdot B + A$$

to simplest form using algebraic identities.

[2.27] Simplify the function

$$K = X + Y \cdot Z + \bar{X} \cdot Y + \bar{X} \cdot \bar{Y}$$

using algebraic identities.

[2.28] Construct a logic network for the function

$$W(A, B, C, D) = \overline{A \cdot B} \cdot \overline{C \cdot D}$$

using only NOR gates.

[2.29] Construct a logic network for the function

$$H = \overline{(x \cdot y)} \cdot \bar{z}$$

using only NOR gates

[2.30] Construct the function

$$X(a, b, c) = (\bar{a} + \bar{b}) \cdot c$$

using only NAND gates.

[2.31] Construct a logic network for the function

$$d = (a + b) \cdot \bar{b}$$

using only NAND gates.

[2.32] Construct a logic diagram for the function

$$R = (x + y) \cdot (z + w)$$

using IEEE logic symbols.

[2.33] Construct a logic diagram for the function

$$f = \bar{a} \cdot \overline{(b + c)}$$

using IEEE logic symbols.

CHAPTER 3

$A \cdot B$

Combinational Logic Design

Logic design deals with the creation of a digital network that performs a particular task. In terms of Boolean logic, this is equivalent to implementing a particular function. Digital logic design can be summarized as consisting of the following steps:

- Specification of the problem.
- Creation of the logic network that implements the logic functions.
- Building of the network.
- Testing and verification of the logic circuit.

In this chapter we will examine the foundations of logic design. This will be applied in later chapters when we discuss the design of large systems.

3.1 Specifying the Problem

Proper specification of the problem is critical since it provides the basis for the entire design process. There are several ways to specify the function that is needed. Depending on how the information is presented, some approaches may be more useful than others.

Combinational logic deals with networks that use logic gates to combine the input variables as needed to produce logic functions. In a combinational circuit, the value of the output is determined by the current value of the inputs. If any of the inputs are changed, then the value of the output may change as specified by the function. To design a combinational logic network, we usually start with a specified set of input variables that are to produce one or more outputs. In the context of Figure 3.1, this means that our program is to find the logic network inside of the box in terms of basic logic operations (NOT, OR, AND) or a set of logic gates.

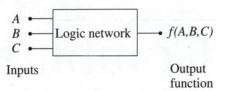

Inputs

Output
function

Figure 3.1 Generic logic block symbol

Problem specification deals with the approach that we use to state the problem. We must put it in a form that provides us with the information needed to create a digital logic network solution. Since we do not live in a binary world, it is often necessary to start with a nondigital basis to which we introduce an encoding scheme. Once the encoding is chosen, the problem is in the digital domain and our work can proceed from that point. If, on the other hand, we are already working with binary numbers, then we can specify the problem using 0s and 1s.

Let us assume for now that we have already worked our way down into the binary number system and that we are interested in building a digital logic network in the world of 0s and 1s. We will examine how to interface to the real world later on in the book. There are two primary techniques that we can use to specify the design problem.

Function Tables

A function table is just a truth table that has been extended to an arbitrary function. An example is shown in Figure 3.2 for the case of three input variables A, B, and C, which give the output function $f(A, B, C)$. There are $2^3 = 8$ possible binary combinations, so a complete specification of the problem requires that we have eight entries. For each entry, the value of output is either $f = 0$ or $f = 1$.

Inputs			Output
A	B	C	f
0	0	0	0
0	0	1	1
0	1	0	1
0	1	1	0
1	0	0	1
1	0	1	0
1	1	0	0
1	1	1	1

Figure 3.2
Problem specification using a
function table

Boolean Expressions

It is often possible to specify the problem by using one or more Boolean equations as the starting point. Once we have the logic expressions, we can proceed to apply the laws of Boolean algebra to manipulate the functions as needed. Gates can be chosen, and the logic network can be constructed directly from the equations.

It is important to note that, for a specific problem, the two approaches are completely equivalent. They are simply different ways of viewing the same information. Logic designers use both types of problem specifications as starting points. As we will learn in this chapter, the statement of the problem is often given in general terms that must be translated into an efficient binary design.

3.2 Canonical Logic Forms

Structured logic is based on the ability to write Boolean equations in a manner that uses various types of regular and repeated forms. This is often a useful starting point in the analysis as structured equations provide a uniform view of the problem specifications. In modern technology, the structured equations themselves are sufficient to create the logic network in certain types of electronic circuits.

Two types of structured forms are especially useful in logic design. These are known as the **sum-of-products** (**SOP**) and **product-of-sums** (**POS**) forms, respectively, and both are discussed below. Their utility arises from the fact that *any* logic function can be expressed in SOP or POS form. Once we understand how to create and deal with these types of equations, then the results can be used to build any combinational logic network.

3.2.1 Sum-of-Products (SOP) Form

A sum-of-products expression consists of AND terms that are ORed together. The terminology SOP arises from viewing an AND operation such as $A \cdot B$ as resembling a product (from multiplication) and viewing the OR operation $(X + Y)$ as looking like a sum (from addition). Equivalently, an SOP equation is said to have AO (AND/OR) form. For a function to be in **canonical SOP** structure, every variable must appear in each term in either normal or complemented form. Otherwise, the function is simply in SOP form.

Example 3-1

Suppose that we have the variables A, B, and C. Then the following functions are in canonical SOP form:

$$
\begin{aligned}
f &= A \cdot B \cdot C + \overline{A} \cdot B \cdot C + A \cdot \overline{B} \cdot C + A \cdot B \cdot \overline{C} \\
g &= \overline{A} \cdot \overline{B} \cdot \overline{C} + \overline{A} \cdot B \cdot \overline{C} + A \cdot \overline{B} \cdot \overline{C}
\end{aligned}
\tag{3.1}
$$

The canonical structure is due to the fact that every term has factors of A, B, and C in it.

Example 3-2

Consider the Boolean expression $F(A, B, C)$ written as

$$
F = A \cdot B + \overline{A} \cdot C + B \cdot \overline{C}
\tag{3.2}
$$

This is in sum-of-products form, but it does not have canonical structure because each term has only two of the three possible variables.

Example 3-3

Consider the expression

$$G(a, b, c) = a \cdot \bar{b} \cdot c + \bar{a} \cdot b \cdot c + a \cdot c \qquad (3.3)$$

Even through two of the three terms satisfy the criteria for canonical form, the last term only has a and c in it, so that G is not in canonical form.

Formation of a Canonical SOP Function

Consider the function

$$h(x, y, z) = x \cdot y + y \cdot z \qquad (3.4)$$

Although this has SOP structure, it is not classified as being in canonical SOP form since the first term is missing a z entry, and the second term has no factor of x. This can be cast into canonical form by using the identities

$$\begin{aligned} (x + \bar{x}) &= 1 \\ (y + \bar{y}) &= 1 \end{aligned} \qquad (3.5)$$

and distributing the terms. To see how this is accomplished, let us write explicitly

$$\begin{aligned} h(x, y, z) &= x \cdot y \cdot 1 + 1 \cdot y \cdot z \\ &= x \cdot y \cdot (z + \bar{z}) + (x + \bar{x}) \cdot y \cdot z \\ &= x \cdot y \cdot z + x \cdot y \cdot \bar{z} + x \cdot y \cdot z + \bar{x} \cdot y \cdot z \\ &= (x \cdot y \cdot z) + (x \cdot y \cdot \bar{z}) + (\bar{x} \cdot y \cdot z) \end{aligned} \qquad (3.6)$$

The final form has the desired characteristic of canonical SOP form; the parentheses are not really needed but have been added to emphasize the structure of the terms.

This technique can be applied to any SOP expression to work it into canonical form. Note, however, that casting an equation into canonical form usually increases the complexity and adds extra terms. The canonical form will often require a larger number of gates in the final circuit than if the original expression were used. The tradeoff is a more structured approach to designing large, complex logic networks. The choice of approach depends on the type of electronic circuit that actually will be used to build the final network.

3.2.2 Product-of-Sums (POS) Form

This type of expression consists of OR terms that are ANDed together. As with the SOP case, every variable must appear in each term in either normal or complemented form. A POS function implements OA (OR/AND) logic.

Example 3-4

The 2-input function

$$f(x, y) = (\bar{x} + y) \cdot (x + \bar{y}) \qquad (3.7)$$

is in canonical POS form since each term contains both x and y factors. An example of a noncanonical POS expression is

$$g = x \cdot (x + \bar{y}) \tag{3.8}$$

This is designated as having POS structure by treating the variable x as the OR term $x = (x + 0)$.

Example 3-5

Consider the 3-variable equation

$$h = (A + B + C) \cdot (\bar{A} + B + C) \cdot (\bar{A} + \bar{B} + C). \tag{3.9}$$

This is in canonical POS form; it satisfies all of the requirements.

Example 3-6

The function

$$h(A, B, C) = (A + \bar{C}) \cdot (B + \bar{C}) \cdot (\bar{A} + B + \bar{C}) \tag{3.10}$$

is in POS form, but it does not have canonical structure since the first and second factors do not contain all three variables.

3.3 Extracting Canonical Forms

Let us now consider the very important situation wherein a problem is specified by a function table, and we would like to extract a Boolean expression from the data. This is easily accomplished by applying the relationship between each entry and structured logic forms.

Let us investigate this relationship by means of a specific case where we start with the function table shown in Figure 3.3. The input variables are labeled A, B, C, and the output function is $f(A, B, C)$. Our first point of observation is that since f is a binary number, it has values of only 0 or 1. This seemingly trivial comment gains more importance when we recall from Chapter 2 that several special identities include 0 or 1 explicitly. These can be used to help us write a logic expression that contains the same information as that presented in the table.

Inputs			Output	
A	B	C	f	
0	0	0	0	
0	0	1	1	← $\bar{A}\bar{B}C = 1$
0	1	0	1	← $\bar{A}B\bar{C} = 1$
0	1	1	0	
1	0	0	1	← $A\bar{B}\bar{C} = 1$
1	0	1	0	
1	1	0	0	
1	1	1	1	← $ABC = 1$

Figure 3.3
Identification of SOP terms using a function table

Let us now proceed to extract the canonical SOP form from the function table. This is accomplished by first identifying the input combinations that yield an output of $f = 1$, and then remembering the identity

$$1 + Anything = 1 \tag{3.11}$$

where *Anything* means any binary variable or group of variables. Logically, the requirement $f = 1$ means that one of the entries marked with an arrow must evaluate to a value of 1. There are four possible combinations that give this result, so the SOP form for f will have four terms and can be decomposed into four separate functions

$$f = f_1 + f_2 + f_3 + f_4 \tag{3.12}$$

such that if any term is 1, then $f = 1$. Each term will then be constructed as a 3-variable AND group that can be obtained from the input values listed.

Consider the first occurrence of $f = 1$, which is the second entry in the table. This says that

If $(A = 0)$ AND $(B = 0)$ AND $(C = 1)$

then $f = 1$

Let us then construct the AND term

$$f_1 = \bar{A} \cdot \bar{B} \cdot C \tag{3.13}$$

where we have associated 0 entries using complements $\bar{A}$ and $\bar{B}$ and the 1 entry using simply C. By inspection, this gives $f_1 = 1$ for the specified input values.

Using this example as a basis, we can formulate a general approach to constructing every SOP term needed in the complete expression. For a given input combination, we create an AND term that uses every variable. If the specified input value is a 1, then the variable appears in its normal form; if the input is a 0, then we use the complemented form.

To see how this works, consider the second case where $f = 1$ (which is the third entry in the table). This term is described by the statement

If $(A = 0)$ AND $(B = 1)$ AND $(C = 0)$,

then $f = 1$

leading us to construct

$$f_2 = \bar{A} \cdot B \cdot \bar{C} \tag{3.14}$$

to represent this term. Similarly, the remaining terms are given by

$$f_3 = A \cdot \bar{B} \cdot \bar{C}$$
$$f_4 = A \cdot B \cdot C \tag{3.15}$$

which correspond to the input combinations 100 and 111, respectively. Because all of these product (AND) terms give $f = 1$, the complete expression is obtained by taking the sum (OR), giving

$$f = \bar{A} \cdot \bar{B} \cdot C + \bar{A} \cdot B \cdot \bar{C} + A \cdot \bar{B} \cdot \bar{C} + A \cdot B \cdot C \tag{3.16}$$

which is in canonical SOP form.

Example 3-7

Let us apply this technique to obtain the SOP function corresponding to the information in the table below (Figure 3.4). By inspection, there are four AND terms, giving

$$g = \overline{A} \cdot \overline{B} \cdot \overline{C} + \overline{A} \cdot \overline{B} \cdot C + A \cdot B \cdot \overline{C} + A \cdot B \cdot C \qquad (3.17)$$

as the required expression.

A	B	C	g
0	0	0	1
0	0	1	1
0	1	0	0
0	1	1	0
1	0	0	0
1	0	1	0
1	1	0	1
1	1	1	1

Figure 3.4
Example of SOP formation using a function table

3.3.1 Minterms and Maxterms

In SOP expressions, a function is created by summing AND terms. Once we are given the number and names of the variables, every term has the same form. This principle can be formalized by introducing the concept of a **minterm**, which is a term that has every variable in regular or complemented form ANDed together.

To understand the concept of a minterm, let us study the case where the variables A, B, C are used. A typical minterm is $\overline{A} \cdot B \cdot \overline{C}$, which corresponds to a binary grouping of 010. To simplify our notation, note that 010 corresponds to a decimal value of 2. We then define the minterm m_2 by

$$m_2 = \overline{A} \cdot B \cdot \overline{C} \qquad (3.18)$$

Extending this notation to all possible cases gives

$$
\begin{array}{ll}
m_0 = \overline{A} \cdot \overline{B} \cdot \overline{C} & m_4 = A \cdot \overline{B} \cdot \overline{C} \\
m_1 = \overline{A} \cdot \overline{B} \cdot C & m_5 = A \cdot \overline{B} \cdot C \\
m_2 = \overline{A} \cdot B \cdot \overline{C} & m_6 = A \cdot B \cdot \overline{C} \\
m_3 = \overline{A} \cdot B \cdot C & m_7 = A \cdot B \cdot C
\end{array}
\qquad (3.19)
$$

These are easy to remember since each subscript corresponds to the decimal value of the binary word.

Minterm notation can be used to simplify the appearance of SOP expressions. Consider the function

$$f = \overline{A} \cdot \overline{B} \cdot \overline{C} + \overline{A} \cdot B \cdot C + A \cdot \overline{B} \cdot \overline{C} + A \cdot B \cdot C \qquad (3.20)$$

This can be written as a summation over minterms in the form

$$f = m_0 + m_3 + m_4 + m_7$$
$$= \sum m(0, 3, 4, 7) \tag{3.21}$$

where the summation sign (Σ) indicates an ORing of the designated minterms listed in parentheses. The shorthand notation provides a compact way to express SOP functions.

Example 3-8

Let us expand the function

$$g(a, b, c) = \sum m(1, 2, 5) \tag{3.22}$$

by writing

$$g(a, b, c) = m_1 + m_2 + m_5 \tag{3.23}$$

To obtain the proper representation in terms of a, b, and c, we note the decimal-to-binary equivalents

$$1_{10} \to 001$$
$$2_{10} \to 010 \tag{3.24}$$
$$5_{10} \to 101$$

Thus,

$$g = \bar{a} \cdot \bar{b} \cdot c + \bar{a} \cdot b \cdot \bar{c} + a \cdot \bar{b} \cdot c \tag{3.25}$$

is the same function in expanded form.

When we are working with POS functions, it is convenient to introduce the concept of a **maxterm**. A maxterm contains each possible variable ORed with every other. For example, $(\overline{A} + B + \overline{C})$ is a maxterm for a 3-variable function. In general, a maxterm M_i is defined in terms of the corresponding minterm m_i by

$$M_i = \overline{m_i} \tag{3.26}$$

As an example, consider the 3-variable maxterm M_2. Since $m_2 = \overline{A} \cdot B \cdot \overline{C}$ (corresponding to 010), the maxterm is computed as

$$M_2 = \overline{m_2}$$
$$= \overline{(\overline{A} \cdot B \cdot \overline{C})} \tag{3.27}$$
$$= (A + \bar{B} + C)$$

where we have used the DeMorgan reduction in arriving at the final form. A complete listing of 3-variable maxterms follows.

$$M_0 = \bar{m}_0 = A + B + C \qquad M_4 = \bar{m}_4 = \bar{A} + B + C$$

$$M_1 = \bar{m}_1 = A + B + \bar{C} \qquad M_5 = \bar{m}_5 = \bar{A} + B + \bar{C}$$

$$M_2 = \bar{m}_2 = A + \bar{B} + C \qquad M_6 = \bar{m}_6 = \bar{A} + \bar{B} + C \tag{3.28}$$

$$M_3 = \bar{m}_3 = A + \bar{B} + \bar{C} \qquad M_7 = \bar{m}_7 = \bar{A} + \bar{B} + \bar{C}$$

A POS function may be written in simplified form using maxterms. For example,

$$
\begin{aligned}
g &= (A + B + \bar{C}) \cdot (A + \bar{B} + \bar{C}) \cdot (\bar{A} + \bar{B} + \bar{C}) \\
&= M_1 \cdot M_3 \cdot M_7 \\
&= \prod M(1, 3, 7)
\end{aligned}
\tag{3.29}
$$

where the product sign ($\prod$) is used to denote ANDing among the listed maxterms.

3.3.2 Properties of SOP and POS Forms

Now let us examine a few important properties of SOP and POS functions. Consider the minterm SOP equation

$$
\begin{aligned}
F &= m_1 + m_2 + m_5 + m_6 \\
&= \sum m(1, 2, 5, 6)
\end{aligned}
\tag{3.30}
$$

as an example of a typical logic expression with canonical structure. This form provides us with enough information to construct the function table shown in Figure 3.5. Once we have listed all possible input combinations, we assign a value of $F = 1$ to lines that correspond to the minterms m_1, m_2, m_5, and m_6; these are shown with arrows. The remaining possibilities must have $F = 0$, completing the table.

A	B	C	F
0	0	0	0
0	0	1	1	←
0	1	0	1	←
0	1	1	0
1	0	0	0
1	0	1	1	←
1	1	0	1	←
1	1	1	0

Figure 3.5
Extracting a POS form from a function table

Now suppose that we want the POS form of the same function. By the definition of a maxterm, $M_1 = \bar{m}_1$, we should look for entries in the function table where the *minterm*s are 0; these entries correspond to maxterms that have a value of 1. From the function table we thus identify the maxterms

$$M_0 = \overline{m_0} \qquad M_3 = \overline{m_3} \qquad M_4 = \overline{m_4} \qquad M_7 = \overline{m_7} \tag{3.31}$$

as the important terms. In canonical POS form, all maxterms are ANDed together, so if one or more of the maxterms is 0, then the entire function is 0. This says that the POS form should look like

$$F = \prod M(0, 3, 4, 7) \tag{3.32}$$

where we simply use the maxterms corresponding to the entries where $F = 0$. Note that the maxterm list (0, 3, 4, 7) is simply the **complement** of the minterm list (1, 2, 5, 6). That is, the two sets give the complete listing 0 through 7.

Next, let us take the complement of F.

$$
\begin{aligned}
\overline{F} &= \overline{(m_1 + m_2 + m_5 + m_6)} \\
&= \overline{m_1} \cdot \overline{m_2} \cdot \overline{m_5} \cdot \overline{m_6} \\
&= M_1 \cdot M_2 \cdot M_5 \cdot M_6 \\
&= \prod M(1, 2, 5, 6)
\end{aligned}
\tag{3.33}
$$

This shows that $\overline{F}$ is simply the POS form where the maxterms have the same listing as the minterms in the SOP function, and vice versa.

3.4 The Exclusive-OR and Equivalence Operations

Let us apply the techniques discussed above to describe the exclusive-OR (**XOR**) operation. The truth table for the 2-input XOR (XOR2) operation is shown in Figure 3.6(a). By definition, the XOR2 gate provides an output that is identical to that of the OR2 gate except for the case where both inputs are 1; the XOR2 output is 0 in this case. The name "exclusive-OR" arises from this exception in that the output

The XOR function is assigned its own logic gate symbol.

A B	$A \oplus B$
0 0	0
0 1	1
1 0	1
1 1	0

(a) Function table

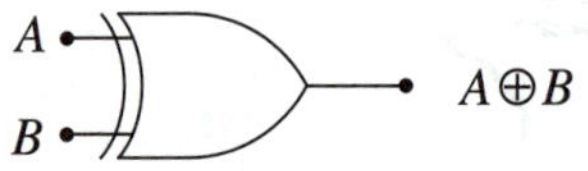

(b) XOR symbol

Figure 3.6 Exclusive-OR (XOR) gate

is a 1 when only a single input is 1 exclusively, that is, by itself.

The logical description of the gate can be extracted from the function table. The two cases that results in a 1 at the output are $AB = 01$ and $AB = 10$, so the function is given by

$$A \oplus B = A \cdot \overline{B} + \overline{A} \cdot B \tag{3.34}$$

via the SOP construction technique. The XOR operation is denoted by the symbol $\oplus$, and the expression is read "A o-plus B." The XOR operation occurs so frequently that it is assigned its own gate symbol as shown in Figure 3.6(b). This is just an OR gate that has been modified by adding another curved line on the input side.

The complement of the XOR function is the exclusive-NOR (**XNOR**) operation $\overline{A \oplus B}$. We can tell from the XNOR2 function table in Figure 3.7(a) that

$$\overline{A \oplus B} = A \cdot B + \overline{A} \cdot \overline{B} \tag{3.35}$$

by reading off the SOP terms where the output is a 1. This shows that

$$\overline{A \oplus B} = 1 \text{ iff } A = B$$

Because of this property, the XNOR is also referred to as the **equivalence function**. It is very useful in many practical applications that require the comparison

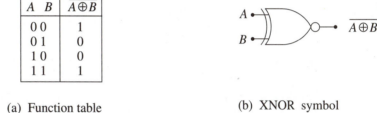

A B	$\overline{A \oplus B}$
0 0	1
0 1	0
1 0	0
1 1	1

(a) Function table

(b) XNOR symbol

Figure 3.7 The Exclusive-NOR (XNOR) or equivalence function

of two quantities. Figure 3.7(b) shows the XNOR2 logic symbol, which is simply the XOR gate with a complemented output. The IEEE symbols are shown in Figure 3.8. In this notation, the exclusive-OR operation is denoted by the greater-than sign > whereas the regular OR function is indicated by the greater-than-or-equal-to sign $\geq$. An XNOR symbol is obtained by simply adding an inversion bubble to the output.

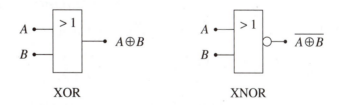

XOR

XNOR

Figure 3.8 IEEE standard symbols for XOR and XNOR gates

The 2-input XOR/XOR gates can be extended to 3-input and higher circuits by cascading the simpler 2-input gates. Figure 3.9(a) shows the function table for an XOR3 gate that has an output of

$$A \oplus B \oplus C \qquad (3.36)$$

Although we can write the SOP form directly, let us examine the table a little more carefully to obtain a qualitative description. For each row, you may verify that the table shows an output of

$$A \oplus B \oplus C = 0 \text{ if there are an even number of 1s at the input}$$

and

$$A \oplus B \oplus C = 1 \text{ if there are an odd number of 1s at the input}$$

The XOR gate is therefore called the **odd function** since it may be used to determine whether there are an odd number of 1s at the input. Figure 3.9(b) and (c) illustrate the evolution of the XOR3 symbol. An XNOR3 gate may be obtained in the

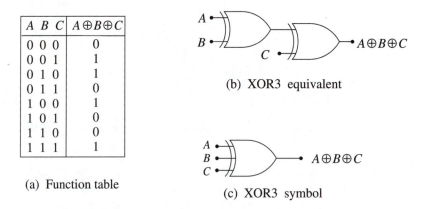

A B C	$A \oplus B \oplus C$
0 0 0	0
0 0 1	1
0 1 0	1
0 1 1	0
1 0 0	1
1 0 1	0
1 1 0	0
1 1 1	1

(a) Function table

(b) XOR3 equivalent

(c) XOR3 symbol

Figure 3.9 Characteristics of a 3-input XNOR logic gate

same manner and provides the **even function** where

$$\overline{A \oplus B \oplus C} = 1 \text{ if there are an even number of 1s at the input}$$

and

$$\overline{A \oplus B \oplus C} = 0 \text{ if there are an odd number of 1s at the input}$$

These arguments can be extended to build gates with a larger number of inputs.

3.5 Logic Arrays

Logic arrays are structured networks that can be configured to produce specific forms of logic expressions. For example, a logic array may be designed to accept inputs and produce specific minterms as outputs. These circuits are often viewed as a "sea of gates" that allows one to map logic forms directly from the equations to the circuit. In this section we will examine the concept of basic logic arrays and show how they can be used to implement canonical SOP and POS functions.

3.5.1 AND and OR Arrays

Logic arrays are classified according to either the function they perform or the form of the Boolean term they produce.

Let us begin by examining the structure of what we will term an **AND array**. By definition, this network accepts input variables and produces minterms at the output. An example of this type of logic array is shown in Figure 3.10. The inputs are A, B, and C, and inverters are used to provide the complements $\overline{A}$, $\overline{B}$, and $\overline{C}$ for use as inputs. The AND3 gates are wired such that each produces a distinct minterm:

$$\begin{aligned}
f_1 &= A \cdot B \cdot C = m_7 \\
f_2 &= A \cdot \overline{B} \cdot \overline{C} = m_4 \\
f_3 &= \overline{A} \cdot B \cdot \overline{C} = m_2 \\
f_4 &= \overline{A} \cdot \overline{B} \cdot \overline{C} = m_0
\end{aligned}$$

(3.37)

Additional minterms can be generated by simply adding AND gates to the existing array. The number of inputs can also be expanded as needed.

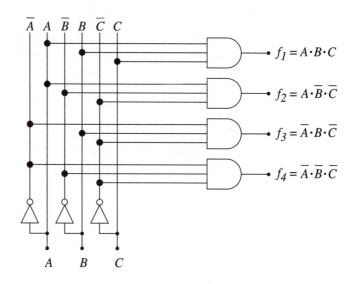

Figure 3.10 A programmed AND logic array

The important idea behind this example is the structured approach to drawing the network. In general, a logic array is created by using an array of gates that can be wired as needed. Figure 3.11 shows an example of an **uncommitted** logic array. An uncommitted array provides the general layout of the logic gates but does not

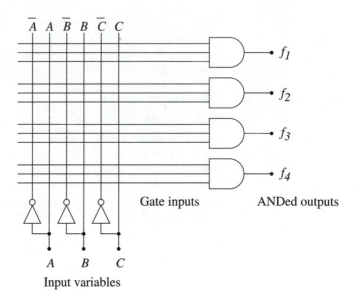

Figure 3.11 An uncommitted AND logic array

indicate any of the gate input connections. Once again, the important concept is the structure of the network. All input variables and their complements are provided as potential inputs. **Programming** of the array is accomplished by specifying the input connections. This feature gives rise to programmable logic arrays (PLAs), which are very useful in modern digital design.

An **OR array** produces maxterms and has the same general structure as the AND array. An uncommitted OR array is shown in Figure 3.12. The inputs are X, Y, and Z, and inverters are used to provide $\overline{X}$, $\overline{Y}$, and $\overline{Z}$. The ORed outputs are denoted by g_i with one output for each gate. Comparing this network to the AND array shows that the only difference is in the use of OR logic.

To program the array, we must specify the inputs to each gate by providing the appropriate connections. Figure 3.13 is an example of a programmed array. The inputs have been wired to produce the maxterms

$$g_1 = X + Y + Z = M_0$$
$$g_2 = \overline{X} + Y + \overline{Z} = M_5$$
$$g_3 = X + \overline{Y} + Z = M_2 \tag{3.38}$$
$$g_4 = \overline{X} + \overline{Y} + \overline{Z} = M_7$$

as can be verified by tracing the inputs to each gate. Note that the programmed wiring diagram is identical to the unprogrammed diagram, except that we have provided the connection dots. Using structured logic arrays of this type thus allows rapid design and implementation of a digital system.

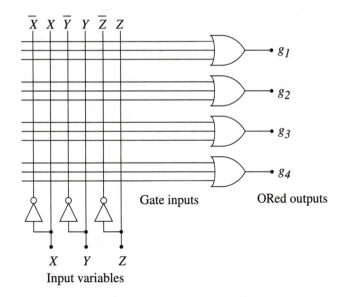

Figure 3.12 An uncommitted OR logic array

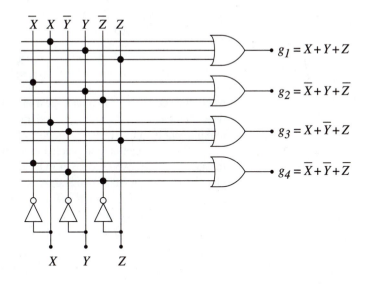

Figure 3.13 Example of a programmed OR array

3.5.2 SOP and POS Arrays

The AND and OR arrays above can be combined to create a structured approach to generating canonical logic forms.

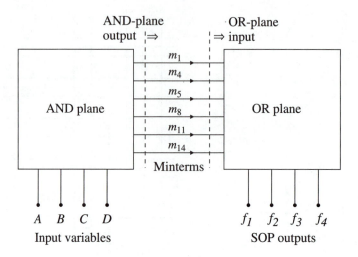

Figure 3.14 General structure of an AND-OR PLA

Consider first a sum-of-products expression. Since this is just the AND-OR logic pattern, we may cascade two logic arrays as shown in Figure 3.14 for SOP to produce SOP functions. The operation is straightforward. The AND-plane uses the input variables A, B, C, and D to produce the minterms m_i. The minterms act as inputs for the OR-plane logic, which results in outputs of the general form

$$f_n = \sum_i m_i \tag{3.39}$$

where the actual terms m_i that appear in f_n are specified by the OR gate wiring. This type of logic array is quite general in form. It can be modified to fit any SOP requirement by adjusting the numbers of input and output lines.

A POS array can be constructed in a similar manner. Since the product-of-sums format is just OR-AND logic, a POS logic array is obtained by cascading an OR-plane array into an AND-plane array as shown in Figure 3.15. The inputs are designated as A, B, and C, which are used (along with their complements) to produce maxterms M_j as OR-plane outputs. The maxterms are fed into the AND-plane, resulting in outputs of the general form

$$g_m = \prod_j M_j \tag{3.40}$$

The actual maxterm included in each is determined by the AND-plane wiring. As in the case of the SOP array, the numbers of inputs and outputs of each plane may be varied as needed.

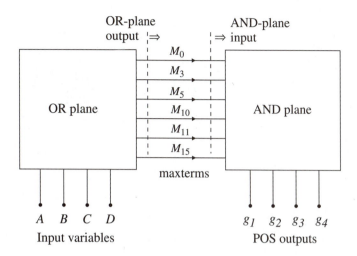

Figure 3.15 An OR-AND PLA

Example 3-9

Let us examine the construction of a PLA in detail. Consider the network shown in Figure 3.16. This is an AND-OR array that has inputs A, B, and C and provides canonical SOP output functions denoted by h, g, r, and y.

To determine the functions, we first find the outputs of the AND array. The first AND output is $A \cdot B \cdot C$, which is denoted by (111) in the diagram. Similarly, the second line is $\overline{A} \cdot \overline{B} \cdot \overline{C} = (000)$, and so on. A total of six product minterms are generated by the AND array. These are then used as inputs into the OR array gates to obtain the sums. Each connection is shown by a "dot" that connects the lines. Working from left to right, the SOP outputs are

$$h = A \cdot B \cdot C + \overline{A} \cdot \overline{B} \cdot \overline{C} + A \cdot \overline{B} \cdot \overline{C}$$

$$g = A \cdot \overline{B} \cdot C + \overline{A} \cdot \overline{B} \cdot C + \overline{A} \cdot B \cdot \overline{C}$$

$$r = A \cdot B \cdot C + \overline{A} \cdot \overline{B} \cdot C + A \cdot \overline{B} \cdot \overline{C}$$

$$y = \overline{A} \cdot \overline{B} \cdot C + A \cdot \overline{B} \cdot \overline{C} + \overline{A} \cdot B \cdot \overline{C}$$

(3.41)

Each function is defined independently. Any function can be changed by simply altering the connections to the OR gates. Note that there is no specification that requires every output to have the same number of minterms. OR gates with 2, 4, or more inputs can be used to generate equations as required in the logic design.

Large arrays can become quite tedious to draw, so a shorthand notation is sometimes used. This is shown in Figure 3.17 for the same AND-OR PLA. The gate inputs have been replaced by a single line that represents several independent lines. Connections are still shown by dots, but a dot must be interpreted as connecting two of the individual lines, not the entire group. The number of inputs to a gate is equal to the number of connection dots in the input line.

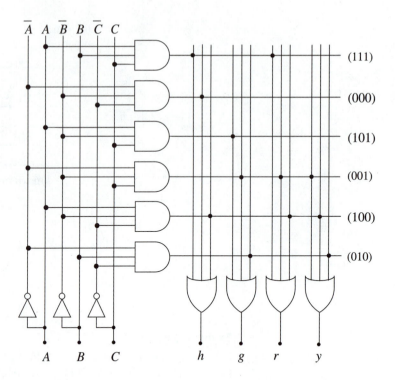

Figure 3.16 PLA example

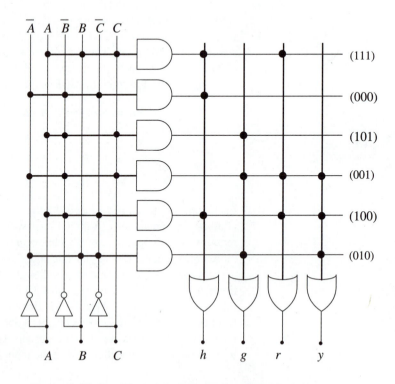

Figure 3.17 Simplified wiring notation for arrayed logic

Example 3-10

Let us examine the characteristics of the array in the previous example if we interchange the AND and OR gates to yield the network in Figure 3.18. This creates an OR-AND array with inputs A, B, and C and outputs G, H, J, and K.

The OR array uses the inputs to generate maxterms. For example, the first line is $(A + B + C)$, the second line is $(\overline{A} + \overline{B} + \overline{C})$, and so forth. The AND array produces the products of these maxterms, resulting in a POS function. Every output has the general form

$$OUT = M_a \cdot M_b \cdot M_c \tag{3.42}$$

where the maxterms depend on the connections. Reading the outputs from left to right gives

$$G = (A + B + C) \cdot (\overline{A} + \overline{B} + \overline{C}) \cdot (A + \overline{B} + \overline{C})$$
$$H = (A + \overline{B} + C) \cdot (\overline{A} + \overline{B} + C) \cdot (\overline{A} + B + \overline{C})$$
$$J = (A + B + C) \cdot (\overline{A} + \overline{B} + C) \cdot (A + \overline{B} + \overline{C})$$
$$K = (\overline{A} + \overline{B} + C) \cdot (A + \overline{B} + \overline{C}) \cdot (\overline{A} + B + \overline{C})$$

$$\tag{3.43}$$

for the array. As with the POS array, we may alter any function by changing the connection dots.

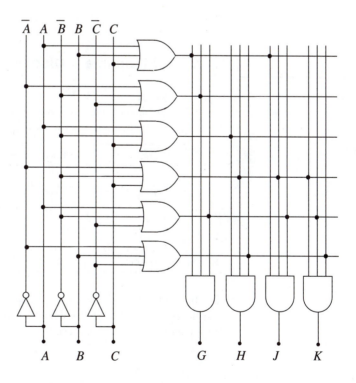

Figure 3.18 OR-AND PLA example

3.5.3 Application of Logic Arrays

Logic arrays are extremely useful for rapid implementation and prototyping of complex digital networks. Turn-around time can be very short, allowing the designer to test the algorithm or try several different ideas. Many companies produce commercial logic arrays that are inexpensive to use; programming software for each family makes the design fast and straightforward. Moreover, recent product offerings allow the designer to design and test very complex systems. This makes the devices very useful for "field designs" and fast prototyping.[1]

The main drawbacks to using logic arrays are that the resulting circuit will probably not be the most efficient of gates and that the design itself will not be the fastest implementation that can be achieved. The first drawback arises from the fact that uncommitted arrays contain gates that may not be needed in the particular implementation but cannot be removed from the circuit. The speed issue is due to the fact that the switching speed of an array is set by the structure of the circuits, which cannot be changed by the user. Regardless of these points, however, PLAs have given rise to the concept of general **programmable logic devices** (PLDs) of many different varieties. In this category are FPGAs (**field-programmable gate arrays**), which are very powerful logic circuits that can be used to implement highly complex logic networks.

In general, the design of programmable logic is based on the use of CAD (computer-aided design) "tools." These are dedicated programs that allow the designer to enter the problem statement into a computer and that then provide possible wiring solutions for the device chosen.

3.6 BCD and 7-Segment Displays

Binary-coded decimal (**BCD**) is a binary counting system for the base-10 digits 0 through 9. A binary-coded decimal word has bits that are denoted by *ABCD* with the assignments shown in the following table.

ABCD	**Decimal**	*ABCD*	**Decimal**
0000	0	0101	5
0001	1	0110	6
0010	2	0111	7
0011	3	1000	8
0100	4	1001	9

The main difference between BCD and "normal" binary-to-decimal conversion is that the remaining binary combinations 1010 through 1111 are not defined in this system and are therefore not used.

---

[1] The term *field design* implies that the project is done not in a laboratory or a factory but in the applications environment.

BCD is useful for many applications in digital design, and it is particularly well suited for driving visual displays that show numerical data, such as the time on a digital clock face. A common type of numerical display is that based on the **7-segment** layout shown in Figure 3.19. In this drawing, the segments are labeled *a*, *b*, *c*, *d*, *e*, *f*, *g*, and each can be individually controlled by an electronic signal to switch ON or OFF. An ON segment is visually different from an OFF segment. For example, displays that use **LED**s (light-emitting diodes) have segments that emit light when they are ON and are dark when they are OFF. By varying the lit segments, the engineer can use the display to form the decimal digits illustrated in Figure 3.20.

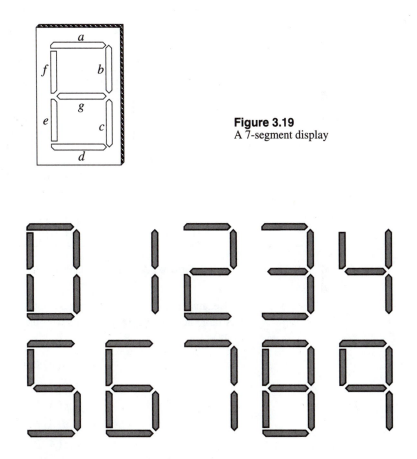

Figure 3.19
A 7-segment display

Figure 3.20 Formation of base-10 digits using a 7-segment display

A useful example in digital design is that of designing the BCD-to-7-segment **decoder** circuit in Figure 3.21 that accepts a BCD input and sends the signals to the display segments *a* through *g* that form the appropriate digit. The basic problem is to determine which segments are to be ON and OFF for a given BCD input. Assuming that a segment turns ON with a logic 1 applied gives the function table in Figure 3.22. Each row is a BCD entry and shows explicitly the segments that are to be lit; these values were obtained by using the graphical display possibilities above.

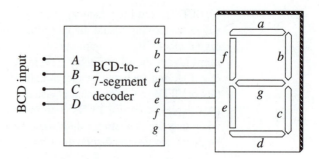

Figure 3.21 BCD-to-7-segment decoder function

ABCD	Digit	*a*	*b*	*c*	*d*	*e*	*f*	*g*
0 0 0 0	0	1	1	1	1	1	1	0
0 0 0 1	1	0	1	1	0	0	0	0
0 0 1 0	2	1	1	0	1	1	0	1
0 0 1 1	3	1	1	1	1	0	0	1
0 1 0 0	4	0	1	1	0	0	1	1
0 1 0 1	5	1	0	1	1	0	1	1
0 1 1 0	6	1	0	1	1	1	1	1
0 1 1 1	7	1	1	1	0	0	0	0
1 0 0 0	8	1	1	1	1	1	1	1
1 0 0 1	9	1	1	1	0	0	1	1

Figure 3.22 Function table for the BCD-to-7-segment decoder function

Because the decoder produces output logic signals $a, b, c, \ldots$ to each segment, this table actually represents the seven different logic functions shown in equation (3.44)

$$
\begin{aligned}
a &= a(A, B, C, D) \\
b &= b(A, B, C, D) \\
c &= c(A, B, C, D) \\
d &= d(A, B, C, D) \\
e &= e(A, B, C, D) \\
f &= f(A, B, C, D) \\
g &= g(A, B, C, D)
\end{aligned}
$$

(3.44)

as clarified by the drawing in Figure 3.23. Each output column in the function table provides the information needed to write the SOP expression directly. Segment a has eight logic 1 entries, so the SOP logic equation is given by

$$
\begin{aligned}
a = {}& \overline{A}\,\overline{B}\,\overline{C}\,\overline{D} + \overline{A}\,\overline{B}C\overline{D} + \overline{A}\,\overline{B}CD + \overline{A}B\overline{C}\overline{D} + \overline{A}BC\overline{D} \\
& + \overline{A}BCD + A\overline{B}\,\overline{C}\,\overline{D} + A\overline{B}\,\overline{C}D
\end{aligned}
$$

(3.45)

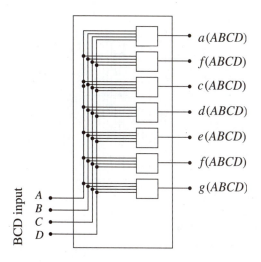

Figure 3.23 Block diagram showing individual segment circuits

as can be verified by inspection. Similarly, the expression for segment b is

$$b = \overline{A}\,\overline{B}\,\overline{C}\,\overline{D} + \overline{A}\,\overline{B}\,\overline{C}D + \overline{A}\,\overline{B}CD + \overline{A}B\overline{C}D + \overline{A}BC\overline{D}$$
$$+ \ \overline{A}BCD + A\overline{B}\,\overline{C}\,\overline{D} + A\overline{B}\,\overline{C}D \tag{3.46}$$

The same procedure can be used to obtain the remaining functions. The details are left as an exercise for the reader.

Let us now turn to the problem of designing in the logic blocks in the decoder. Suppose that we decide to use a logic array to implement the decoder. Since the expressions are already in SOP form, an AND-OR array can be easily programmed as shown in Figure 3.24 to provide all of the desired functions. The usefulness of the PLA is due to its simplicity. The AND-plane is wired to produce every needed minterm combination of $ABCD$. The individual functions are then generated by wiring the inputs to the OR gates as specified by the table; note that the number of inputs for each OR gate is varied as needed.

3.7 Karnaugh Maps

Canonical SOP and POS forms can often be simplified by applying the rules of Boolean algebra. This is preferred if we are trying to build a network using the smallest number of gates. In the simplified function, the types of gates and their placement in the logic network will be "random" in that they cannot be predicted. This characteristic leads to our calling this design technique **random logic**, as compared with the structured logic networks discussed above.

Karnaugh maps allow us to simplify functions using a visual mapping technique that helps us recognize Boolean reductions by their locations on a grid. We will examine the problem where we are given a logic function $f(A, B, C, \ldots)$ that is

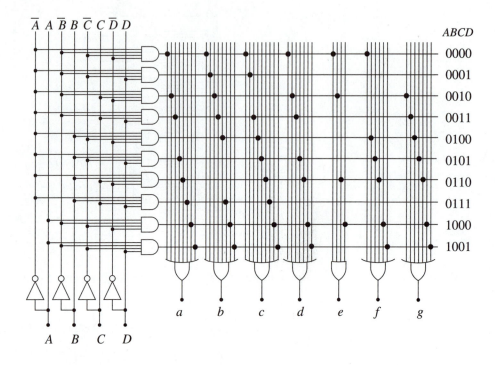

Figure 3.24 PLA implementation for a BCD-to-7-segment decoder network

in canonical SOP form. Our goal is to reduce the complexity to simpler form; this generally means reducing the number of gates needed to implement the function. The technique of Karnaugh maps relies on the following two identities:

$$A + \overline{A} = 1$$
$$1 \cdot X = X$$

(3.47)

where X is any group of logic variables.

Let us first examine how these identities can be used to simply a POS function. Suppose we have the 3-variable expression

$$g(A, B, C) = ABC + AB\overline{C} + A\overline{B}C + ABC$$

(3.48)

The first two terms can be combined and simplified using the sequence

$$ABC + AB\overline{C} = AB(C + \overline{C})$$
$$= AB \cdot 1$$
$$= AB$$

(3.49)

whereas the third and fourth terms can be reduced by

$$A\bar{B}C + ABC = A(\bar{B} + B)C$$
$$= A \cdot 1 \cdot C \qquad \textbf{(3.50)}$$
$$= AC$$

Combining these two results, we have

$$g(A, B, C) = AB + AC$$
$$= A(B + C) \qquad \textbf{(3.51)}$$

This is entirely equivalent to the original expression but only requires two gates in the final form.[2] It is obvious that the simplified form is much more efficient in random logic.

Karnaugh maps allow us to perform reductions of this type using a visual approach. The technique is very straightforward to learn and use. Starting with a function table, we will map[3] the input-output combinations to a rectangular grid array. The structure of the grid and the associated rules enable us to locate easily the terms where the identity $(X + \bar{X}) = 1$ can be used to simplify the function. Although the use of Karnaugh maps can be applied to functions with arbitrary numbers of variables, the technique gets cumbersome when the number of variables becomes too large. We will examine the application of Karnaugh maps for the case of two, three, and four variables. For more complicated functions, computer programs are much more efficient for the designer and often provide additional capabilities at the system level.

3.7.1 2-Variable Karnaugh Maps

Let us begin by considering a function f of two variables A and B. Since these are intrinsically simple, we will use this case to illustrate the technique of creating and interpreting a Karnaugh map that is easily extended to include more variables.

Karnaugh maps (or *K-maps* for short) use the properties of the function as expressed in a grid-like table. To create this table, we start with the list of all possible input minterms and the resulting output value of the function. For the case of two variables, there are four minterms: $\bar{A} \cdot \bar{B}, \bar{A} \cdot B, A \cdot \bar{B}$, and $A \cdot B$. To construct the basic map structure, we form a 2×2 grid which allows for each variable to take on the value of 0 or 1. This is shown by the general 2-variable map in Figure 3.25(a); Figure 3.25(b) shows the same information but uses minterm notation. To apply the map to a particular function, we examine each input combination and fill the box with a 0 or a 1, corresponding to the value of $f(A, B)$ for the given input pair. This yields a map that helps us simplify the function by using the technique discussed below.

Let us start with the AND function to illustrate the general approach. Figure 3.26 shows the function table and the associated map; we note that these are equivalent representations. The truth table shows that the AND function yields only a single

---

[2] One OR gate and one AND gate.

[3] The term *map* is used here in the mathematical sense, where inputs are assigned a relationship to an output set.

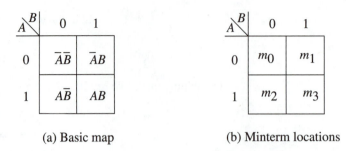

(a) Basic map   (b) Minterm locations

Figure 3.25 General structure for a 2-variable Karnaugh map

case where $f = 1$ corresponding to $A = 1$ and $B = 1$. In this case the function $A \cdot B$ is already in simplest form, as verified by the single entry in the Karnaugh map shown in Figure 3.26(b). No additional work needs to be done.

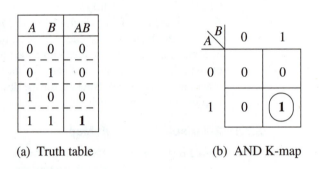

(a) Truth table   (b) AND K-map

Figure 3.26 Karnaugh map for the AND2 operation

The NAND operation gives a different situation. The truth table and associated map are shown in Figure 3.27. Reading the truth table gives us the SOP form

$$f = \overline{A} \cdot \overline{B} + \overline{A} \cdot B + A \cdot \overline{B} \tag{3.52}$$

Although this can be reduced algebraically, let us apply the Karnaugh map technique. First, we locate the 1 entries in the function table in Figure 3.27(a) and enter them on the map in Figure 3.27(b). Then, we search for adjacent squares that contain 1s. (By "adjacent," we mean only vertical or horizontal directions; diagonal squares do not count.) For the NAND function, there are two possible groups of adjacent 1s. These are shown directly on the K-map of Figure 3.27(b). Note that the pairs have been "grouped" by encircling them. As will be seen below, the grouping is the key to simplifying a logic function using K-maps.

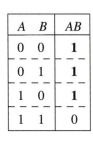

A	B	AB
0	0	1
0	1	1
1	0	1
1	1	0

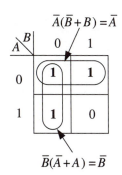

(a) Truth table   (b) NAND K-map

Figure 3.27 Karnaugh map for the NAND2 operation

To illustrate how the K-map reduction technique works, consider first the horizontal grouping corresponding to the sum of terms

$$\overline{A} \cdot \overline{B} + \overline{A} \cdot B = \overline{A}(\overline{B} + B)$$
$$= \overline{A} \qquad (3.53)$$

This reduction can be seen graphically by noting that both entries have the factor $\overline{A}$, leaving a sum of $(B + \overline{B}) = 1$; thus only $\overline{A}$ survives. Next, consider the vertical grouping. We see that $B = 0$ for both terms but that the upper box has $A = 0$ whereas the lower box occurs when $A = 1$. Mathematically, this corresponds to

$$\overline{B}(\overline{A} + A) = \overline{B} \qquad (3.54)$$

Combining the two reduced terms, we have

$$f = \overline{A} + \overline{B} = \overline{A \cdot B} \qquad (3.55)$$

by applying the DeMorgan theorem. The procedure has thus transformed the information in the truth table into the standard NAND equation.

The OR/NOR pair of gates has similar properties when studied using K-maps. These are illustrated in Figure 3.28. The OR-map is given in Figure 3.28(b). There are three 1s in the map, giving the two groupings shown. The vertical group corresponds to $B = 1$ and has both $A = 0$ (upper box) and $A = 1$ (lower box). Thus the grouping reduces to the simple factor B. The horizontal grouping is similar with $A = 1$; the left box has $B = 0$, while the right box has $B = 1$, so the two terms reduce to simply A. Summing yields

$$f = A + B \qquad (3.56)$$

which is the OR expression. The NOR map shown in Figure 3.28(c) has only a single 1 entry for the case where both $A = 0$ and $B = 0$. The map thus represents the function

$$g = \overline{A} \cdot \overline{B} = \overline{A + B} \qquad (3.57)$$

which is the expected result.

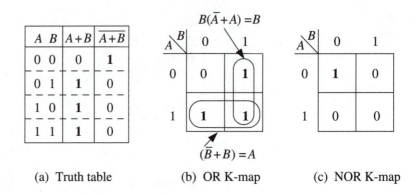

(a) Truth table   (b) OR K-map   (c) NOR K-map

Figure 3.28 Karnaugh maps for the OR2 and NOR2 operations

These basic 2-variable examples illustrate how Karnaugh maps can be used to simplify expressions. The technique can be summarized as follows:

- Fill in the K-map with 1s and 0s as defined by the function table.
- Group adjacent 1 entries in pairs.
- Apply the rule $(a + \overline{a}) = 1$ to eliminate a variable.

The power of using K-maps is that the same approach can be extended to functions of 3 or more variables. The importance of the technique is that it affords a better understanding of how binary variables behave.

3.8 3-Variable Karnaugh Maps

The utility of Karnaugh maps becomes more evident when we apply the technique to functions of three variables. Suppose we have a function $f(A, B, C)$ in canonical SOP form; this means that f consists of a sum of minterms (such as $m_1 = \overline{A} \cdot \overline{B} \cdot C$). Since we have three variables to deal with, we will create a two-dimensional mapping by grouping two variables together for one axis. This can be achieved by creating a map using the variable A with $B \cdot C$ or, alternatively, grouping $A \cdot B$ and keeping C independent; either choice is acceptable.

Figure 3.29 shows the map for the first choice where A is taken to be independent, while $B \cdot C$ are treated as a group. The map in Figure 3.29(a) shows the entries for each box in terms of the variables and their complements; the equivalent minterms are shown in Figure 3.29(b). There is an important property of this and every Karnaugh map that gives the graphical technique the proper characteristics. It is simply that

- Adjacent boxes differ by only one bit entry.

This means, for example, that the box with $\overline{A} \cdot \overline{B} \cdot C$ is adjacent to boxes in which two of the factors are the same but one is complemented. This is why the $B \cdot C$ values are sequenced as 00, 01, 11, 10.

The procedure for reducing a function using a 3-variable Karnaugh map is based on the same principle as that used for the 2-variable map. In general, this means that

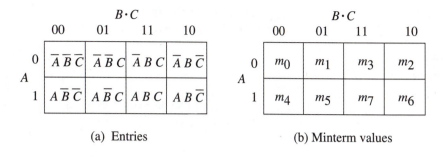

(a) Entries

(b) Minterm values

Figure 3.29 General structure for a 3-variable Karnaugh map

we look for adjacent "1" entries and group them together. Each grouping cor-resonds to a simplification of the expression. The main difference is that the additional variable allows us to choose groups consisting of either two adjacent boxes or four adjacent boxes for reduction. Moreover, we will find that the left and right edges are actually adjacent, allowing us to "wrap" the map into a cylinder.

Example 3-11

Let us start with the 3-variable function table in Figure 3.30 to illustrate the general technique. By inspection, we see that the output function has a value of $f = 1$ for the four input combinations: $\overline{A} \cdot B \cdot C$, $\overline{A} \cdot B \cdot \overline{C}$, $A \cdot \overline{B} \cdot \overline{C}$, and $A \cdot B \cdot C$. These are shown as

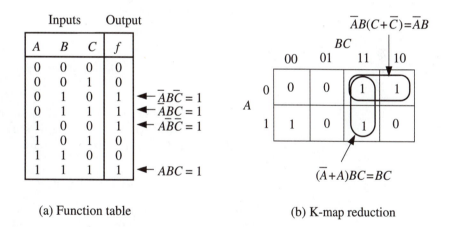

(a) Function table

(b) K-map reduction

Figure 3.30 A 3-variable reduction example

logic 1 entries in the K-map. The two possible groupings are shown. The top horizontal group evaluates to

$$\overline{A}B(C + \overline{C}) = \overline{A}B \tag{3.58}$$

and the second (vertical) group gives

$$(\overline{A} + A)BC = BC \tag{3.59}$$

The term $A \cdot \overline{B} \cdot \overline{C}$ corresponding to the 100 entry cannot be reduced any further. Thus, ORing this term with the simplified terms gives the function as

$$f = A\overline{B}\overline{C} + \overline{A}B + BC \tag{3.60}$$
$$= A\overline{B}\overline{C} + B(\overline{A} + C)$$

where the second step has been used to eliminate one AND function. This completes the problem.

Example 3-12

Let us examine one more reduction starting from the function table given in Figure 3.31(a). This listing shows an output of $g = 1$ for inputs of $A \cdot B \cdot C = 000, 010, 100,$

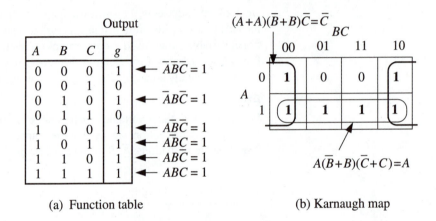

(a) Function table   (b) Karnaugh map

Figure 3.31 3-variable K-map with adjacent edges

101, 110, and 111; these combinations tell us where to place logic 1 entries in the map in Figure 3.31(b). The first grouping we see is the horizontal string of four 1s along the bottom row. The row itself is the variable A, and the grouping reduces the entire row to

$$A(\overline{B} + B)(\overline{C} + C) = A \tag{3.61}$$

The second grouping combines the first and fourth column and deserves a bit more comment. Even though these are on opposite edges, they are considered adjacent and yield the reduced term

$$(\overline{A} + A)(\overline{B} + B)\overline{C} = \overline{C} \tag{3.62}$$

since only the case where $C = 0$ is invariant. Combining these two terms shows

$$g = A + \overline{C} \tag{3.63}$$

as the simplified result.

It is easy to justify that the edges are adjacent by noting that the $B \cdot C$ numbering order of 00, 01, 11, 10 has an arbitrary starting point. For example, we could have chosen 01, 11, 10, 00 as the sequence, and all of the reduction techniques would still have been valid. This argument can be seen using the illustrations in Figure 3.32. The original map is shown in drawing Figure 3.32(a). The equivalent map in Figure 3.32(b) uses the $B \cdot C$ sequence 10, 00, 01, 11, illustrating the point. As an alternative viewpoint, the left and right sides may be visualized as touching by wrapping the map into the cylinder implied by Figure 3.32(c).

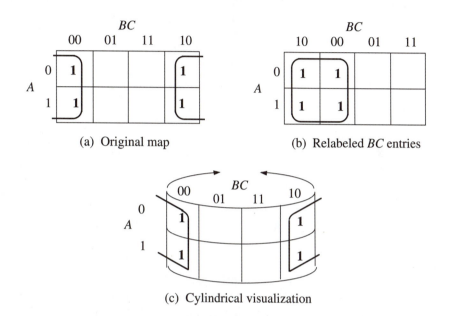

(a) Original map

(b) Relabeled *BC* entries

(c) Cylindrical visualization

Figure 3.32 Visualization of adjacent edges in a 3-variable K-map

The simplification using a 3-variable map may be summarized as follows:

- A group of one minterm gives a term with all three factors $A \cdot B \cdot C$.
- A group of two minterms reduces to a term with two factors.
- A grouping of four entries reduces to a term with one factor.
- A group of all eight minterms is equivalent to a logic 1 result.

The last possibility is the special case where every minterm is a 1, so that the function itself is a constant value of 1 regardless of the inputs. It is important to note that only groups of two, four, or eight minterms are permitted. This is due to the fact that the grouping is equivalent to reductions using the identity $(X + \overline{X}) = 1$, which requires that pairs of boxes be used.

3.8.1 "Don't Care" Conditions

Occasionally we have the situation where the output produced by a particular set of inputs can be specified as *either* a 0 or a 1 without affecting the application of the function. Such an occurrence is called a **don't care** condition and is denoted using an output of X as shown in Figure 3.33(a) for the inputs $(A \cdot B \cdot C) = (010)$ and (110). In performing a map reduction, an X may be treated as either a 0 or a 1. The choice is based on which value gives a simpler reduction.

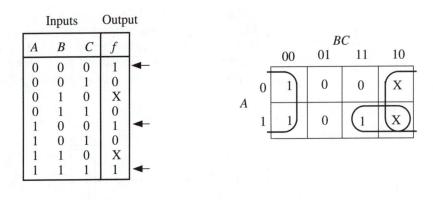

(a) Function table

(b) K-map reduction

Figure 3.33 Example of "don't care" entries

The usefulness of a don't care can be understood by analyzing the example. Consider the entries in the map of Figure 3.33(b). If we assume that the X values shown in the map are both 0, then the function is

$$f = \bar{B} \cdot \bar{C} + A \cdot B \cdot C \qquad (3.64)$$

since only one group (the left column) exists. On the other hand, if we assume that both don't care X entries are 1s and group as shown in Figure 3.33(b), then the left and right columns may be combined, in addition to the lower right covering shown. This yields

$$f = A \cdot B + \bar{C} \qquad (3.65)$$

after simplification. It is important to note that we must specify the don't care conditions as such before performing a reduction. In practice, do not assume that an unknown or unspecified output is the same as a don't care; it may just be a case of insufficient information.

3.8.2 Alternative 3-Variable Map Layout

The previous maps were constructed using two rows and four columns. An alternative approach is to group $A \cdot B$ together while keeping C separate. The map for this choice is shown in Figure 3.34; it consists of two columns and four rows. It has the same characteristics as the other approach, so it may be used if desired. The result of a map reduction remains the same regardless of the geometry chosen.

Example 3-13

Consider the 3-variable map shown in Figure 3.35 using the vertical format. To reduce the function (which we'll call g), we look for groups of 1s. Two are shown, where we note that the top and bottom are considered touching. This gives the function

$$g = \bar{A} \cdot B + \bar{B} \cdot \bar{C} \qquad (3.66)$$

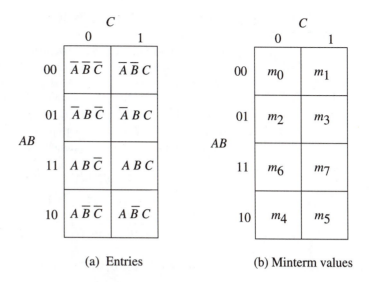

(a) Entries   (b) Minterm values

Figure 3.34 Alternative layout for a 3-variable Karnaugh map

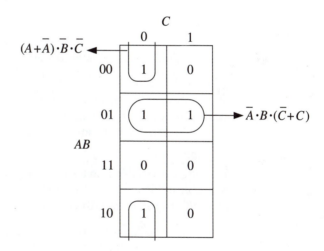

Figure 3.35 Example of a 3-variable vertical map

by ORing the two terms. It is simple to verify this result using either a Boolean algebraic reduction or the alternative layout used earlier for the 3-variable K-map. Since the results are equivalent, the choice of map layouts is really a matter of personal taste.

3.9 4-Variable Karnaugh Maps

The general technique can be extended to four variables A, B, C, D by choosing two groups of two variables. For example, Figure 3.36 provides the basic layout for a map created using the groups $A \cdot B$ and $C \cdot D$. As with the 3-variable map, the entries in adjacent boxes differ in only a single bit entry. In the present example, we have implied the ordering $A \cdot B \cdot C \cdot D$ so that the equivalent minterm entries are those shown in Figure 3.36(b). It is important to remember that once an order is chosen for the variables, it must be maintained throughout the problem.

AB \ CD	00	01	11	10
00	$\overline{A}\,\overline{B}\,\overline{C}\,\overline{D}$	$\overline{A}\,\overline{B}\,\overline{C}D$	$\overline{A}\,\overline{B}CD$	$\overline{A}\,\overline{B}C\overline{D}$
01	$\overline{A}B\overline{C}\,\overline{D}$	$\overline{A}B\overline{C}D$	$\overline{A}BCD$	$\overline{A}BC\overline{D}$
11	$AB\overline{C}\,\overline{D}$	$AB\overline{C}D$	$ABCD$	$ABC\overline{D}$
10	$A\overline{B}\,\overline{C}\,\overline{D}$	$A\overline{B}\,\overline{C}D$	$A\overline{B}CD$	$A\overline{B}C\overline{D}$

(a) General map

AB \ CD	00	01	11	10
00	m_0	m_1	m_3	m_2
01	m_4	m_5	m_7	m_6
11	m_{12}	m_{13}	m_{15}	m_{14}
10	m_8	m_9	m_{11}	m_{10}

(b) Minterm entries

Figure 3.36 General structure of a 4-variable Karnaugh map

In a 4-variable K-map, we search for groups of adjacent 1s. The grouping may consists of two, four, or eight entries. As seen in the examples below, the larger the number in a group, the simpler the resulting term.

Example 3-14

Let's use a 4-variable Karnaugh map to reduce the function described by the function table in Figure 3.37. The grouping in the second column made up of the factors $ABCD = 0101$ and 1101 is easily seen to reduce to

$$(A + \overline{A})B\overline{C}D = B\overline{C}D \tag{3.67}$$

since both $A = 0$ and $A = 1$ are covered. The second grouping shown in the K-map, made up of the terms $ABCD = 1011$ and 1010, simplifies to

$$A\overline{B}C(D + \overline{D}) = A\overline{B}C \tag{3.68}$$

by eliminating D in this case.

The third grouping shown in the map consists of the four corner terms $ABCD =$ 0000, 0010, 1000, and 1010. These are classified as being adjacent since they differ by only a single bit value; this can also be verified by relabeling both axes as shown in Figure 3.38. This grouping of four adjacent entries simplifies to

$$(A + \overline{A})\overline{B}(C + \overline{C})\overline{D} = \overline{B}\,\overline{D} \tag{3.69}$$

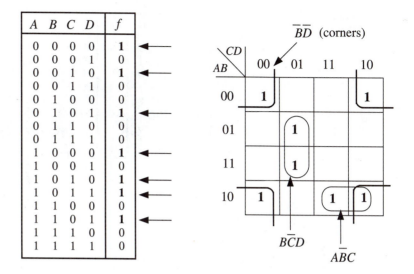

A	B	C	D	f
0	0	0	0	**1**
0	0	0	1	0
0	0	1	0	**1**
0	0	1	1	0
0	1	0	0	0
0	1	0	1	**1**
0	1	1	0	0
0	1	1	1	0
1	0	0	0	**1**
1	0	0	1	0
1	0	1	0	**1**
1	0	1	1	**1**
1	1	0	0	0
1	1	0	1	**1**
1	1	1	0	0
1	1	1	1	0

Figure 3.37 4-variable reduction example

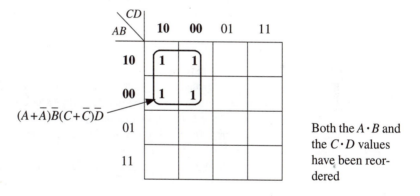

Figure 3.38 Reordering entries to illustrate adjacent entries

since both $(A + \overline{A}) = 1$ and $(C + \overline{C}) = 1$. ORing these together gives

$$f = B\overline{C}D + A\overline{B}C + \overline{B}\,\overline{D} \tag{3.70}$$

as the final expression. Note that one more grouping with $\overline{B}$ can be made if desired.

Example 3-15

Another characteristic of a 4-variable K-map is illustrated by the example in Figure 3.39. In this case, the top and bottom rows all have entries of 1 and have been grouped together. The reduction for this group of minterms is

$$(A + \overline{A})\overline{B}(C + \overline{C})(D + \overline{D}) = \overline{B} \tag{3.71}$$

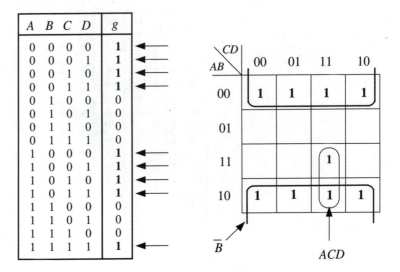

Figure 3.39 A 4-variable map with adjacent top and bottom rows

so that only the factor of $B = 0$ remains. Combining this with the other 2-entry group gives

$$g = \bar{B} + ACD \tag{3.72}$$

as the final form.

The grouping of the top and bottom rows is allowed since we can view them as being adjacent by relabeling the order as in Figure 3.40. Alternatively, the map can be wrapped around so that the top and bottom meet to give the views in Figure 3.41. It is important to keep in mind that all of these maps are equivalent since they provide the same information.

The characteristics of a 4-variable map may be summarized as follows:

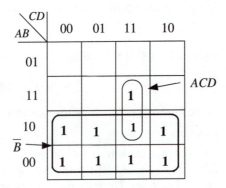

Figure 3.40 Reordering of row entries

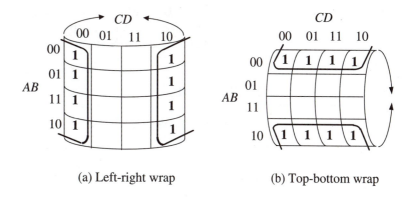

(a) Left-right wrap   (b) Top-bottom wrap

Figure 3.41 Visualization of wrapping in a 4-variable map

- A group of one minterm gives a term with all four factors $A \cdot B \cdot C \cdot D$.
- A grouping of two squares yields a term with three factors.
- A grouping of four entries reduces to a term with two factors.
- A grouping of eight minterms reduced to a term with only one factor.
- A grouping of all 16 minterms is a logical 1.

Note that only groups with specific numbers of minterms (two, four, eight, or sixteen) are permitted. This is due to the fact that the reduction identity $(X + \overline{X})$ requires pairs.

3.10 The Role of the Logic Designer

Logic design as presented in this and the previous chapter provides the basis for all digital networks. The laws of Boolean switching algebra define the operations that can be performed among binary variables, and general combinational logic design techniques allow us to construct logic functions of arbitrary complexity as required.

Logic designers are generally charged with designing the logic networks that perform specific tasks, usually in the context of a large system. This may include digital units such as encoders and decoders, circuits for arithmetic operations, data communication techniques and networks—the list seems to go on forever! In the modern world of engineering, however, no person works alone. Project groups provide the basic work unit. Each individual has an area of specialization but must have knowledge that extends into other areas. Understanding more than one field helps you interact with the other engineers in the group and also gives you a deeper insight into what your work affects. Depending on the specific makeup of a project group, the logic designer may have to interact with system architects who deal with the large-scale properties of the network, "hardware" designers who provide the digital electronic circuits, circuit- and VLSI-oriented personnel, and perhaps even customers who may use the network.

Now that we have seen the foundations of basic logic, we can progress into the other areas that constitute modern digital systems. As we will see, the lessons and techniques derived from Boolean algebra permeate all levels of the design hierarchy.

3.11 Problems

[3.1] Use Boolean algebra to place in canonical SOP form the function

$$f(a, b, c) = a \cdot c$$

[3.2] Structure in canonical SOP form the expression

$$g(x, y) = x + y$$

[3.3] Suppose that we are given the binary expression

$$f(u, v, w) = u \cdot v \cdot \overline{w} + u \cdot v$$

Find the simplest canonical SOP form for f.

[3.4] Write the SOP form for the function $g(A, B, C)$ defined by the following function table.

A	B	C	g
0	0	0	0
0	0	1	0
0	1	0	1
0	1	1	1
1	0	0	1
1	0	1	0
1	1	0	1
1	1	1	1

Inputs / Output

[3.5] Use the function table in Problem [3.4] to obtain the POS form of $g(A, B, C)$.

[3.6] Construct the function table for the following expression.

$$g(a, b, c) = a \cdot b \cdot \overline{c} + \overline{a} \cdot b \cdot c + a \cdot \overline{b} \cdot c + a \cdot \overline{b} \cdot \overline{c}$$

[3.7] Construct the function table for the following expression.

$$F(x, y, z) = x \cdot \overline{y} \cdot \overline{z} + \overline{x} \cdot \overline{y} \cdot \overline{z} + \overline{x} \cdot y \cdot \overline{z} + x \cdot y \cdot z$$

[3.8] Construct the function table that describes

$$f(x, y, z) = x \cdot \overline{y} + x \cdot y \cdot \overline{z} + y \cdot z$$

and then use the table to place f in canonical SOP form.

[3.9] Write the SOP form for the function $f(x, y, z)$ defined by the following table.

Inputs			Output
x	y	z	f
0	0	0	1
0	0	1	1
0	1	0	1
0	1	1	0
1	0	0	0
1	0	1	0
1	1	0	0
1	1	1	1

[3.10] Write the SOP form for the function $g(a, b, c)$ defined by the following function table.

Inputs			Output
a	b	c	G
0	0	0	0
0	0	1	1
0	1	0	1
0	1	1	0
1	0	0	1
1	0	1	1
1	1	0	0
1	1	1	1

[3.11] Write the SOP form for the functions defined by the following function tables.

(a)

Inputs			Output
u	v	w	B
0	0	0	1
0	0	1	1
0	1	0	0
0	1	1	0
1	0	0	1
1	0	1	1
1	1	0	0
1	1	1	1

(b)

Inputs			Output
a	b	c	R
0	0	0	1
0	0	1	0
0	1	0	0
0	1	1	1
1	0	0	1
1	0	1	1
1	1	0	1
1	1	1	1

[3.12] Construct the function table for the expression

$$r(u, v, w) = u \cdot v \cdot w + (\bar{u} + v) \cdot w$$

[3.13] Construct the function table for the 4-variable function

$$f(a, b, c, d) = (a + \bar{b}) \cdot c + b \cdot \bar{c} \cdot d + b \cdot c$$

[3.14] A logic function is written using minterms in the form

$$g(a, b) = m_0 + m_2$$

Write g explicitly as a function of a and b.

[3.15] A logic function is written using minterms in the form

$$f(x, y, z) = m_0 + m_1 + m_5 + m_6$$

Write f explicitly as a function of x, y, and z.

[3.16] A logic function $f(x, y, z)$ is written in minterms as

$$f(x, y, z) = m_0 + m_2 + m_4 + m_5$$

(a) Write f in terms of the variables x, y, and z.
(b) Use Boolean algebra to simplify f to simplest form.

[3.17] Reduce the function

$$h(a, b, c) = \sum m(0, 2, 4, 5, 6)$$

to simplest form. You may use either an algebraic reduction or a Karnaugh map.

[3.18] Expand the function

$$h(x, y) = \sum m(1, 2, 4)$$

to canonical SOP form.

[3.19] A logic function is written using maxterms in the form

$$G(a, b) = M_0 \cdot M_2$$

Write G explicitly as a function of a and b.

[3.20] Expand the function

$$G(A, B, C) = \prod M(1, 3, 4, 7)$$

into an explicit function of A, B, and C.

[3.21] Design an AND-OR PLA that implements the functions

$$f(a, b, c) = \sum m(0, 1, 4, 7)$$
$$g(a, b, c) = \sum m(0, 2, 5, 7)$$

[3.22] Use an AND-OR PLA to implement the expressions

$$f_1(u, v, w) = u \cdot v \cdot \bar{w} + u \cdot \bar{v} \cdot w + u \cdot v \cdot w + \bar{u} \cdot v \cdot \bar{w}$$
$$f_2(u, v, w) = u \cdot \bar{v} \cdot \bar{w} + u \cdot v \cdot \bar{w} + \bar{u} \cdot v \cdot w + u \cdot \bar{v} \cdot w$$

[3.23] Design an AND-OR PLA that implements the function set

$$F(x, y, z, w) = \sum m(0, 1, 5, 8, 9, 12, 14)$$

$$G(x, y, z) = \sum m(1, 2, 4, 7)$$

$$H(x, y, z, w) = \sum m(4, 5, 11, 13, 15)$$

[3.24] Design an OR-AND PLA for the functions

$$F_1(x, y, z) = \prod M(0, 1, 3, 4, 7)$$

$$F_2(x, y, z) = \prod M(3, 5)$$

$$F_3(x, y, z) = \prod M(4, 5, 6)$$

[3.25] Design an OR-AND PLA that provides the function

$$g_0(x, y, z) = (x + \bar{y} + z + \bar{w}) \cdot (x + \bar{y} + z + w) \cdot (\bar{x} + y + z + w)$$

$$g_1(x, y, z) = (\bar{x} + y + z + w) \cdot (\bar{x} + \bar{y} + \bar{z} + w) \cdot (x + \bar{y} + z + w)$$

$$g_2(x, y, z) = (x + y + z + w) \cdot (x + y + z + w) \cdot (x + y + \bar{z} + w)$$

$$g_3(x, y, z) = (x + y + \bar{z} + w) \cdot (x + y + \bar{z} + w) \cdot (x + y + z + w)$$

[3.26] Consider the function

$$f = a \cdot b \cdot c + a \cdot c + a \cdot \bar{b} + \bar{a} \cdot b$$

Cast this into canonical SOP form where every term has all three variables by using the identity $(x + \bar{x}) = 1$ as needed. Then use your SOP form to construct the K-map entries.

[3.27] Simplify the logic expression

$$h(x, y, z) = x \cdot y + x \cdot (\bar{y} + z) + (x + y) \cdot \bar{z}$$

using a 3-variable Karnaugh map.

[3.28] Simplify the logic expression

$$g(a, b, c) = \sum m(0, 1, 2, 3, 6)$$

using a 3-variable Karnaugh map.

[3.29] Use the Karnaugh map technique to simplify the expressions described by the function tables in the following problems.

    (a) Problem [3.4]
    (b) Problem [3.9]
    (c) Problem [3.10]
    (d) Problem [3.11]

[3.30] Simplify the logic expression

$$F = A \cdot B \cdot C + \bar{A} \cdot B \cdot \bar{C} + A \cdot \bar{B} \cdot C + A \cdot B \cdot C + \bar{A} \cdot \bar{B} \cdot C$$

using a 3-variable Karnaugh map.

[3.31] Consider the function

$$f(x, y, z) = \sum m(0, 1, 3, 4, 7)$$

Create the Karnaugh map for f and then simplify.

[3.32] Simplify the function

$$F(x, y, z, w) = \sum m(2, 3, 4, 7, 8, 14, 15)$$

using a 4-variable Karnaugh map.

[3.33] Find the simplest form for the functions described by the following K-maps.

x \ yz	00	01	11	10
0	0	1	0	0
1	1	1	1	1

(a)

x \ yz	00	01	11	10
0	1	0	0	1
1	1	0	1	1

(b)

[3.34] Find the simplest form for the functions described by the following K-maps.

u \ vw	01	11	10	00
0	0	1	1	0
1	1	0	0	1

(a)

u \ vw	11	10	00	01
0	0	1	1	0
1	0	1	1	1

(b)

[3.35] Find the simplest form for the functions described by the following K-maps.

u \ vw	0	1
00	0	1
01	1	0
11	1	1
10	0	1

(a)

u \ vw	0	1
00	1	1
01	1	0
11	1	0
10	0	0

(b)

u \ vw	0	1
10	0	1
11	1	1
01	1	0
00	0	1

(c)

[3.36] Find the simplest form for the function described by this 4-variable K-map.

xy \ zw	00	01	11	10
00	0	1	1	1
01	1	0	1	1
11	1	1	0	0
10	0	1	0	0

[3.37] Construct the function table for the expression

$$h(x, y, z, w) = x \cdot y \cdot (z + \overline{w}) + (\overline{x} + y) \cdot z$$

Then cast it into canonical SOP form.

[3.38] Use a 4-variable K-map to simplify the function

$$F(x, y, z, w) = \sum m(0, 1, 3, 4, 6, 7, 12, 13, 14)$$

to the simplest form.

[3.39] Find the simplest form for the function F described by the following 4-variable K-map.

xy \ zw	11	10	00	01
01	1	1	0	1
11	1	1	0	0
10	1	0	0	0
00	0	1	0	1

[3.40] Find the simplest form for the function g described by the following 4-variable K-map.

uv \ de	00	01	11	10
11	0	1	1	1
10	1	0	0	0
00	1	0	0	0
01	1	1	1	1

CHAPTER 4   Digital Hardware

The previous chapters have dealt with the theoretical foundations of digital logic systems. In order to apply these principles to the real world, we need to study how a digital network is constructed using electronic circuits. The physical realization of a digital element or system is termed **hardware**. In everyday usage, *hardware* refers to the many electronic items that actually are used to build a digital system, such as logic devices and circuit boards. In this chapter we will begin our study of what is involved in translating logic theory into a working system.

4.1 Voltages as Logic Variables

In order to build an electronic circuit that can perform logic operations, we must first find an electrical parameter that can be used to represent logic 0 and logic 1 states. The two obvious choices are the voltage *V*, which has units of *volts* (abbreviated as *v*) or the electrical current *I*, with units of *amperes* (which may be abbreviated as *A* or *amps*). You will recall from basic physics that the two are related: a voltage causes electrical current to flow. Most digital logic chips use two different voltage ranges to define logic 0 and logic 1 conditions.

All electronic networks require a **power supply** to operate. For example, laptop computers use a battery pack to allow for flexible usage. Power supplies come in many shapes and sizes, but they all have the same function: they provide electrical energy to the circuit. In digital circuits, the power supply is usually modeled as a **voltage source**, which is equivalent to a battery, with a value that we will denote by V_{DD}. The more common values of V_{DD} are 5 *v* and 3.3 *v*, but some chips are designed to use smaller values. When the battery is connected to an electronic circuit, it can supply the energy needed for proper operation. This occurs by forcing current flow through the circuit.

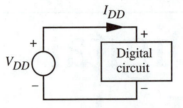

Figure 4.1
Power supply connection to a digital circuit

Electronic circuits are generally described using **schematic diagrams**. A schematic diagram uses graphical symbols to represent each type of electrical component and shows how the components are wired together to form the circuit.[1] This is, of course, identical in concept to creating a logic diagram using logic symbols. A simple schematic diagram is shown in Figure 4.1. It illustrates how a voltage source with a value V_{DD} (the circle-shaped symbol) is connected to a digital circuit (represented by the rectangle). Note that the polarity of the voltage source, i.e., the placement of the positive "+" and the negative "−" signs, is important. This is because the positive side is at a *higher* voltage than the negative side. In particular, the "+" terminal has a voltage that is V_{DD} volts higher than the "−" terminal.

Electrical circuits control the flow of electric current. In the drawing, the current delivered to the circuit from the power supply is denoted by I_{DD} and is measured in amperes. The amount of electrical **power** supplied to the circuit is calculated from

$$P = I_{DD}V_{DD} \tag{4.1}$$

and has units of *watts* (*W*), such that 1 *watt* is equal to the transfer of 1 *joule* of energy per second. When power is absorbed, the temperature of the chip increases because of the process of *Joule heating*. Although the chip is designed to operated at elevated temperatures, usually around 65°C, excessive heat can damage the device and must be avoided. Fans may be used to circulate air around the chip and carry away the heat energy. This is also a major factor in using a reduced value for the power supply voltage; reducing V_{DD} gives a smaller P if we can keep the current I_{DD} the same.

Every digital circuit we will study requires a power supply. Thus we will introduce simplified drawings where the presence of the voltage source is implied but is not shown explicitly in our diagrams. Figure 4.2 illustrates the idea. In Figure 4.2(a), the power supply is shown attached to the top and bottom of the circuit; the simplified drawing in Figure 4.2(b) is entirely equivalent. The positive side of the power supply is shown by the line labeled V_{DD}, and the negative side of the source is denoted by the symbol for a **ground**. In practice, the ground is used as reference for all other voltage in the circuit. Since a voltage can exist only between two points, the reference voltage of the ground connection is taken to be 0 *v*.

---

[1] The name *schematic diagram* arises from the fact that the drawing shows the "scheme" used to construct the circuit.

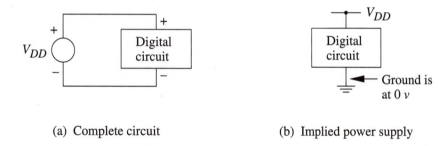

(a) Complete circuit   (b) Implied power supply

Figure 4.2 Simplified representation of the power supply voltage

4.1.1 Logic Levels

We are now in a position to define what is meant by logic 0 and logic 1 values in a digital circuit. If a power supply with a voltage of value V_{DD} is applied to a circuit, then the range of possible voltages inside the network is $[0, V_{DD}]$. Thus a natural definition is where

$$\text{Logic } 0 \Rightarrow 0 \, v$$

$$\text{Logic } 1 \Rightarrow V_{DD}$$

give the **ideal** voltages that represent logic 0 and logic 1 values, respectively. For example, if we use a power supply of value $V_{DD} = 5 \, v$, then

$$\text{Logic } 0 \Rightarrow 0 \, v$$

$$\text{Logic } 1 \Rightarrow 5 \, V$$

gives the association. These definitions are identical in concept to the encoding process used for defining binary words.

In practice, it is not necessary to define logic 0 and logic 1 voltage levels so precisely. Instead, we may represent 0s and 1s as defined over ranges of values, as shown in Figure 4.3. The convention wherein low voltages represent logic 0s, and high voltages represent logic 1s, is called **positive logic**. We will assume positive logic in all of our discussions. The opposite convention, called **negative logic**, uses

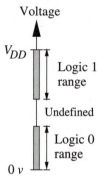

Figure 4.3
Voltage ranges corresponding to logic 0 and logic 1 binary values

high voltages to represent logic 0s and low voltages to represent logic 1 states. Negative logic is useful in some types of high-speed circuits.

4.2 Digital Integrated Circuits

Modern digital systems are built using devices called integrated circuits (**ICs**). These marvels of modern technology are also called **computer chips** in the popular literature. They provide all of the logic operations using time-varying voltages that correspond to changing Boolean variables.

Externally, an integrated circuit appears as a plastic or ceramic package with tiny metal strips or pins providing the needed electrical connections. One of the simpler types of packages, called the **dual in-line package** (**DIP**), is shown in Figure 4.4(a); the number of metal "legs" depends on the enclosed circuit and can vary from 8 pins to 40 or more pins. The end and top views of the DIP package are shown in Figures 4.4(b) and (c), respectively. The pins are numbered relative to a reference mark molded on the top of the plastic case.

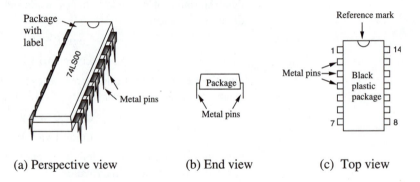

(a) Perspective view   (b) End view   (c) Top view

Figure 4.4 Integrated circuit dual in-line package

Integrated circuits are mounted on **printed circuit** (PC) **boards** and the pins are wired according to metal lines, or "traces," on the board itself. Figure 4.5(a) illustrates the general idea. The printed circuit board is a thin piece of insulating material (such as epoxy) that has metal lines on at least one side. Holes are drilled into the board to accept the IC pins. The chip is inserted into the board, and the pins are soldered to the metal lines as in Figure 4.5(b). The circuit connections are created by patterning the metal lines as required by the circuit diagram; Figure 4.6 shows that the pins of the IC are connected to the other components on the circuit board by routing the metallic paths. The case illustrated, where only one side of the board has a metal interconnect pattern, is the simplest possible structure. Complex networks often require the use of multilayer boards that provide several interconnection planes separated from each other by thin insulating layers.

The heart of the device cannot be seen without removing some of the package material. It is a tiny piece of silicon that is enclosed inside the plastic as shown in Figure 4.7. The silicon chip is actually *the* integrated circuit in that it provides the

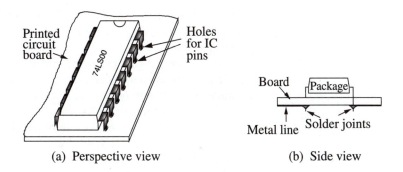

(a) Perspective view   (b) Side view

Figure 4.5 Placement of a DIP integrated circuit on a PC board

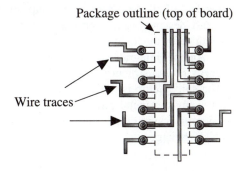

Figure 4.6 Wiring on a printed circuit board

electronic circuitry; the package is just for protection and ease of use. The silicon integrated circuit is a very complex device created by adding elements such as boron and phosphorus to the silicon itself and then applying additional layers of materials such as silicon dioxide (quartz glass) and metals on top of the silicon. Each new element or layer must be created with a precisely defined shape and size using a process called **optical lithography**. This and other aspects of the silicon

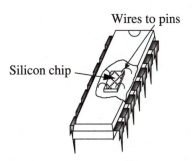

Figure 4.7
Cutaway view of DIP, showing silicon integrated circuit

integrated circuit fabrication process are examined in more detail in Chapter 7. For the moment, let us examine the IC as just a hardware device that provides electronic logic operations.

To the user, a digital integrated circuit may be viewed as a type of "black box" that has inputs and outputs that are specified by the manufacturer. There are literally thousands of different types of integrated circuits, and the actual functions provided by a particular device are not known until one consults the **data sheets**. A data sheet is exactly what its name implies: information on the IC, and the important specifications that are of interest to the user. Chips are identified by a part number written on top, which allows one to find the required data sheet in the manufacturer's **data book**.[2] The logic functions that a particular chip can implement are usually shown by embedding equivalent logic diagrams in package outline drawings, as shown in Figure 4.8. These are called **pin-out diagrams**, since they define how each pin is connected to the internal circuitry. In the example drawings, Figure 4.8(a) shows a hex (6) inverter chip, and Figure 4.8(b) is a quad (4) AND chip. Each gate is independent of all the others on the chip. Armed with this information, the electronic board designer can determine how to wire the integrated circuits together as required by the logic diagram.[3]

Another integrated circuit package you will probably encounter is the **pin grid array** (PGA) structure. A photograph of a PGA package is shown in Figure 4.9. The silicon chip is in the center of the package, and electrical access is attained by using the metal pins that are arranged in an ordered grid array. Most large microprocessor chips, such as the Intel Pentium® shown in Figure 4.9, use the PGA package.

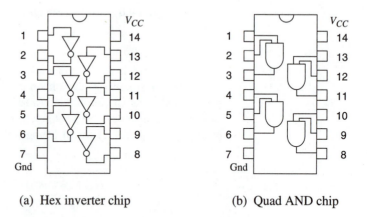

(a) Hex inverter chip   (b) Quad AND chip

Figure 4.8 Example of pin-out diagrams for a DIP IC

---

[2] Sometimes the chips have cryptic markings called **house numbers** specifying devices that are not listed in data books. These are specialized devices that are not usually available for general designs and can be very difficult to identify.

[3] V_{CC} is an alternative notation for the power supply voltage.

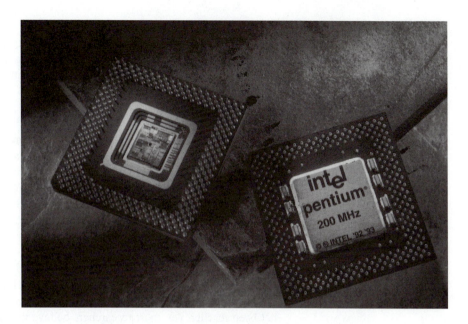

Figure 4.9 The pin grid array (PGA) integrated circuit package

4.2.1 Integration Levels

Integrated circuit technology has advanced to a very sophisticated level that allows extremely complex digital systems to be placed on a single piece of silicon 2 or 3 centimeters on a side. The **integration level** is a designator that is used to provide a measure of the number of gates (or switching devices known as transistors) that are on a chip. The standard integration levels are listed below. Note, however, that the separation between two adjacent levels is somewhat arbitrary, and no strict measures are used.

SSI

Small-scale integration (SSI) refers to integrated circuits that have a few gates, such as the hex inverter and quad AND examples presented above. SSI-level circuits are useful for building circuits from basic gates and also for use as **glue logic** networks that interface (or glue) complex networks together. They are inexpensive and readily available, which makes working with them an ideal way to learn digital logic and hardware concepts.

MSI

Medium-scale integration (MSI) describes more complex digital logic blocks. These are usually specialized circuits that are widely used in a large number of different systems. An example is a network that can add two binary words together. **Adder circuits** (which are examined in Chapter 8) can require on the order of 10 to 50 gates per bit, and an MSI adder circuit may consist of a few hundred (10^2) gates.

LSI

Circuits made in **large-scale integration** can be quite complex and generally contain as few as a thousand gates to as many as one hundred thousand gates. Examples of LSI circuits are calculator chips, electronic control units, and small microprocessors.

VLSI

This acronym stands for **very large-scale integration** and is one level above LSI. At this point, the number of switching devices (**transistors**) is often quoted instead of the number of gates. Typically, a VLSI circuit contains a few million transistors (10^6), which is roughly equivalent to several hundred thousand gates. Modern 16b and 32b microprocessors fall into this category. For example, the Intel Pentium® shown earlier contains about 3.3 million transistors. More recent designs have as many as 10–50 million transistors on them.

ULSI

The next step beyond VLSI is usually called **ultra large-scale integration**. It refers to systems that employ around one billion (10^9) transistors. Although true ULSI has not been reached yet, most researchers believe that it will be achieved in the near future.

The various levels of integration listed here are similar to the hierarchical view of digital logic design that we will adopt in this book. At the lowest (SSI) level, we deal with logic gates and simple functions as in Chapters 2 and 3. The MSI level is the next step. It consists of moderately complex logic blocks such as those introduced in Chapter 8. The most complex systems are those built in LSI and above. Although it is possible to use this technology to build enormous and powerful systems, all networks are based on the same set of fundamental circuits. This observation is true at all levels. The same comment holds for logic design in general. Regardless of the complexity, all digital networks are designed using the same basic logic operations.

Now that we have seen the basic ideas involved in integrated circuit technology, let us examine how the ICs actually implement the theoretical logic principles covered in Chapters 2 and 3. Once we understand the theory-to-practice interface, we will be better prepared to design a practical, working system.

4.3 Logic Delay Times

Consider the inverter shown in Figure 4.10(a); the input is A, and the output is $\overline{A}$. Suppose that the input is changed from a value $A = 0$ to $A = 1$ at a time $t = 0$, as shown in Figure 4.10(b). Ideally, the output of the logic gate would change from $\overline{A} = 1$ to $\overline{A} = 0$ at the same time as shown in Figure 4.10(b). In the real world, however, because of limitations imposed by the laws of physics, a physical parameter such as a voltage in an electronic switching network cannot change instantaneously. The voltage is being used to represent the Boolean variable A, so we must modify our viewpoint such that our theory corresponds to what actually occurs in a real gate. The **waveform** (a plot of a variable as a function of time) measured in the laboratory will be very different from that portrayed in the drawing. In particular, the output response will be delayed because it takes a finite amount of time for the signal to change. This is a characteristic of every physical logic gate and is a basic limitation on how fast a digital network can be operated.

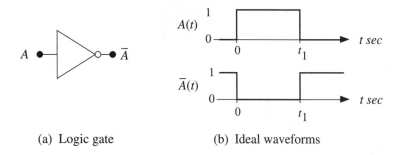

(a) Logic gate   (b) Ideal waveforms

Figure 4.10 Ideal inverter response

Let us briefly examine some of the important parameters that are used to describe the properties of an electronic logic gate.

4.3.1 Output Switching Times

Consider first the circuit shown in Figure 4.11(a), where we acknowledge that the logic gate is an electronic circuit by including the power supply voltage V_{DD} and ground connections. The output voltage $V_{out}(t)$ varies with time t and is used to represent a binary variable. Associated with it are predefined voltage ranges that represent "0" and "1" logic states. In an ideal circuit, these would correspond to 0 V and V_{DD}, respectively, but the actual values depend on the electronic circuit.

Two important switching times can be seen by assuming that the output voltage $V_{out}(t)$ changes like the waveform in Figure 4.11(b). The rate of change of the output voltage allows us to define the time intervals

- t_{LH}, the output **low-to-high time,** also called the **rise time,** t_r

- t_{HL}, the output **high-to-low time,** also referred to as the **fall time,** t_f

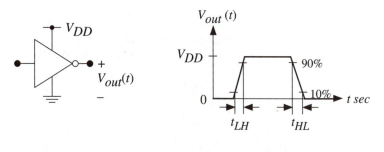

(a) Logic gate   (b) Low-to-high and high-to-low times

Figure 4.11 Output transition times

By convention, these time intervals are not measured between 0 v and V_{DD}, but represent the transition required between the 10% and 90% voltage levels, as shown.[4] For the output waveform shown, the signal swings between 0 v and V_{DD}. This gives the 10% voltage as $0.1V_{DD}$ and the 90% voltage as $0.9V_{DD}$. These delay times illustrate the fact that a physical gate requires a finite amount of time to change the output voltage levels. Both t_{LH} and t_{HL} are determined by the circuit used to create the inverter operation, and both are also affected by the next gates in the logic chain.

The absolute minimum amount of time needed for the gate to switch from a logic 0 to a logic 1 voltage level and back again is given by

$$t_{min} = t_{LH} + t_{HL} \tag{4.2}$$

This allows us enough time to let the output rise and fall to well-defined values. The **maximum switching frequency** f_{max} is then given by

$$f_{max} = \frac{1}{t_{LH} + t_{HL}} \quad [in \ hertz \ (Hz)] \tag{4.3}$$

This represents the maximum number of logic transitions that the gate can make in 1 second. The smaller the value of the minimum switching time t_{min}, the faster the gate evaluates logic inputs. Obviously, high-speed digital logic circuits depend on making t_{min} as small as possible. This is equivalent to saying that a fast digital gate can change its output voltage very quickly. In modern design, t_{LH} and t_{LH} are on the order of **nanoseconds** (*ns*), where 1 $ns = 10^{-9}$ *sec*.

Example 4-1

Consider an inverter that is calculated to have

$$t_{LH} = 7.2 \ ns \text{ and } t_{HL} = 3.9 \ ns$$

The maximum signal frequency is given by

$$
\begin{aligned}
f_{max} &= \frac{1}{(7.2 \times 10^{-9}) + (3.9 \times 10^{-9})} \\
&= 90.09 \ MHz
\end{aligned}
\tag{4.4}
$$

where 1 *MHz* is 10^6 (one million) *Hz*.

It is important to note that complex logic networks are built using many gates that are cascaded into a logic chain. The actual signal frequency for these types of systems is determined by the total delay through the chain, not by one gate.

---

[4] The actual range of the output voltage from a gate depends on the type of electronic circuitry used. The particular range from 0 V to V_{DD} is valid for CMOS logic gates.

4.3.2   The Propagation Delay

Although the time intervals t_{LH} and t_{HL} characterize the rising and falling edges of an output waveform from a gate, it can be cumbersome to have to keep track of both delay times for every logic gate. At the logic design level, it is simpler to introduce a single delay time that represents an average switching time, called the **propagation delay**, from the input to the output. This is used to include the physical delay of a logic signal as it "propagates" through a chain of gates.

Consider the logic voltages shown in Figure 4.12. The input in denoted by $V_{in}(t)$, and the output is $V_{out}(t)$. It is important to remember that both the input and the output voltages can have values between 0 v and V_{DD}, but only certain ranges actually represent logic 0 and logic 1 states in the Boolean sense. An electronic logic gate is designed so that it reacts to an input voltage and tries to produce a reasonably well-defined low or high voltage at the output. This feature is shown by the waveforms in Figure 4.13. Note that since the input voltage to the stage was obtained as an output voltage from a previous logic circuit, both the input and the output have the same general shape.

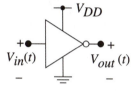

Figure 4.12
Inverter input and output voltages

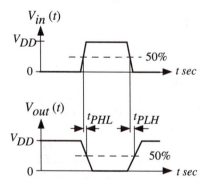

Figure 4.13
Switching waveforms

The (logic) propagation delay from the input to the output is defined by using two basic time intervals. The first is t_{PHL}, which is the propagation delay for an output transition from a high to a low state. As seen in Figure 4.13, this represents the delay time from the input 50% voltage (when V_{in} reaches a value of $V_{DD}/2$) to the output 50% voltage (when V_{out} falls to a value of $V_{DD}/2$). Similarly, we use t_{PLH} as the propagation delay for an output transition from a low to a high state. The propagation delay t_p is then taken to be the simple average of the two:

$$t_p = \frac{1}{2}(t_{PHL} + t_{PLH}) \tag{4.5}$$

This can be used as a reasonable estimate for the logic delay time for the electrical network. Sometimes we use the alternative definition

$$t_p = \max(t_{PHL}, t_{PLH}) \tag{4.6}$$

that is, the larger of the two.

Even though the timing delay is a characteristic of the electronic nature of a gate, it is important to include it in our logic models so that we can account for it when designing and analyzing real networks. This is accomplished as shown in Figure 4.14, where we include the propagation delay as a feature when describing the logic variable $A(t)$. Care must be exercised when using extrapolations of this type. If A is defined to be a Boolean variable, then by definition it must be a 0 or a 1; it cannot take on an intermediate value as Figure 4.14 implies. This is not a problem so long as we remember that this type of drawing is used to include physical effects.

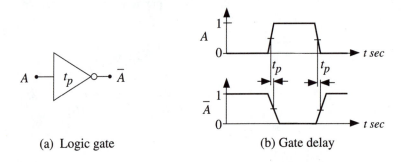

(a) Logic gate   (b) Gate delay

Figure 4.14 Effect of the propagation delay through a gate

4.3.3 Fan-Out and Fan-In

The switching time of an electronic gate depends on the number of gates that are driven at the output. The **fan-out** of a gate is the number of gates that are driven by the output, and it depends on how the gate is used in the logic chain. In general, increasing the fan-out slows down the logic flow through the gate.

Fan-out considerations can be understood by means of the circuits in Figure 4.15. The propagation delay t_p of a gate is due to two factors. The first is the intrinsic delay of the gate itself. This delay that the gate exhibits even without any additional gates at the output is called the "no load" condition. In Figure 4.15(a), it is called the internal delay t_{p0} and represents the case where the fan-out is 0. Connect-

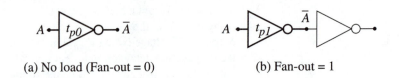

(a) No load (Fan-out = 0)   (b) Fan-out = 1

Figure 4.15 Calculation of the propagation delay

ing a single inverter to the output as illustrated in Figure 4.15(b) yields a circuit where the first gate has a fan-out of 1. The delay through the first gate for this case is given by

$$t_{p1} = t_{p0} + 1t_{pL} \tag{4.7}$$

where t_{pL} is the additional delay time needed to drive the load (which is the next inverter). Each load added to the circuit will increase the delay. For example, the circuit in Figure 4.16 has a fan-out of 4, so we find that the propagation delay would be approximately

$$t_{p4} = t_{p0} + 4t_{pL} \tag{4.8}$$

by assuming that each gate adds a factor of t_{pL} to the overall delay. In general, for a fan-out of N, the delay is approximately

$$t_{pN} = t_{p0} + Nt_{pL} \tag{4.9}$$

Although logic design is sometimes easier with large fan-out networks, the delay at the physical level usually restricts the design to relatively small fan-out values.

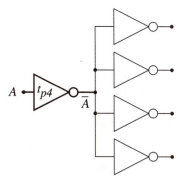

Figure 4.16
The propagation delay circuit for a fan-out of 4

Example 4-2

Suppose that we have an inverter that is characterized by

$$t_{p0} = 1ns \text{ and } t_{pL} = 0.25 \ ns$$

The propagation delay is plotted as a function of the fan-out N in Figure 4.17. The extrapolated dashed line illustrates that the propagation delay increases in a linear manner with N. This implies that a design that uses a large fan-out will exhibit a long delay time.

The **fan-in** of a digital logic gate refers to the number of inputs. For example, an inverter has a fan-in of 1, a NAND2 has a fan-in of 2, and so on. To a logic designer, the fan-in of a gate is chosen to accommodate the number of inputs. At the hardware level, however, the fan-in provides information about the intrinsic speed of the gate itself. In general,

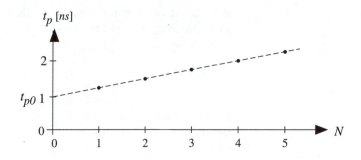

Figure 4.17 Increase in propagation delay with fan-out

The propagation delay increases with the fan-in.

This means, for example, that a NOR2 gate is faster than a NOR3 gate. This is because a larger number of inputs requires a more complicated electronic circuit, and every electronic device used in a logic gate can potentially slow down the switching response.

4.3.4 Extension to Other Logic Gates

The foregoing discussion illustrates an important point. Every digital logic gate is characterized by a set of intrinsic delay times; in our approach, we have emphasized the propagation delay t_p and the output transition times t_{HL} and t_{LH}. The actual numerical values of these time intervals depend on (1) the gate itself and (2) the load that it is driving. When designing a realistic logic network, we must consider both the logic formation (from Boolean algebra) and the timing behavior (from the hardware).

To understand the implications of hardware, let us extend the discussion from a simple inverter to multiple-input gates. Figure 4.18 shows a 2-input AND gate with inputs $A(t)$ and $B(t)$ plotted in the timing diagrams; the output is designated as $f(t)$. The output remains at 0 for the first three combinations $AB = 00, 10, 01$. In the fourth time interval, $B = 1$ and A makes a transition from 0 to 1. The output f responds after a propagation delay t_p as shown. When the inputs both fall back down to $AB = 00$, it again requires a delay of t_p before the output responds.

Another example is shown in Figure 4.19 for an XOR2 gate. Using the same inputs as for the AND2 gate gives the output $g(t)$ shown. Since the XOR2 operation gives a 1 when only one input is a 1, switching A from a 0 to a 1 when $B = 0$ induces the same transition at the output as shown in the second time interval. For the third set of inputs where $A = 0$ and $B = 1$, the g remains at a 1. The output falls back to a 0 after a propagation delay when A changes from a 0 to a 1 in the fourth time interval.

These simple examples serve to illustrate that the propagation delay occurs whenever the output changes from a 0 to a 1 or from a 1 to a 0, regardless of the input combination that induced the switching event. Although timing diagrams based on the propagation delay are sufficient for basic designs, it is important to remember that the output switching delays t_{HL} and t_{LH} are not generally equal. The

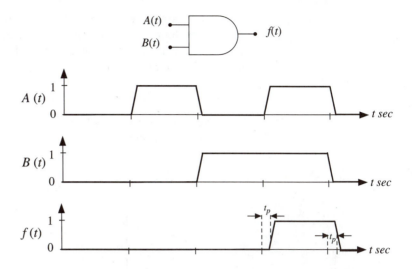

Figure 4.18 Effects of propagation delay in an AND2 gate

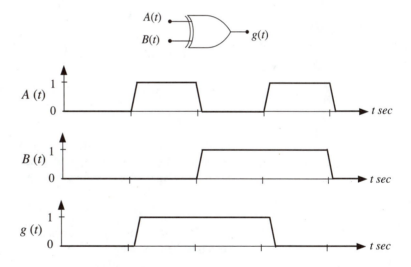

Figure 4.19 Propagation delay in an exclusive-OR gate

difference between the two values usually increases with the fan-in. The asymmetry of the output waveform is important when one is designing critical data paths.

4.3.5 Logic Cascades

The propagation delay of each gate contributes to the delay of a logic signal through a cascaded logic chain. This important characteristic limits the conversion of logic equations to real-world hardware since the timing may become critical.

Let us examine how to model the logic delay through a cascade using the simple equations we have examined. Figure 4.20 portrays a linear chain of inverters. Each gate has a delay time t_{dn} associated with it. The total delay through the chain can be estimated by adding these together to obtain

$$t_d = t_{d1} + t_{d2} + t_{d3} + t_{d4} \tag{4.10}$$

This represents the total delay between the signal $A(t)$ entering the chain at the left and the output $B(t)$ on the right, as shown in the timing diagrams. The first three

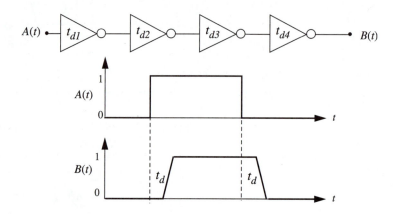

Figure 4.20 Signal delay through an inverter chain

gates have a fan-out of 1 into an identical inverter, so these gates have delays of

$$t_{dn} = t_{p0, NOT} + t_{pL, NOT} \qquad (n = 1, 2, 3) \tag{4.11}$$

where $t_{p0, NOT}$ is the internal delay of an inverter gate, and $t_{pL, NOT}$ represents the delay induced by driving the input of an identical NOT gate. The load at the output of the fourth stage is not specified, so we write

$$t_{d4} = t_{p0, NOT} + t_L \tag{4.12}$$

where t_L is the delay due to the load at B. Our expression for the total delay through the chain is then given by

$$t_d = 4t_{p0, NOT} + 3t_{pL, NOT} + t_L \tag{4.13}$$

It is important to note that the fan-out of each gate is important in this sum.

To understand the significance of this analysis, consider the modified chain shown in Figure 4.21. Two additional gates have been added at the output of stage 1, and stage 3 has one additional load to drive. The new delays for these gates are now given by

$$t_{d1} = t_{p0, NOT} + 3t_{pL, NOT}$$
$$t_{d3} = t_{p0, NOT} + 2t_{pL, NOT} \tag{4.14}$$

so the total delay through the chain in this case is

$$t_d' = 4t_{p0,NOT} + 6t_{pL,NOT} + t_L \qquad (4.15)$$

which is larger than t_d by a factor of $3t_{pL,NOT}$. The important point here is that the delay through the chain has increased even though there are still the same number of gates between $A(t)$ and $B(t)$. The slowdown is due to the fact that both gates 1 and 3 exhibit longer delays because of the larger loads.

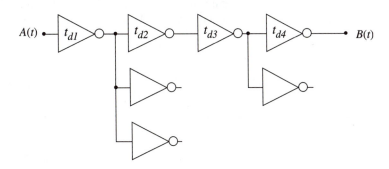

Figure 4.21 Inverter chain with increased internal delay

Although inverter cascades are used in logic design, most logic chains use different gate operations to perform the necessary operations. Each type of gate is characterized by a particular set of internal and load characteristics, so it is necessary to modify the analysis to account for these differences. Let us examine the situation where we use only the NOT, NAND2, and NOR2 gates in our designs. For each gate, we introduce the internal characteristic delay t_{p0} and the load delay t_{pL} seen by a driving gate, as summarized in Figure 4.22. The total propagation delay through a chain of N gates is still obtained by summing the individual delays via

$$t_d = \sum_{n=1}^{N} t_{pn} \qquad (4.16)$$

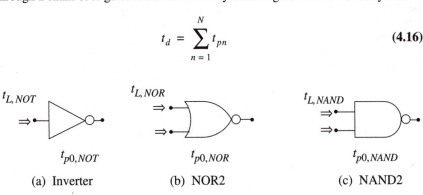

(a) Inverter   (b) NOR2   (c) NAND2

Figure 4.22 Basic contributions to gate delay

Now, however, each gate must be characterized by the proper numbers.[5]

---

[5] This approximation assumes that every gate has about the same drive capability so that the load delay times used in the fan-out terms are identical.

An example of a basic cascade is shown in Figure 4.23(a). The total delay from the input to the output is given by

$$t_d = t_{d1} + t_{d2} + t_{d3}$$
$$= (t_{p0,NOT} + t_{pL,NOR}) + (t_{p0,NOR} + t_{pL,NOT}) + (t_{p0,NOT} + t_L) \tag{4.17}$$

where we have once again introduced a load time t_L at the output. The first term is the internal delay of the NOT gate plus the load of the NOR2 gate, and the second term is the internal NOR delay added to the NOT load delay. The third term is one more internal NOT delay in addition to the output load time. This technique can be

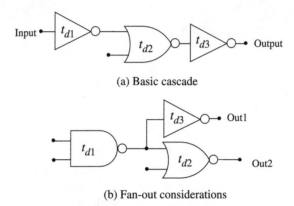

(a) Basic cascade

(b) Fan-out considerations

Figure 4.23 Signal delay in logic cascades

extended to more complex schemes, such as that shown in Figure 4.23(b) for the case where the first gate has a of FO = 2. The delay between the inputs of the NAND gate and Out1 is

$$t_{d,Out1} = t_{d1} + t_{d3}$$
$$= (t_{p0,NAND} + t_{pL,NOR} + t_{pL,NOT}) + (t_{p0,NOT} + t_L) \tag{4.18}$$

and the delay to Out2 is given by

$$t_{d,Out2} = t_{d1} + t_{d2}$$
$$= (t_{p0,NAND} + t_{pL,NOR} + t_{pL,NOT}) + (t_{p0,NOR} + t_L) \tag{4.19}$$

Both logic paths are affected by the dual fan-out of the first stage.

4.4 Basic Electric Circuits

Electronic circuits are used to build computing systems because they can be switched much faster than, say, a mechanical apparatus.

Let us first examine the concept of electric current flowing through a wire as depicted in Figure 4.24. The current I is measure in amperes (A) and represents the flow of charge across a reference area as shown. When we denote the charge by Q, which is measured in units of coulombs (C), the current is the time rate of change of the charge across the area, as given by the derivative

$$I = \frac{dQ}{dt} \tag{4.20}$$

This defines the unit of ampere as 1 coulomb/second crossing the reference plane.

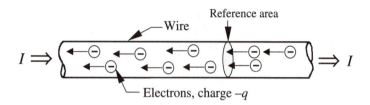

Figure 4.24 Electric current flow in a wire

In a wire, the charge flow is due to electrons that experience a force from the electric field established by the voltage applied to the wire. Electrons are negatively charged with a value $Q = -q$, where $q = 1.602 \times 10^{-19}$ coulombs [C] is the fundamental charge unit. To obtain the charge motion from right to left as shown, the voltage on the left side must be higher than the voltage on the right side. It is important to note that we will use conventional current flow throughout this book. Conventional current is defined to flow in the direction of positive charge, so it flows in a direction opposite to that of electron motion. In the present example, the current I flows from left to right.

4.4.1 Resistance

Current flow can be viewed as the motion of charges in response to an applied voltage difference. However, when we force current through any material, internal atomic or molecular structure results in a natural tendency to "resist" the flow of charge. Since electronic circuits rely on the flow of current, we formalize this concept, introducing a parameter R called the **resistance** and assigning a value of R to every electronic element we use.

The general problem is shown in Figure 4.25(a). Given a voltage V across the device, the current I depends on the internal characteristics. The symbol used for a resistor is provided in Figure 4.25(b); the "jagged, rough" current path implied by the symbol is a good reminder that this represents a resistor. The resistance is defined as the ratio of voltage V to current I

$$R = \frac{V}{I} \tag{4.21}$$

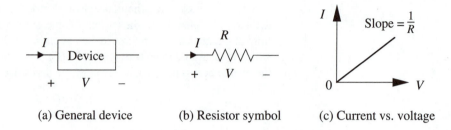

(a) General device   (b) Resistor symbol   (c) Current vs. voltage

Figure 4.25 Resistor characteristics

The unit of resistance is the **ohm,** and resistance is denoted by the Greek letter omega Ω. It is defined such that a 1-ohm resistor admits 1 ampere of current with 1 volt applied across it. This is also a statement of **Ohm's Law**, which is usually written as

$$V = IR \qquad (4.22)$$

This says that the voltage across a resistor is proportional to the current (and vice versa), R being the proportionality constant. In terms of circuit theory, we usually write

$$I = \frac{1}{R}V \qquad (4.23)$$

and plot the current I as a function of V as shown in Figure 4.25(c). In this case, the quantity $(1/R)$ is the slope.

The above equations tell us that the behavior of a resistor is very simple. Consider the case where the voltage V is known. A small value of R implies that the current will be large (it doesn't resist the flow much), and a large value of R means that the current will be small.

The power dissipation P of a resistor in units of **watts** (W) is given by

$$\begin{aligned} P &= VI \\ &= (IR)I \\ &= I^2R \end{aligned} \qquad (4.24)$$

where we used Ohm's Law from equation (4.22) in the second step. In integrated circuit design, this type of power dissipation leads to heating and can cause thermal instabilities and failures. This illustrates a tradeoff, since a small value of R gives large current flow but is accompanied by large power dissipation.

Analyzing digital networks often leads to the case where we find two resistors that are arranged as shown in Figure 4.26(a). This is called a **series** connection and is defined by the fact that both resistors R_1 and R_2 have the *same* current I flowing through them. Two resistors in series increase the total resistance between the left and right sides such that the two elements may be modeled as a single resistor with a value of

$$R = R_1 + R_2 \qquad (4.25)$$

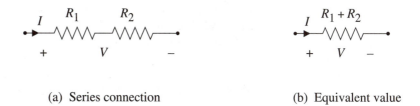

(a) Series connection   (b) Equivalent value

Figure 4.26 Resistors in series add to give a single resistor

as shown in Figure 4.26(b). This may be extended to any number of series-connected resistors.

4.4.2 Capacitance

Capacitance is a parameter that describes how a particular device can store electric charge. Figure 4.27(a) shows the parallel-plate capacitor that is familiar from your studies in physics. This device consists of two metal plates that are separated from each other by an insulating layer such as glass. If a voltage V is applied to the plates as shown, then a charge of $+Q$ will be induced on the top plate and a charge of $-Q$ will be on the bottom plate. The circuit symbol for a capacitor is shown in Figure 4.27(b); it is easy to remember since one can visualize the two plates.

The amount of charge Q stored on the capacitor and the applied voltage V are related by the equation

$$Q = CV \tag{4.26}$$

where C is a proportionality constant that is called the **capacitance**. In the charge-voltage plot of Figure 4.27(c), C is the slope of the line. The capacitance gets its name from the fact that C tells us the charge storage "capacity" of the device; the larger C, the greater the amount of charge that can be stored for a given voltage. The unit of capacitance is the **farad** (F) such that

$$C = \frac{Q}{V} \tag{4.27}$$

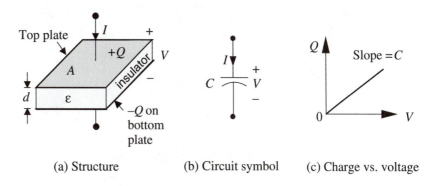

(a) Structure   (b) Circuit symbol   (c) Charge vs. voltage

Figure 4.27 Characteristics of a capacitor

says that a 1-farad capacitor can store 1 coulomb of charge with 1 volt applied. In the world of microelectronics, realistic capacitors have very small values of C, typically on the order of $10^{-12} F$ or smaller.

To move charge onto or off the plates of the capacitor requires that a current flow be established. The current into the positive plate of a capacitor can be computed by using the definition of current in equation (4.20). This gives

$$
\begin{aligned}
I &= \frac{dQ}{dt} \\
&= \frac{d}{dt}(CV) \qquad\qquad\qquad \textbf{(4.28)} \\
&= C\frac{dV}{dt}
\end{aligned}
$$

where we have substituted for the charge $Q = CV$ from the definition. This tells us that the current I is proportional to the time rate of change of the voltage (dV/dt). One important consequence of this result is that the voltage $V(t)$ cannot change in an instantaneous manner (in zero time). This can be seen by noting that an instantaneous change would take place in a time interval $dt \approx \Delta t \to 0$, but this would make ($\Delta V/\Delta t$) infinite. Since the current is proportional to $\Delta V/\Delta t$, this could take place only if I were also infinite, which is not physically possible.

Figure 4.28(a) shows two capacitors C_1 and C_2 that are connected in a **parallel** arrangement. The parallel connection is defined by the fact that both capacitors have the same voltage V across them. For a given value of V, the total charge stored on the two devices is

$$
\begin{aligned}
Q &= Q_1 + Q_2 \\
&= C_1 V + C_2 V \qquad\qquad\qquad \textbf{(4.29)} \\
&= (C_1 + C_2)V
\end{aligned}
$$

This leads us to write that the equivalent value of the two capacitors is given by the sum

$$
C = C_1 + C_2 \qquad\qquad\qquad \textbf{(4.30)}
$$

This can be extended to any number of capacitors that are connected in parallel.

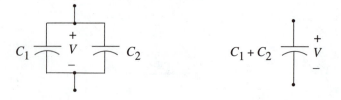

(a) Parallel connection   (b) Equivalent value

Figure 4.28 Capacitors in parallel add values

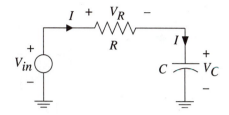

Figure 4.29 An RC circuit

4.4.3 The RC Circuit

Digital switching networks can often be modelled by using the series-connected RC circuit shown in Figure 4.29. This simple network consists of an input voltage source with its positive side connected to a resistor R; the other side of the resistor is connected to a capacitor C, and the capacitor is connected back to the voltage source through the ground connection. Since the three components are in series, they all carry the same current I. However, three different voltages are shown in the circuit, one across each element. These are the input voltage V_{in}, the resistor voltage V_R, and the capacitor voltage V_C. Following the circuit in a clockwise direction starting from the ground connection on the left-hand side shows that there is a voltage rise [from negative (–) to positive (+)] of V_{in} due to the voltage source. Continuing around the circuit, we see that this is followed by a voltage drop [from positive (+) to negative (–)] of V_R across the resistor and then by another voltage drop of V_C across the capacitor. The relationship among these voltages is from **Kirchhoff's Voltage Law** (KVL), which states that around a closed circuit,

Sum of voltage rises = sum of voltage drops

In mathematical terms, this allows us to write

$$V_{in} = V_R + V_C \tag{4.31}$$

which shows that the resistor voltage V_R and the capacitor voltage V_C respond to changes in the input voltage V_{in}.

Let us examine the behavior of the circuit when the input voltage is switched from 0 v to a high voltage V_H at time t_1 and then back to 0 v at time t_2, as shown in the upper waveform of Figure 4.30; this is the voltage equivalent of logic pulse. Initially, all of the voltages are 0. In particular, there is no charge on the capacitor so V_C starts at 0 v.

When V_{in} is changed to a high voltage V_H, current starts flowing through the resistor from the top (+ terminal) of the voltage source to the top of the capacitor. The current is given by

$$I = \frac{dQ}{dt} \tag{4.32}$$

so the current "charges the capacitor," and the capacitor charge Q_C increases with time. Since the capacitor voltage is given by

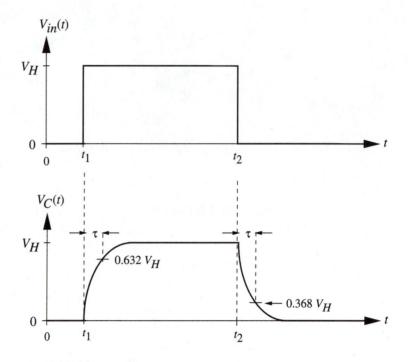

Figure 4.30 Voltage waveforms for the RC circuit

$$V_C = \frac{Q_C}{C} \qquad (4.33)$$

V_C increases with time, as shown in the lower waveform of Figure 4.30. Eventually, the capacitor voltage reaches a value $V_C = V_H$, and the current flow ceases. When the input voltage switches from V_H back down to 0 v at time t_2, the reverse situation takes place. The charge on the capacitor now flows through the resistor and the voltage source to ground, and the capacitor voltage V_C eventually decays to 0 v.

Let us first examine the charging event in more detail. The time delay as the capacitor voltage changes from 0 v to V_H is due to the resistance R in the circuit, which impedes the current flow. Once the current flow starts, the voltage across the resistor increases since

$$V_R = IR \qquad (4.34)$$

where

$$V_R = V_{in} - V_C \qquad (4.35)$$

by KVL. This automatically limits the level of the current and the rate of charge transfer, since we may write

$$I = \frac{V_{in} - V_C}{R} \qquad (4.36)$$

by combining the two equations. The important lesson from this analysis is that resistance always slows the flow of charge in an electrical network. Since resistance cannot be eliminated in the real world,[6] it is important to understand the nature of the delay.

A mathematical analysis of the series RC circuit gives a simple result for the capacitor voltage $V_C(t)$ for both events. When the capacitor is charged, the voltage increases according to the formula

$$V_C(t) = V_H[1 - e^{-(t-t_1)/\tau}] \qquad t \geq t_1 \qquad \textbf{(4.37)}$$

which is called an **exponential dependence**. In this equation, $e = 2.71828\ldots$ is the "Euler e" familiar from the Calculus, while the Greek letter τ (*tau*) is called the **time constant** and is computed from

$$\tau = RC \quad \text{seconds} \qquad \textbf{(4.38)}$$

Note that the time constant has units of seconds. The significance of this can be seen in the plot of $V_C(t)$: τ represents the delay for the capacitor voltage to change from 0 V to a high value that is chosen at time τ seconds from t_1. Mathematically, the voltage at this time is given by

$$V_C(t = t_1 + \tau) = V_H[1 - e^{-1}]$$
$$\approx 0.632 V_H \qquad \textbf{(4.39)}$$

This is often approximated as $(2/3)V_H$, which is easier to remember.

A small τ means a fast transition and a large τ describes a slow transition. The discharge event is described in a similar manner. The capacitor voltage is given by the exponential dependence

$$V_C(t) = V_H e^{-(t-t_2)/\tau} \qquad (t \geq t_2) \qquad \textbf{(4.40)}$$

so that the value of τ also describes the transition from a high voltage V_H to a low value of

$$V_C(t = t_2 + T) = V_H e^{-1}$$
$$\approx 0.368 V_H \qquad \textbf{(4.41)}$$

This is often simply approximated as $(1/3)V_H$ when a calculator is not handy.

Example 4-3

Consider an RC circuit with values of $R = 1000\Omega$ and $C = 10^{-12}$ F, as shown in Figure 4.31(a). The time constant for this circuit is given by

---

[6] The exception to this statement is in a **superconductor** material that exhibits zero resistance when the temperature is reduced to very low temperatures. Modern ceramic superconductors must be cooled to around 80–90 K for this to occur.

$$\tau = RC$$

$$= 10^3 \times 10^{-12} \tag{4.42}$$

$$= 10^{-9} sec$$

which is equal to 1 nanosecond (*ns*) such that $1\ ns = 10^{-9}\ sec$. If we take the voltage source to have a value of $V_H = 5\ v$, then a low-to-high transition in which the capacitor voltage swings from $V_c = 0\ v$ to $V_c = (0.632)(5) = 3.16\ v$ takes a time $t = \tau = 1$ *ns*. Similarly, if the capacitor voltage is initially at $V_C = 5\ v$, then it takes the high-to-low time $t = \tau = 1\ ns$ for the voltage to fall to a value of $V_C = (0.368)(5) = 1.84\ v$. Now then, suppose that we use the same circuit but reduce the resistance to a value $R = 500\Omega$ and $C = 5 \times 10^{-13}\ F$, as shown in Figure 4.31(b). Then the time constant is reduced to

$$\tau = (500) \times (5 \times 10^{-13}) \tag{4.43}$$

$$= 0.25\ ns$$

which is 1/4 the value of the first circuit. This means that the capacitor voltage can change 4 times faster. The difference in the response of the two circuits is seen in

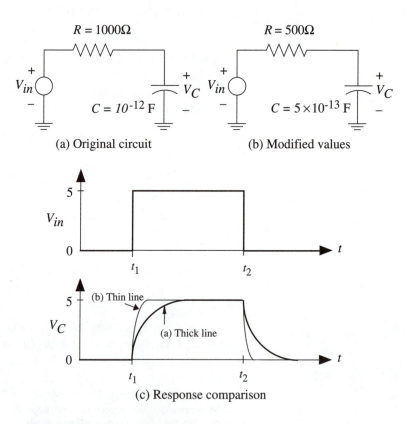

(a) Original circuit   (b) Modified values

(c) Response comparison

Figure 4.31 RC circuit example

Figure 4.31(c). Note that the same result could have been achieved by keeping $R = 1000\Omega$ but reducing the capacitance to $C = 2.5 \times 10^{-13} F$.

4.4.4 Application to Digital Circuits

Electronic networks provide the building blocks for computational systems. When we use a changing voltage $V(t)$ to represent a Boolean variable, we must allow sufficient time for the voltage to stabilize to a low "0" or high "1" value. The switching time of every digital electronic network is governed by any resistance and capacitance in the network. In the physical world, it is not possible to eliminate either resistance or capacitance. **Parasitic** elements are unwanted resistance and capacitance contributions that cannot be eliminated and that act to slow down the network response.

Because the switching speed of a digital logic gate depends on the parasitic elements in the circuit, much of the design effort is directed toward reducing their values. This is important at every level of hardware design, from the integrated circuit all the way up to the completed system.

4.5 Transmission Lines

Another type of logic delay arises when we analyze the physics of transmitting a voltage along a wire. Consider the situation shown in Figure 4.32, where the Transmitter circuit produces an output voltage pulse $V_T(t)$. This is sent to the Receiver unit through a wire that is called a **transmission line** and is eventually picked up as $V_R(t)$ there. Ideally, there would not be any time delay between the Transmitter and the Receiver, and the Receiver would detect the transmitted pulse as it was being generated. This, however, would violate the laws of physics, which do not allow an instantaneous transmission of energy from one point to another.

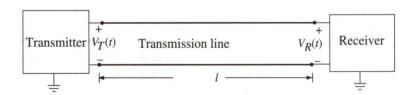

Figure 4.32 Transmission line connecting two digital units

The delay in this circuit arises from the fact that energy is being sent along the transmission line as an **electromagnetic wavefront** with a finite velocity v_w. Consider the close-up of the pair of wires shown in Figure 4.33. The voltage across the wires is due to a difference in electric potential between them, and the electromagnetic energy is stored in the energy wavefront as it moves along the wire. At this level we see that this is just an electromagnetic wave that is guided along the wire. The velocity of the wavefront is given by

$$v_w = \frac{c}{n} \tag{4.44}$$

Figure 4.33 A "twin-lead" transmission line

where $c \approx 3 \times 10^{10}$ *cm/sec* is the speed of light in a vacuum, and $n \geq 1$ is a factor that accounts for the energy storage properties of the insulating region between the two wires.[7] The value of n also depends on the geometry of the line. Figure 4.34 shows the cross-sectional view of a **coaxial cable,** which is used for very high data transmission rates. In this type of transmission line, the signal is carried on the wire in the center of the cable. This is surrounded by an insulating plastic, which itself is encased in a braided metal layer that is connected to ground and acts as a shield against any electromagnetic disturbances.

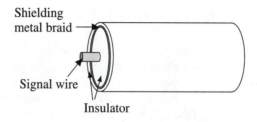

Figure 4.34 Structure of a coaxial cable

Now let us return to the problem of signal delay. If we measure the transmitter and receiver voltages $V_T(t)$ and $V_R(t)$, respectively, we obtain plots similar to those shown in Figure 4.35. The total delay is shown as a time t_d in the drawings and is due to two distinct factors. First, there is the transmission line signal delay t_s due to the finite speed of propagation along the wire. With a wavefront velocity v_w on a wire of length x, this is given by the time

$$t_s = \frac{x}{v_w} \tag{4.45}$$

Although this delay is usually quite small, it can be significant in systems where long wires are required to connect units together.

The remaining portions of the time delay t_d are due to the fact that the voltage at the end of a transmission line usually takes a small amount of time to "build up" to the final value. This is due to a problem known as a mismatch, wherein the Receiver cannot absorb all of the energy at one time but actually reflects some of the energy

---

[7] In optics, n is called the index of refraction.

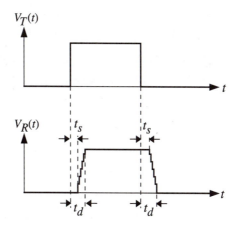

Figure 4.35 Effects of transmission line delay

back to the transmitter. The transmitter then retransmits some of the energy back to the receiver, and so on. This leads to the situation where a small amount of energy literally "bounces" between the transmitter and the receiver, delaying the transmission of the total signal. This is an example of a phenomenon called **ringing,** in which a physical parameter such as voltage oscillates in time. You may have seen the effect on a television receiver that uses a wire antenna; sometimes, multiple images appear on the screen (called ghosting); these images are due to reflections. Although the details of this type of problem are beyond the scope of this book, it is important to keep the effects in mind when designing the wiring between logic units.

4.5.1 Crosstalk

Another problem that occurs in data transmission is called **crosstalk**. This situation occurs when a portion of the signal energy on one wire is transferred to a neighboring line.

Figure 4.36 illustrates the origin of the problem. When two independent conductors are close to each other, capacitance always is present. In the case of two interconnect line such as those on a printed circuit board, this induces a coupling capacitance C_c that couples the interconnect lines via electric field interactions. If one line is at a voltage V_1 and the other at a voltage V_2, then the voltage across the

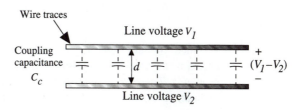

Figure 4.36 Coupling between closely spaced transmission lines

capacitance is $(V_1 - V_2)$. If one of the voltages changes in time, then the electric coupling causes a change on the other line. The coupling increases with the value of the coupling capacitance C_c. This, in turn, depends on the separation distance d between the two lines as shown in Figure 4.37. Closely space lines with small values of d lead to large values of C_c, which in turn increase the coupling strength.

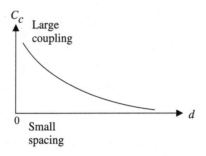

Figure 4.37 Coupling capacitance as a function of separation distance d

Crosstalk can lead to logic errors in hardware. This can be illustrated by considering Figure 4.38(a) where the input line to an inverter is at a voltage $V_{in} = 0$ V. Suppose that a neighboring line has a voltage pulse traveling along it. The coupling capacitance C_c allows an energy transfer between the two and may cause a "blip" in the input voltage, as shown in Figure 4.38(b). If the change is too large, it may cause the inverter to switch, resulting in an error in the output.

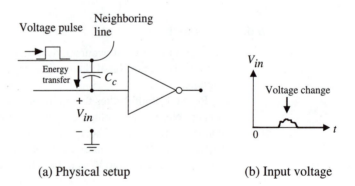

(a) Physical setup   (b) Input voltage

Figure 4.38 Signal transfer due to crosstalk

Since the coupling capacitance increases with the distance of separation between the lines, high-density interconnect networks can be subject to crosstalk problems. These are not restricted to printed circuit boards but occur at all levels, including wiring between boxes and micro wiring on the chip itself.

4.5.2 Electromagnetic Interference

A problem similar to crosstalk is electromagnetic interference (**EMI**). EMI is due to the fact that all electrical systems emit energy in the form of electric and magnetic fields. For example, both electric motors and the spark plugs in gasoline engines emit large levels of stray electric fields. EMI occurs when this radiation is "picked up" by a circuit, causing a change in the voltage level.

The problem is illustrated in Figure 4.39. The electromagnetic energy emitted from the source is detected by the circuit wiring, which can be viewed as acting like a basic antenna. This induces a random voltage fluctuation on the wire, which may in turn cause a false trigger at the input of the logic gate. This type of interference is particularly troublesome in environments that are noisy (in the electromagnetic sense), such as a factory that uses high-current electric motors. However, even automobiles and clocks emit stray electromagnetic fields that may cause problems. Because it is necessary to design digital hardware that can operate in ordinary locations where one has no control over the EMI levels, a substantial amount of system design is dedicated to blocking the unwanted energy. The most basic approach is to **shield** the circuits by enclosing them inside metallic containers that are grounded. This is usually accompanied by other techniques, such as strategic placement of the circuit boards and use of special electronic circuits.

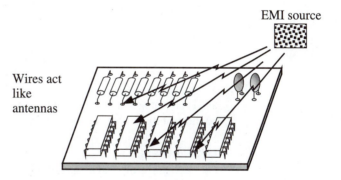

Figure 4.39 EMI noise problems

The inverse problem is that of **radiation,** where changing currents in the circuits cause electromagnetic energy to move away from the network into the space around it. In other words, the circuits themselves can act as antennas and cause problems in other electronics devices. Shielding helps diminish the radiated energy in the same way that it protects the circuits from outside influences. The problem is so important that the government limits the amount of radiation that devices are allowed to emit.

4.6 Logic Families

Electronics provides several different types of technologies and design techniques that can be used to build digital logic circuits. This is accomplished by starting with a simple logic gate, such as a NOT or a NAND, that is built from transistors that are used as electronic switches. Once a basic gate is created, it may be possible to use the same ideas to create other useful logic gates. The logic gates that are created using the same basic circuit design are known as a **logic family**. Gates that belong to a particular logic family are similar in parameters such as power supply value, output voltage levels, propagation delay times, and other characterisics. This similarity allows for easy interfacing for building large systems.

In this section, we will briefly describe the logic families that are most commonly used to build digital systems.

4.6.1 CMOS

CMOS is an acronym that stands for **C**omplementary **M**etal **O**xide **S**emiconductor. This is a type of technology that allows for extremely high-density integrated circuits and forms the basis of modern chip design. The transistors used to create the switching circuits in CMOS are called **MOSFET**s (**MOS** **F**ield-**E**ffect **T**ransistors). MOSFETs can be made extremely small: devices smaller than about $0.3 \ \mu m \times 1 \ \mu m$ are typical.[8] Since the transistors are so small, several million individual devices can be created on a single "chip" of silicon that is only 1 or 2 cm on a side. This allows for complex digital networks, such as microcomputers, to be integrated into a single low-cost unit.

Because of its importance to modern digital system design, CMOS is discussed in detail in later portions of this book, starting with Chapter 6. For now, we will just summarize some of the important characteristics.

Power Supply

CMOS integrated circuits commonly use a power supply value of $V_{DD} = 5$ volts. However, some CMOS chips can operate over a range of values (typically 5 v to 15 v), and more advanced designs use 3.3 v and lower. A small power supply voltage is advantageous for reducing the power consumption and hence the heating. In addition, it makes battery-operated systems more practical.

Logic Levels

One advantage of CMOS is that it provides output logic voltages that range from 0 v to the value of the power supply V_{DD}. This provides ideal voltage levels for logic 0 and logic 1 states.

Propagation Delay

The propagation delay t_p through a simple inverter (NOT) circuit is on the order of 0.1 nanoseconds or smaller if measured inside the integrated circuit. More complex logic networks have propagation delays of several nanoseconds.

4.6.2 TTL Integrated Circuits

TTL (**T**ransistor-**T**ransistor **L**ogic) is another popular logic gate family. In the early days of microcomputing, TTL chips were used almost exclusively to provide "glue logic" that allowed various chips in the computer to communicate with each other.

TTL is based on a different type of transistor from those used in CMOS chips. These devices, known as bipolar transistors,[9] can be used to design very fast

---

[8] 1 μm is read "1 micrometer" or "1 micron," where 1 $\mu m = 10^{-6} \ m$. The micrometer is a basic unit of measurement in designing integrated circuits.

[9] *Bipolar* means "two polarities." The term is applied because a bipolar transistor uses both negative and positive charges in creating a switch-type action.

switching networks. However, bipolar transistors are much larger than MOSFETs, they are more difficult to connect together, and bipolar circuits can exhibit extreme heating problems. Owing to these considerations, they are not usually the first choice for high-density chip designs. However, bipolar logic circuits and TTL ICs are still important in hundreds of applications, because they provide reliable, low-cost logic chips that are easy to use.

Power Supply

TTL integrated circuits are designed for a power supply voltage of $V_{CC} = 5$ volts only. If a different value is used, then the circuit will not operate properly and may be destroyed.

Logic Levels

Even though TTL circuits use a power supply voltage of $V_{DD} = 5$ v, the output levels measured from a typical gate cannot attain this level. Typically, $V \approx 0.3$ v is the lowest value of the output voltage, and $V \approx 3.6$ v is the highest value. This reduced logic swing is due to electrical characteristics of the bipolar transistors. In general, the reduced logic voltage range is not a problem if TTL is used throughout the design.

Propagation Delay

Bipolar transistors can be switched very quickly. A TTL integrated circuit usually has propagation delays on the order of a few nanoseconds as measured outside the circuit package.

4.6.3 Emitter-Coupled Logic (ECL)

ECL is another logic family that is based on bipolar junction transistors. The fastest of all the silicon logic circuits, it was used for many years to build the large mainframe computers. ECL circuits do dissipate a relatively large amount of heat, and they have been replaced by CMOS circuits in most computers. However, ECL is still very common in high-speed digital networks such as those used in fiber-optic networks and other communication systems. Another recent example is the 900-MHz communications devices that are so popular.

Power Supply

ECL integrated circuits use negative logic with a power supply providing ground and a negative voltage called V_{EE}. $V_{EE} = -5.2$ v is a typical value. This consideration makes it more natural to use negative logic in ECL logic gates.

Logic Levels

ECL achieves its speed through many means, one of which is the use of small voltage swings. The difference between logic 0 and a logic 1 voltages is on the order of 0.2 v, so the time needed to switch between the levels is very short.

Propagation Delay

Bipolar transistors can be switched very quickly, and typical on-chip propagation delays in ECL circuits are on the order of $t_p = 50$ picoseconds (0.05 nanoseconds) or less. When combined with the small voltage swings used, ECL allows for high signal switching frequencies.

Power Consumption

One problem with ECL integrated circuits is that they consume large amounts of power, which in turn generates a substantial amount of heat. Older mainframe computers had to be placed in special air-conditioned rooms to cool the systems down.

Logic design can usually be accomplished using any logic family, although some modifications to the equations may be necessary to adhere to the available gates. In practice, the logic family is chosen before the design is started to ensure hardware compatibility.

4.7 The Hardware Designer

Hardware considerations are critical for building a digital system that behaves as expected. Basic signal delays originate from the hardware aspects of logic design and establish basic limits on the speed of any physical network. Logic and hardware design go hand in hand; one cannot design a complex logic network without considering issues such as propagation delays and interconnect wiring. These may limit the usefulness of a particular algorithm or logic design technique. Indeed, a system design is not completed until it can be demonstrated with actual functioning hardware.

Hardware designers work closely with logic designers to create working networks. The distinction between the two becomes blurred in practice, and an engineer with a solid foundation may perform both the design and the hardware implementation.

4.8 Problems

[**4.1**] Consider an integrated circuit that uses a power supply voltage of $V_{DD} = 5$ v. The average power supply current is measured to be 125 mA, where 1 mA (a milliampere) is equal to 10^{-3} A.

Calculate the power P delivered to the circuit.

[**4.2**] An integrated circuit operates with a power supply voltage of 3.0 volts and dissipates 10 watts of power. What is the value of the current I in amperes?

[**4.3**] Compute the unknown quantities for each situation.

(a) $I = 210$ mA, $V = 5$ v, $P = ?$
(b) $V = 3.1$ v, $I = 1.4$ A, $P = ?$
(c) $P = 15W$, $V = 2.6$ v, $I = ?$

[4.4] An inverter (NOT) gate is designed with $t_{HL}= 0.5$ *ns* and $t_{LH}= 1.0$ *ns*. The input voltage V_{in} is shown in the accompanying figure, and it is known that the output voltage ranges from $V_{out}= 0$ *v* to $V_{out}= 5$ *v*.

(a) Draw $V_{out}(t)$.

(b) What is the maximum switching frequency for this circuit? Sketch what the output voltage V_{out} would look like at this frequency.

(c) Suppose that the input is driven at twice the maximum frequency. Sketch what the output voltage would look like in this case, and explain why the gate will not operate properly.

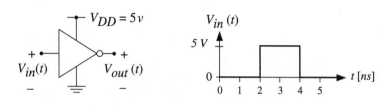

[4.5] Consider the NOT cascade that follows. Calculate the delay from left to right if $t_{p0,\ NOT}= 0.5$ *ns* and $t_{pL,\ NOT}= 0.5$ *ns*.

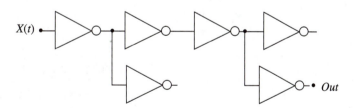

[4.6] Calculate the delay between the input signal $X(t)$ and *Out* for the NOT cascade shown. Use $t_{p0,\ NOT}= 0.5$ *ns* and $t_{pL,\ NOT}= 0.4$ *ns* and assume that the output node is connected to the inputs of two inverters.

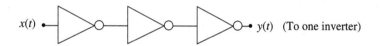

[4.7] Calculate the gate delay from *Input* to *Output* if $t_{p0,\ NOT}= 0.5$ ns, $t_{pL,\ NOT}= 0.5$ ns, $t_{p0,\ NOR}= 0.75$ ns and $t_{pL,\ NOR}= 0.9$ ns. Assume that the output drives the input to one inverter.

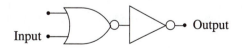

[4.8] An inverter is characterized by the propagation delay times $t_{p0} = 1$ *ns* and t_{pL} = 0.25 *ns*. Suppose that six identical inverters are cascaded in a linear chain. Find the total delay time t_d from the input to the output.

[4.9] The current through a 500-Ω resistor is measured to be 25 *mA*. Find the voltage across the device.

[4.10] Consider a 1500-Ω resistor. The voltage is known to be 0.86 V. Find the current through the device.

[4.11] Calculate the delay between the input signal $a(t)$ and the output $f(t)$, using the following parameters:

$t_{p0, NOT} = 0.5$ *ns*, $t_{pL, NOT} = 0.4$ *ns*, $t_{p0, NOR} = 0.75$ *ns*, $t_{pL, NOR} = 0.9$ *ns*,
$t_{p0, NAND} = 0.85$ *ns*, $t_{pL, NAND} = 0.95$ *ns*.

Assume that the output drives the input to one inverter.

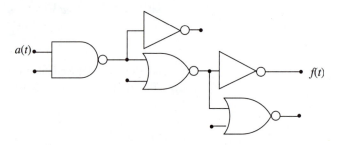

[4.12] A 1200-Ω resistor has a voltage of 0.35 *V* across it. How much power does the device dissipate?

[4.13] Consider the series-connected resistors that follow. Calculate the total resistance seen between the listed nodes.

(a) x and z
(b) y and w
(c) x and w

$$R_1 = 200 \, \Omega \qquad R_2 = 450 \, \Omega \qquad R_3 = 80 \, \Omega$$

$x \qquad\qquad\qquad y \qquad\qquad\qquad z \qquad\qquad\qquad w$

[4.14] The voltage across a 1200-Ω resistor is found to be given by

$$V_R(t) = 5e^{-t/\tau} \, v$$

where $\tau = 4$ *ns*. Find the resistor current I in units of milliamperes (*mA*) for the following times:

(a) $t = 2$ *ns*
(b) $t = 4$ *ns*
(c) $t = 6$ *ns*

Then plot the current $I(t)$ as a function of time.

[4.15] A capacitor of value $C = 10$ μF (microfarads; where 1 $\mu F = 10^{-6}$ F) has a voltage of 2 volts across it. Calculate the amount of charge Q, in coulombs, on each plate.

[4.16] Consider a 100 fF capacitor; 1 femtofarad (fF) is equal to 10^{-15} farads. The charge Q is known to have a value of $Q = 2.8$ nC (1 $nC = 10^{-9}$ C). What is the voltage across the device?

[4.17] What is the total capacitance between nodes x and y in the capacitor circuit shown?

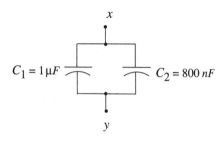

[4.18] Refer to your calculus book to answer the following question: What is the significance of the Euler e? Hint: Consider the function

$$f(x) = e^x$$

of the variable x with respect to the main operations of calculus.

[4.19] Consider the following RC network. What is the time constant τ?

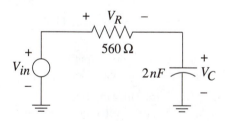

[4.20] The capacitor in an RC circuit has a value of $C = 20$ pF. The time constant for the circuit is given as $\tau = 14$ ns. Find the value R of the resistor in ohms.

[4.21] The resistor in an RC circuit has a value of 240 Ω. What is the largest capacitance that can be connected in the circuit if the maximum acceptable time constant is $\tau = 3$ ns?

[4.22] The voltage across a capacitor with a value of $C = 140$ fF is given by the time function

$$V_C(t) = -4e^{-t/\tau}\ v$$

where the time constant is given by $\tau = 2$ *ns*. What is the amount of store charge on the capacitor at a time $t = 3$ *ns*?

[4.23] An RC circuit is built using a resistor with a value of $R = 1200$ Ω and a capacitor that has $C = 4.7$ *pF*, where 1 picofarad (*pF*) equals 10^{-12} *F*.

(a) Calculate the time constant for the circuit in units of nanoseconds.

(b) Suppose that the applied voltage pulse makes a transition from 0 *v* to 3 *v*. What is the value of the capacitor voltage V_C when $t = \tau$?

(c) Now consider the case where the capacitor voltage is at a value of 3 *v*, and then the applied voltage is dropped to 0 *v*. What is V_C when $t = \tau$ for this case?

[4.24] The time constant for the circuit shown is $\tau = 2.55$ *ns*. What is the value C of the capacitor?

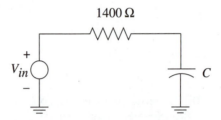

[4.25] A resistor of value $R = 1500$ Ω is used in an RC circuit. A capacitor with a value C initially has a voltage of $V_C = 5$ *v* when the applied voltage is switched to zero volts at time $t = 0$. The capacitor voltage decays according to the equation

$$V_C(t) = 5e^{-(t/\tau)}$$

where τ is the time constant. Find the value of the capacitance C if $V_C = 2$ *v* when $t = 2$ *ns*.

[4.26] Examine the output jacks in a video cassette recorder (VCR). What type of cable is needed for the video input/output?

[4.27] A transmission line that is 4 meters long causes a signal delay of 23 *ns*. What is the signal velocity on the line?

[4.28] A transmission line in a computer network induces a signal delay of 1.25 *ns*. The signal wavefront velocity is characterized by equation (4.44) with $n = 1.7$. Find the length of the line in centimeters.

[4.29] Two sections of a large computer are connected with a transmission line that is 162 *cm* long. The signal is transmitted at the wavefront velocity given in equation (4.44) with $n = 1.65$. What is the value of the signal delay time that is caused by the transmission line?

[4.30] A salesperson tries to tell you that there is no signal delay between the output of your computer and the input to an auxiliary (external) CD player. What basic law of physics contradicts that statement? Hint: Look in your modern physics textbook.

[4.31] A transmission line is characterized by an effective signal wavefront velocity of

$$v_w = \frac{c}{3}$$

Calculate the time delay in picoseconds required for a signal to propagate end to end on a line that is 30 *cm* long.

[4.32] Two interconnect wires are spaced a distance of $d = 4$ *mm* from one another. What is the increase in coupling capacitance C if the distance is reduced to

(a) 3 *mm*
(b) 1 *mm*

[4.33] Electromagnetic radiation is emitted from any device that has currents of voltages that change values with time. Can you list five sources of electromagnetic interference (EMI) in your apartment or dorm room?

4.9 Appendix

This supplement is provided to derive the solutions for the capacitor voltage to the RC circuit. The starting point is the current I given by

$$I = \frac{V_{in} - V_C}{R} \tag{4.46}$$

Since the current flows into the capacitor, it is also given by

$$I = C\,\frac{dV_C}{dt} \tag{4.47}$$

Equating yields

$$I = \frac{V_{in} - V_C}{R} = C\,\frac{dV_C}{dt} \tag{4.48}$$

which may be rearranged to give the differential equation

$$\frac{dV_C}{dt} + \frac{1}{RC}V_C(t) = \frac{1}{RC}V_{in}(t) \tag{4.49}$$

for $V_C(t)$. Defining the time constant by

$$\tau = RC \tag{4.50}$$

then gives the form

$$\frac{dV_C}{dt} + \frac{1}{\tau}V_C(t) = \frac{1}{\tau}V_{in}(t) \tag{4.51}$$

In this equation, the input voltage $V_{in}(t)$ acts as the *driving term* that changes in time and induces changes in the capacitor voltage.

In general, the solution $V_C(t)$ can be constructed by superposing two terms that are called the homogeneous and the particular solutions. The homogeneous solution is obtained by solving the equation when there is no driving force:

$$\frac{dV_h}{dt} + \frac{1}{\tau}V_h(t) = 0 \tag{4.52}$$

This has solutions

$$V_h = Ae^{-(t/\tau)} \tag{4.53}$$

where A is a constant that needs to be determined. To obtain the particular solution, we must specify the value of the driving term. Let us suppose that we have a discharging event where $V_{in} = 0$ v. In this case, the particular solution satisfies the same equation as the homogeneous solution, so that

$$V_C(t) = Ae^{-(t/\tau)} \tag{4.54}$$

is now the capacitor voltage. To find the value of A, we must apply the initial conditions on V_C. For a discharge event, this states that $V_C(0) = V_{DD}$. To ensure that our solution satisfies this value at time $t = 0$, we write

$$\begin{aligned} V_C(t = 0) &= V_{DD} \\ &= Ae^{-(t/\tau)}\Big|_{t=0} \\ &= A \end{aligned} \tag{4.55}$$

where the last line is obtained by noting that $e^0 = 1$. Thus, $A = V_{DD}$ and

$$V_C(t) = V_{DD}e^{-(t/\tau)} \tag{4.56}$$

as we saw in the discussion.

The equation for charging the capacitor,

$$V_C(t) = V_{DD}[1 - e^{-(t/\tau)}] \tag{4.57}$$

is obtained in the same manner, except that $V_{in} = V_{DD}$ is used in the equation for the particular solution. Solving

$$\frac{dV_p}{dt} + \frac{1}{\tau}V_p(t) = \frac{1}{\tau}V_{DD} \tag{4.58}$$

gives the particular solution

$$V_p(t) = V_{DD} \tag{4.59}$$

which is combined with the homogeneous equation to yield

$$V_C(t) = Be^{-(t/\tau)} + V_{DD} \tag{4.60}$$

where we have denoted the constant as B. In the case of the charging event, the initial condition is that $V_C(0) = 0$ v, so

$$V_C(0) = 0$$
$$= B + V_{DD}$$

$$(4.61)$$

shows that $B = -V_{DD}$. Substituting gives the desired form.

CHAPTER 5

First Concepts in VHDL

Hardware description languages (HDLs) provide an effective means for the computer-aided design (CAD) of digital logic networks. An HDL is simply a type of high-level computer language that allows us to describe all of the important characteristics of a logic network of arbitrary complexity. Included in the listing are individual functions, properties of the signals, and details of how the blocks are connected to form a complete system. Once a system is characterized using the appropriate format and syntax, it is compiled by an HDL program. This results in a set of outputs that can be used to simulate and verify the behavior of the logic network.

In this chapter, we will learn how to describe simple digital circuits using a hardware description language called VHDL. You will find that it is a very natural language to learn and use.

5.1 Introduction

In hierarchical design, every system can be viewed as being made up of smaller blocks. Each block provides a specific function and may itself be made up of even smaller blocks. If we start with the simplest functions and then build more complex blocks, we are using what is called the **bottom-up** approach to system design. Conversely, we can start at the highest (system) level, then break up the system into smaller and smaller logic blocks; this is called the **top-down** approach.

Hardware description languages help us to manage various levels of complexity by providing us with a description of a digital system at any desired level. Once we create an HDL file listing for a logic network, it can be used to verify the operation and performance of the design. Trouble spots can be identified and fixed, and we obtain a reasonable degree of confidence in how well the system will work if it is built in hardware.

In this book, we have chosen a specific hardware description language known as **VHDL** as a vehicle for the presentation of the main concepts. The "V" stands for **VHSIC**, another acronym used by the Department of Defense (DoD) for an area of work known as Very High-Speed Integrated Circuits.[1] Although other excellent HDLs exist, VHDL has a set of established standards that are well documented in literature. VHDL compilers are available as stand-alone programs, and they can be found in several commercial CAD tool sets. Most importantly, VHDL is representative of the features found in other hardware description languages.

5.1.1 Basic Concepts

Consider a general logic function block as shown in Figure 5.1. The unit is characterized by input ports that accept logic variables A, B, C and an output port where the function $f(A,B,C)$ is taken. This type of description can be used regardless of the internal complexity of the block. It does not matter if the logic function is a simple pair of gates, or a very complicated logic function with many inputs, outputs, and logic operations.

Now suppose that you want to describe the operation of the block to another person with whom you are speaking on the telephone. Logic diagrams require that you transmit a graphical image, or attempt to describe the diagram in words. While the verbal description of the logic circuit may be adequate for a small unit, it is not practical for describing a complex network. Moreover, there is a high probability for error.

Another problem exists even if one has a logic diagram available: how do we test the logic design to insure that it operates as it should? A good designer always performs many levels of testing to avoid costly changes if an error is found in the later stages of the design cycle. While we can perform **logic verification** by tracing every possible sequence through the logic diagram, this is very tedious and also introduces a large margin for error.

A hardware description language provides the solution to both of these problems. It allows one to describe the operation of a digital network using precisely defined statements in a form that resembles a generic high-level computer language such as C++. Conversely, one can construct the logic diagram directly from the HDL description. However, HDLs have a much more important application, which is the ability to simulate the logic network using a computer. Testing and redesign can be performed

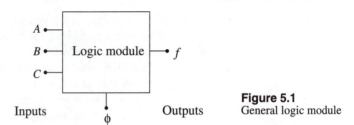

Figure 5.1
General logic module

---

[1] VHDL is pronounced by just reading the letters. VHSIC, on the other hand, is pronounced as *vis-hic*.

at the HDL level, and any changes or corrections can be translated back to the original design. As implied by its name, an HDL allows us to include hardware features in the simulation. Physical problems such as the propagation delays through logic gates can be included to insure that the circuit will be working as intended by the designer. Moreover, HDLs allow one to simulate the detailed behavior of a complex logic network, find trouble spots, and improve the design.

Another very useful aspect of VHDL is that it can be used to describe any digital network. Moreover, it is standardized and can be used on a wide variety of platforms, including standard PC and workstation environments.

5.1.2   Using a Hardware Description Language

To use a hardware description language, we first create a description of the network using the specified syntax and constructs that are defined within the language itself. These are entered into a text file and saved. Once the description is completed, the file is used as the input into the HDL compiler itself. This is illustrated schematically in Figure 5.2. The compiler accepts the file and produces an output file that contains all of the information on the network. The output file is then used in a simulation environment with another program that allows the user to specify input signals and view the corresponding results. This is usually a graphical interface that portrays the signals as they change in time.

In this chapter we will concentrate on learning how to write VHDL input files that can be run on a compiler. Emphasis will be placed on the details and interpretation of the language statements. If you have access to a VHDL compiler, then you should consult the instruction manual[2] to learn how to use its features in the actual simulation environment. Many excellent packages are available for use on desktop PCs, which make it easy to become proficient in this important language.

5.2   Defining Modules in VHDL

A logic unit like that shown in Figure 5.1 will be referred to as a **module** or a **unit**. To describe a given module in VHDL we must do two things. First, we must define the block itself by giving it a name, and by specifying the input and output lines (which

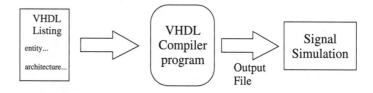

Figure 5.2 Using a VHDL compiler

---

[2] The *instruction manual* is the information booklet that usually comes with a software package, but is often ignored!

are called **ports**). This is called an **entity** statement. Once an entity is introduced, we must specify what the module actually does, i.e., how the outputs are related to the inputs. This is called the **architecture** statement for the module. The architecture can be described in different ways depending upon the level of modeling desired.

As with any computer language, VHDL requires that we adhere to a set of rules that define **keywords** and **syntax**. A "keyword" (also called a reserved word) is a word that has a special meaning in the language, and cannot be used for any other purpose. In our discussion, keywords will be denoted by using a "block-type" font to help you to identify and learn them; for example, entity would indicate that "entity" is a keyword in VHDL. Of course, you do not have to change fonts when you type in a VHDL listing; only the words themselves are important, not how they look on screen. An alphabetical listing of VHDL reserved words is shown in Figure 5.3. For the moment, this should be used as a reference of words that should not be used for naming quantities. The meaning of many of these words will be discussed in this and later chapters.

abs	file	of	sra
access	for	on	srl
after	function	open	subtype
alias		or	
all	generate	others	then
and	generic	out	to
architecture	group		transport
array	guarded	package	type
assert		port	
attribute	if	postponed	unaffected
	impure	procedure	units
begin	in	process	until
block	inertial	pure	use
body	inout		
buffer	is	range	variable
bus		record	
	label	register	wait
case	library	reject	when
component	linkage	rem	while
configuration	literal	report	with
constant	loop	return	
		rol	xor
disconnect	map	ror	xnor
downto	mod		
	nand	select	
else	new	severity	
elseif	next	shared	
end	nor	signal	
entity	not	sla	
exit	null	sll	

Figure 5.3 Reserved words in VHDL

"Syntax" refers to the word usage and order that must be used to write statements, and also to the punctuation requirements that define special points in the statements, such as the end of a command.[3] Figure 5.4 provides the list of special VHDL notation symbols and syntax. Some of these will be introduced and discussed in this chapter, but this listing is primarily for future reference. Although we will examine both in our discussion, it is important that you not lose sight of the language itself. Keywords and syntax are the formal details, and are an integral and necessary part of VHDL. However, the language provides a very powerful approach to digital logic simulation and verification that transcends the details of the written code. As with any computer language, familiarity comes with usage. Practice is therefore highly desirable, and the reader is encouraged to apply the techniques shown here on as many networks as possible.

Symbol	Meaning
+	Addition, or positive number
−	Subtraction, or negative number
/	Division
=	Equality
<	Less than
>	Greater than
&	Concatenator
|	Vertical bar
;	Terminator
#	Enclosing based literals
(	Left parenthesis
)	Right parenthesis
.	Dot notation
:	Separates data object from type
"	Double quote
'	Single quote or tick mark
**	Exponentiation
=>	Arrow meaning "then"
=>	Arrow meaning "gets"
:=	Variable assignment
/=	Inequality
>=	Greater than or equal to
<=	Less than or equal to
<=	Signal assignment
<>	Box
--	Comment

Figure 5.4 Defined symbols in VHDL

---

[3] This is an example of a **delimiter** that defines the end of a statement.

Let us now return to the VHDL description of a module. The first step is to write an **entity statement** that identifies the module as a distinct logic block, and also defines the input and output ports. Consider the module portrayed in Figure 5.5, which we will name *Simple_ gate* with input bits *a*, *b*, and *c*; the output is denoted by *f*, which is also a binary number. The simplest entity statement for this module is

entity *Simple_ gate* is
    port(*a*, *b*, *c* : in *bit*;
            f : out *bit*);
end *Simple_ gate*;

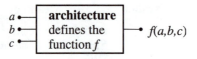

entity describes the unit

Figure 5.5 Generic VHDL module

Let us study this declaration to understand the meaning. The first line identifies this block of statements as an entity declaration for the module *Simple_ gate*. The input and outputs are listed in the port listing that identifies the quantities *a*, *b*, and *c* as binary inputs using the keyword grouping in; the additional **type** identifier *bit* has been added to clarify the nature of the quantities. A *bit* is a predefined quantity that can only have a value of '0' or '1'; the quotation marks are used to distinguish *bit* value from the literal values 0 and 1. The result *f* is specified to be a binary output through out *bit*. Note use of the colon " : " in the port listing form

identifier : in

to associate the *identifier* with the type of port (in or out). The last line completes the entity statement by the designation end *name*.

Once we have an entity defined, we must describe exactly what the block does by using an **architecture declaration**. In the case of *Simple_ gate*, this means that we need to define what *f*(*a*,*b*,*c*) is. VHDL provides various ways to accomplish this task; each is characterized by the manner in which the information is written. The major classifications for architectural declaration are

- **behavioral** descriptions where we provide an explicit relationship between the inputs and the output(s);

- **structural** listings that construct logic functions by combining more primitive elements, such as gates; and

- **dataflow** models that describe modules by defining the "flow" of the data signals. This can be viewed as a type of behavioral model.

In general, a VHDL architectural declaration assumes the form

> **architecture** *type_of_description* **of** *gate_name* **is**
> > *Declaration_1*;
> > *Declaration_2*;
> >
> > . . .
>
> **begin**
> > *Statement_1*;
> > *Statement_2*;
> >
> > . . .
>
> **end** *type_of_description* ;

In this listing, *type_of_description* is the user-defined name for the architecture of the module. Typical choices might be one of the three approaches (behavioral, structural, or dataflow), or some other identifier that is easy to remember and understand. The lines denoted by *Declaration_1*, etc. are used to define any necessary parameters or "building blocks" that are needed in the description. The actual description starts with the keyword **begin**, with the details provided by the listings denoted as *Statement_1*, etc. that give the specifics of the logical function or operation. The **end** line indicates that the description is complete. Once we have defined a module using **entity** and **architecture** listing, then the module can be connected to other modules using other constructs.

Consider the network *Gate_1* shown in Figure 5.6 as our first example of how VHDL descriptions are written. The entity statement can be constructed as

> **-- First Example**
> **entity** *Gate_1* **is**
> > **port**(*a,b* : **in** *bit*;
> > > *f* : **out** *bit*);
> >
> **end** *Gate_1*;

by analogy with our previous **entity** listing. Other than modifying the name and parameters, the only difference is the **comment line**

> **-- First Example**

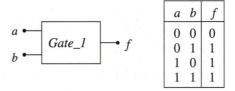

a	*b*	*f*
0	0	0
0	1	1
1	0	1
1	1	1

Figure 5.6 Gate description example

Comments are identified by the two dashes " -- " such that all entries to the right are ignored by the VHDL compiler. We will often emphasize comment lines in our listings using a boldface font to call attention to them, as they are useful for following the structure of a listing. A conscientious VHDL writer will add many comments to explain the details of the code; this is particularly useful for the inevitable debugging that follows. Remember: what is obvious now may be a total mystery in two weeks!

Next, let us introduce two direct approaches to writing the **architecture** declaration. A quick examination of the function table shows that *Gate_1* is just the OR function

$$f = a + b \tag{5.1}$$

VHDL provides for several predefined logic operations that can be used in our descriptions; these are listed in Figure 5.7 with examples of usage and their standard interpretation. To describe this in the **architecture** declaration, we construct

> architecture *Logic* of *Gate_1* is
> begin
> $f <= a$ or b ;
> end *Logic*;

where the description has been named *Logic* and the actual operation is defined by the entry

$$f <= a \text{ or } b$$

in the listing. The notational syntax

$$Left <= Right$$

is read as "*Left* is assigned the value of *Right*." Combining this with the **architecture** declaration provides a complete description of the module.

Keyword	Example	Meaning
not	not a	$\overline{a}$
and	c and d	$c{\cdot}d$
or	u or v	$u+v$
xor	x xor y	$x{\oplus}y$
nand	a nand b	$\overline{a{\cdot}b}$
nor	c nor d	$\overline{c+d}$
xnor	x xnor y	$\overline{x{\oplus}y}$

Figure 5.7 Concurrent operations in VHDL

To gain more practice, let us construct similar VHDL descriptions for the two modules shown in Figure 5.8. In Figure 5.8(a), the output is the NAND operation

$$g = \overline{a \cdot b} \tag{5.2}$$

which is a VHDL-defined operation. Thus, *Gate_2* can be described by the listing

> -- **Second Example**
> -- Entity declaration is first
> entity *Gate_2* is
>   port(*a*, *b* : in *bit*;
>   *g* : out *bit*);
> end *Gate_2*;
> -- **Architecture declaration follows**
> architecture *Logic* of *Gate_2* is
> begin
>   *g* <= *a* nand *b* ;
> end *Logic*;

Similarly, we see from the truth table that *Gate_3* in Figure 5.8(b) is the equivalence (XNOR) function such that

$$h = \overline{x \oplus y} \tag{5.3}$$

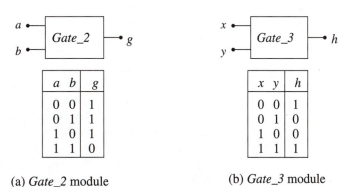

a	b	g
0	0	1
0	1	1
1	0	1
1	1	0

x	y	h
0	0	1
0	1	0
1	0	0
1	1	1

(a) *Gate_2* module   (b) *Gate_3* module

Figure 5.8 Additional examples of VHDL modules

Thus, we may describe it using

> -- **Third Example**
> entity *Gate_3* is
>   port(*x*, *y* : in *bit*;
>   *h* : out *bit*);

end *Gate_3*;

architecture *Logic* of *Gate_3* is

begin

$\quad$ $h <= x$ xnor y;

end *Logic*;

These simple examples illustrate the basic ideas of the **entity** and **architecture** declarations, and provide the basis for describing more complex systems.

At this point, you would probably agree with the observation that the basic VHDL statements above are straightforward to read and interpret. Learning how to write VHDL code is simply a matter of learning the keywords and syntax and becoming familiar with how the statements are combined to describe the logic network. The most important aspect of VHDL to note is that the language defines modules in a unique manner. With these facts in mind, let us now elevate the power of our VHDL statements by studying more advanced concepts.

5.2.1 Concurrent Operations

The keyword operators **not**, **and**, **or**, **xor**, **nand**, **nor**, and **xnor** are used to describe what are called **concurrent** operations. This mean that the operations are performed without regard to any other timing constraint. For example, when we use the AND3 statement

$$f <= x \text{ and } y \text{ and } z;$$

in an architecture declaration, the value of f is updated whenever there is a change in the status of the inputs x, y, z.

Concurrent operations can be attractive to use for describing modules in VHDL because they have a one-to-one correspondence with standard Boolean operations. However, when they are used to construct more complex logic functions in VHDL, it is important to note that the precedence rules are slightly different. The **not** operation has the highest precedence level; it is evaluated first. The remaining functions have a lower precedence level than the **not** but are otherwise equal. In particular, **or** and **and** have the same level. When operators of equal precedence are encountered in a VHDL statement, they are implemented in order from left to right. Parentheses may be used to increase the level of precedence of a group of operations.

To understand the consequences of this, suppose that we have the Boolean operation

$$q = a + b \cdot c \qquad \qquad \textbf{(5.4)}$$

Our convention in writing Boolean expressions says that the AND operation has a higher precedence level than the OR operation. Thus, the operational sequence implied in evaluating this equation would be $(b \cdot c)$ first, followed by $a + (b \cdot c)$. In VHDL, however, the equivalent statement would have to be written as

$$q <= a \text{ or } (b \text{ and } c);$$

where we have added parentheses to ensure that the AND function is calculated first. If we had instead expressed the operation using

> -- **This does not give the desired function**
>
> $q <= a$ or b and c;

then the function would have been incorrectly evaluated as

$$q = (a + b) \cdot c \tag{5.5}$$

Once the precedence level is clarified, concurrent statements are straightforward to construct and use in VHDL descriptions.

Example 5-1

Let us write the concurrent VHDL statement for the Boolean expression

$$h = \overline{a + b \cdot \bar{c} + d} \tag{5.6}$$

Since the AND function is implied to have precedence over the OR, we write

> $h <=$ not (a or (b and not c) or d)

as the required assignment. Note that the not c entry for $\bar{c}$ does not need parentheses since the not is always evaluated before the and operation.

Example 5-2

From the discussion above, we may translate the VHDL statement

> $G <= (x$ or y) and (z or not (w and v))

into the logic expression

$$G = (x + y) \cdot (z + \overline{(w \cdot v)}) \tag{5.7}$$

by a straightforward translation.

Let us use concurrent statements to write an alternative description of *Gate_3* in Figure 5.8(b). Reading the truth table directly gives the SOP form

$$h = x \cdot y + \bar{x} \cdot \bar{y} \tag{5.8}$$

so that we can construct the equivalent alternative listing

> -- **Third Example SOP Form**
> entity *Gate_3* is
>   port(x, y : in *bit*;
> h : out *bit*);

end *Gate_3*;
architecture *Logic _Alt* of *Gate_3* is
begin
        $h <=$ (x and y) or (not x and not y);
end *Logic_Alt*;

if desired. This is equivalent to the previous form and did not require that we make the correlation with the XNOR operation.

Example 5-3

Consider the logic cascade shown in Figure 5.9. Logically, we can construct the output as

$$g = \overline{(x \cdot y)} + \bar{z} \tag{5.9}$$

by tracing the inputs through the gates. Concurrent statements can be used to construct the VHDL listing as

```
-- Logic Gate Example
entity Gate_Ex is
        port(x, y, z : in bit;
                g : out bit);
end Gate_Ex ;

architecture Concurrent of Gate_Ex is
begin
        g <= (x nand y ) or not z ;
end Concurrent ;
```

The procedure is easily extended to an arbitrary gate configuration.

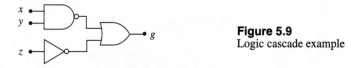

Figure 5.9
Logic cascade example

Example 5-4

Let us examine one more straightforward example of this approach to writing VHDL descriptions. Consider the logic diagram in Figure 5.10. This can be described by using concurrent functions as in the following code.

```
-- Logic Gate Example
entity Gate_Ex is
        port(a, b, c, d: in bit;
                result : out bit);
end Gate_Ex ;
```

architecture *Concurrent* of *Gate_Ex* is
begin

    result <= (*a* or *b*) xnor (*c* and *d*);

end *Concurrent* ;

The important idea to remember is that the structure of the logic flow provides the concurrent statements in a direct manner.

Figure 5.10
Logic cascade

5.2.2 Identifiers

We have introduced several identifiers such as *Gate_1* and *logic* in our examples. It is useful to choose identifiers that are easy to remember, or have direct significance in the logic network.

VHDL specifies a few rules that must be followed when creating identifiers. The basic rules that apply to all standard versions of VHDL can be summarized in the following statements.

- Identifiers may consist of upper-case and lower-case letters (*a* through *z* and *A* through *Z*), the digits 0 through 9, and the underscore " _ " symbol.

- Every identifier is a single string of characters, and it may be of any length desired.

- Identifiers are **case-insensitive.** This means that upper-case and lower-case letters are interpreted as being identical. For example, the identifier *Or_gate_3* is the same as *OR_GATE_3*, which is the same as *OR-Gate_3*, and so on. Care must be taken to insure that you do not use equivalent identifiers for different quantities.

- The underscore " _ " cannot be used as the first or last character in the identifier. Also, letters and numbers must be used on both sides of the underscore, i.e., it is not permissible to use two or more underscores in a row.

An expanded set of allowed characters (such as the dash "-" and accent marks) are allowed in the latest implementations. However, the above rules will be sufficient for our uses.

Example 5-5

The following are examples of legal identifiers:

Temporary_read_signal
AOI4
b3
CPU_IO_Activate

Logic_construction_14
My_special_circuit

An example of an illegal identifier is

_Basic_logic_function

since it starts with an underscore. Similarly,

*restart_2_*

cannot be used since it ends with an underscore.

5.2.3 Propagation Delay

Our VHDL examples up to this point have only dealt with the description of logic gates. The connection between these constructs and hardware (remember the *H* in VHDL) can be made in various ways. One of the easiest to understand is the concept of including the propagation delay t_p introduced in the previous chapter.

Consider the simple NOT gate shown in Figure 5.11. The propagation delay t_p originates from the electronic components and circuits that are used to create the physical realization of the logic function. This can be included in a VHDL architecture listing by using the keyword **after** along with the value of the delay in the concurrent operation statement. Let us assume that t_p has a value of 2 nanoseconds (2 *ns*). Then, the module may be described by

```
-- NOT with Propagation Delay
entity Invert is
        port ( x : in bit;
                    Result : out bit);
end Invert;


architecture Prop_delay of Invert is
begin
        Result <= not x after 2ns;
end Prop_delay;
```

The effect of the line

$$Result <= not\ x\ after\ 2ns;$$

is first to evaluate the complement $\bar{x}$ and then to assign that value to *Result* after a 2-nanosecond delay.

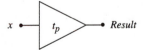

Figure 5.11
Inverter with propagation delay

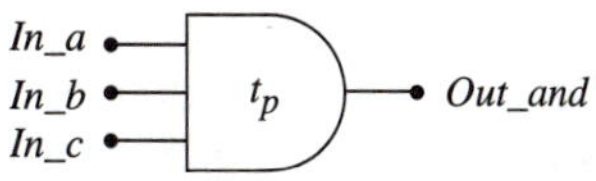

Figure 5.12
AND3 propagation delay

This type of delay statement can be used to describe multiple input modules such as the AND3 gate shown in Figure 5.12. Suppose that t_p has a value of 5 *ns*. We can include the propagation delay as shown by the listing

-- AND3 with Propagation Delay
entity *AND_3* is
    port(*In_a, In_b, In_c:* in *bit*;
        Out_and : out *bit*);
end *AND_3*;
architecture *Delay* of *AND_3* is
begin
    Out_and <= *In_a* and *In_b* and *In_c* after *5*ns;
end *Delay*;

As with the NOT gate, any change in the value of the output *Out_and* due to changes at the inputs is delayed in time by 5 nanoseconds. It is important to remember that the actual value of the propagation delay depends upon the technology (such as the IC family, e.g., CMOS or TTL) and other considerations as discussed in the previous chapter.

Example 5-6

Let us practice by interpreting a couple of VHDL statements. Consider the description

    result <= *X* xor *Y* after 1ns;

This is simply an XOR gate that has a 1-nanosecond delay. Similarly,

    F <= not ((*a* and *b*) or *c*) after 2.75ns;

is a "complex" gate that implements the function

$$F = \overline{(a \cdot b) + c}$$

with the output delayed by 2.75 nanoseconds. This is a special type of gate that occurs in some electronics circuits; there is only one "gate" but it is capable of providing more than a single primitive operation. These and others are presented in Chapter 6.

The delay introduced above is called the **inertial delay** model in VHDL. The name arises from the fact that electronic gates require a short amount of time to respond to the movement of energy within the circuit. Inertial delays are very common in VHDL and are the default models for time delays in logic gates.

VHDL is a very rich and powerful language. It provides many variations from basic situations that allow the digital designer to construct more accurate computer simulations. In digital electronics, other types of signal delays exist that affect the overall performance of the network. VHDL recognizes this situation and provides other types of delay models.

Transport delays are associated with the time delay along an interconnect wire; the specific value of the delay increases with the length of the wire in a nonlinear manner.[4] Hardware designers strive to minimize wire transport delays to the point where they are negligible when compared to the gate propagation times. However, as the size of electronic systems shrinks from generation to generation, transport delays are increasingly important in modern design. The VHDL statement for a transport delay assumes the form

$$output <= \text{transport } (X) \text{ after } 10\text{ps};$$

This describes a signal X that is moved to the output after a delay of 10 picoseconds (ps), where $1\ ps = 10^{-12}\ sec$.

5.3 Structural Modeling

Complex logic networks are built by combining logic gates and other basic units. A VHDL structural description of a logic module is based on the same concept. Larger networks are created by connecting simpler modules together. This approach is attractive since it corresponds to the philosophy we have used to learn digital logic.

Let us examine how structural modeling can be applied to the AOI circuit shown in Figure 5.13. A schematic diagram of this type exhibits the structure by its very nature: the AOI function is created by cascading two AND gates ($G1$ and $G2$) into an OR gate ($G3$). Since we are used to tracing the flow of logic signals through diagrams of this type, we can write by inspection that the output is given by

$$f = a \cdot b + c \cdot d \tag{5.10}$$

This is, of course, obtained by noting that the AND gates produce outputs

$$\begin{aligned} X1 &= a \cdot b \\ X2 &= c \cdot d \end{aligned} \tag{5.11}$$

which are used as inputs to the OR gate such that

$$f = X1 + X2 \tag{5.12}$$

---

[4] Interconnect delays are discussed from a physical standpoint in Chapter 7.

The three equations for *X1*, *X2*, and *f* provide the basic structure of the logic network in that they provide the information needed to reconstruct the logic diagram exactly as it was originally presented. Structural models are written using the same concept.

VHDL structural models are created by first introducing the concept of a building block called a **component**. This is a logic module that has been defined by the usual **entity** and **architecture** listings and is used to create another logic network. A component may be as simple as a logic gate, or it may itself represent a complicated network. We always view a component as a simple building block regardless of its internal complexity. To create a structural model, we first choose the components and then we describe how the components are wired together. The wiring is specified by the formal listing that maps the ports of the modules.

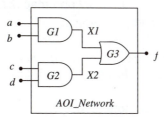

Figure 5.13
Structural modeling example

Let us use these concepts to write a structural description of the module *AOI_Network* shown in Figure 5.13. This requires that we introduce the keywords **component**, **signal**, and **port map** into our VHDL vocabulary. Since the concepts are relatively straightforward, let us present the declaration listing first and then study the details of the new entries. Boldfaced comment lines are used to separate the main sections.

> **-- Structural Description**
> entity *AOI_Network* is
> > port(*a, b, c, d* : in *bit*;
> > > *f* : out *bit*);
> end *AOI_Network* ;
> **-- The listing below builds the network using components**
> architecture *Structural* of *AOI_Network* is
> **-- Define an AND gate as a component**
> component *AND2*
> > port(*x, y* : in *bit*;
> > > *z* : out *bit*);
> end component;

```
-- Define an OR gate as a component
component OR2
        port( x, y : in bit;
                z : out bit);
end component;
-- The next line declares the internal module signals
signal X1, X2 : bit;
-- The port maps specify the internal wiring
begin
        G1: AND2 port map (a, b, X1);
        G2: AND2 port map (c, d, X2);
        G3: OR2 port map (X1, X2, f);
end Structural;
```

Now then, let us address the new concepts that have been introduced in the above statement listing. The first is that of a **component** declaration such as

```
component AND2
        port( x, y : in bit;
                z : out bit);
end component;
```

To use a module as a component in this network, it must have been previously defined by **entity** and **architecture** statements. This means that we have included declarations such as

```
entity AND2 is
        port(u, v : in bit;
                q : out bit);
end AND2;
architecture Logic of AND2 is
begin
        q <= u and v;
end Logic;
```

and

```
entity OR2 is
        port(u, v : in bit;
                q : out bit);
end OR2;
architecture Logic of OR2 is
```

begin

$q <= u$ **or** v ;

end *Logic*;

in the complete VHDL listing of the network. The concept of a component can be understood using the concept of a **design library**. A design library is a collection of different modules, each defined by entity and architecture statements. Once the cells are defined in the library, we may use "copies" of the cells in our design via the component command. This is called "instancing" the cell, and the component itself is called an **instance** of the original.

Figure 5.14 portrays the concept of a library and instancing. Note that an instance is not the same as "withdrawing" the cell from the library. Instead, an instance should be viewed as a literal copy of the cell that is currently stored in the library. This means that if we change the cell in the library, then all instances change accordingly. A cell may be instanced as many times as desired in the network. It is easy to see that this saves time and work in creating complex networks since the library entries only need to be defined once.

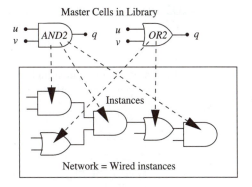

Figure 5.14 Instancing from a cell library

Next, let us examine the use of the signal keyword in the line

signal *X1, X2* : *bit*;

This is used to define the internal identifiers *X1* and *X2* that define the outputs of the AND2 gates *G1* and *G2* in the original logic diagram. The signal definitions are different from the identifiers used in the port statements in that they exist inside the module. Internal identifiers are then used to wire the gates together to form the port connections.

Once the components have been declared, the network is described by the use of the port map statements. These provide the "wiring diagram" information by specifying how component signals interface. Let's examine the entries

-- **The port maps specify the internal wiring**
begin
        G1: *AND2* port map (*a*, *b*, *X1*);
        G2: *AND2* port map (*c*, *d*, *X2*);
        G3: *OR2* port map (*X1*, *X2*, *f*);

in detail. The first map

        G1: *AND2* port map (*a*, *b*, *X1*);

tells us that gate *G1* is the AND2 component and is wired to the port identifiers *a*, *b*, and internal signal *X1*. The order of the signals in the AND2 component statement was given as (*u*, *v*, *q*) corresponding to the usage input 1, input 2, and output. The port map (*a*, *b*, *X1*) maintains this implied order. Similarly,

        G2: *AND2* port map (*c*, *d*, *X2*);

is another AND2 instance that maps the AOI module inputs *c* and *d* to the inputs of the AND2 gate and produces an output of *X2*. Finally,

        G3: *OR2* port map (*X1*, *X2*, *f*);

says that *G3* uses the internal signals *X1* and *X2* as inputs to the OR2 instance and provides the output *f* of the module *AOI_Network*. This completes the structural listing.

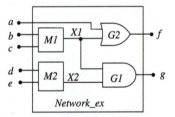

Figure 5.15
Port mapping example

Let us examine another example of structural modeling by writing the VHDL listing for *Network_ex* shown in Figure 5.15. This has five inputs *a*, *b*, *c*, *d*, and *e*, and two outputs *f1* and *g2*. The structural description can be written as follows.

-- **Structural Description**
entity *Network _ex* is
        port(*a*, *b*, *c*, *d*, *e* : in *bit*;
                f, *g*: out *bit*);
end *Network_ex*;
architecture *Structural* of *AOI_Network* is

-- Component definitions
component *Gen_1*
    port(*x, y* : in *bit*;
        g : out *bit*);
end component;
component *Gen_2*
    port(*x, y* : in *bit*;
        g : out *bit*);
end component;
component *OR_2*
    port(*x, y* : in *bit*;
        g : out *bit*);
end component;
component *AND_2*
    port(*x, y* : in *bit*;
        g : out *bit*);
end component;

-- Define internal signals
signal *X1, X2* : *bit*;

-- Port Maps
begin
    M1: *Gen_1* port map (*b, c, X1*);
    M2: *Gen_2* port map (*d, e, X2*);
    G2: *OR_2* port map (*a, X1, f*);
    G1: *AND_2* port map (*X1, X2, g*);
end *Structural*;

Note that the general components *Gen_1* and *Gen_2* must be defined elsewhere in the listing using the entity construct so that their input-output relations are known.

    These examples illustrate how to build up a network component-by-component, making a direct connection to standard digital logic. Given a network, it is a straightforward task to write the VHDL description. Similarly, the logic diagram can be extracted directly from a VHDL description. This provides a critical link between VHDL and the concepts already learned in our earlier studies. Note that it is a simple matter to introduce propagation (inertial) delays into the modeling by modifying the component definitions.

5.4 Conditional Models

Anyone who is familiar with a high-level programming language such as Pascal or C understands the usefulness of constructs that allow conditional branching. VHDL allows the use of various conditional statements to construct architectural declarations that are based solely on the behavior of the module itself.

Consider the module named *Equals* shown in Figure 5.16. This is defined by the function table such that the output *same* equals 1 when the inputs *a* and *b* are equal, but *same* = 0 if they are not equal. A conditional statement in VHDL allows us to assign a value to a signal if certain conditions are satisfied. For this case, we may write the *Equals* declarations as

> entity *Equals* is
>   port(*a, b,* : in *bit*;
>   *same* : out *bit*);
> end *Equals*;
> architecture *Dataflow* of *Equals* is
> begin
>   *same* <= '1' when *a = b* else
>   '0' ;
> end *Dataflow*;

a b	same
0 0	1
0 1	0
1 0	0
1 1	1

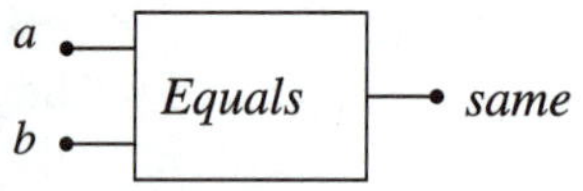

Figure 5.16 Equality detector module

In writing the architectural declaration, we have introduced a decision line using the general construct

> *identifier* <= '1' when *Condition* else
>   '0' ;

This gives *identifier* the value of 1 if *Condition* is true and assigns *identifier* = 0 if the condition is not satisfied. In the present example, the command

> *same* <= '1' when *a = b* else
>   '0' ;

assigns the value *same* = 1 if the inputs are equal, and 0 otherwise. This is exactly what is specified in the function table.

This type of construct allows us to write VHDL descriptors without having to know the internal details of a module. For example, suppose that we have a module that is defined solely by the function table shown in Figure 5.17. Although we can determine the function $g(x, y, z)$ using standard switching algebras, a VHDL description can be created using the when - else construct as follows:

entity *Table_Example* is
    port(*x, y, z*: in *bit*;
        g : out *bit*);
end *Table_Example*;
architecture *Dataflow* of *Table_Example* is
begin
    g <= '1' when (x = '0' and y = '0' and z = '1') else
        '1' when (x = '1' and y = '0' and z = '1') else
        '1' when (x = '1' and y = '1' and z = '0') else
        '1' when (x = '1' and y = '1' and z = '1') else
        '0' ;
end *Table_Dataflow*;

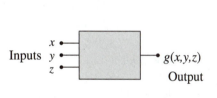

x	y	z	g
0	0	0	0
0	0	1	1
0	1	0	0
0	1	1	0
1	0	0	0
1	0	1	1
1	1	0	1
1	1	1	1

Figure 5.17 Module defined by a function table

This approach is "brute-force" in that we have just listed every input condition that gives an output $g = 1$. While this does not have the elegance of a closed-form Boolean equation, it is easy to construct and use. This is often the quickest way to verify the operation of a network.

Example 5-7

Consider the function shown below. This unit is called Add because (as we shall see in Chapter 8) it performs binary addition. For now, we will just treat it as a 3-input (a, b, and ci), 2-output (s and co) module described by the function table.

A dataflow description can be constructed as follows.

entity *Add* is
    port(*a, b, ci*: in *bit*;
        s, co: out *bit*);

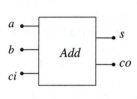

a	b	ci	s	co
0	0	0	0	0
0	1	0	1	0
1	0	0	1	0
1	1	0	0	1
0	0	1	1	0
0	1	1	0	1
1	0	1	0	1
1	1	1	1	1

end *Add*;
architecture *Dataflow* of *Add* is
begin
s <= '1' when (a = '0' and b = '1' and ci = '0') else
'1' when (a = '1' and b = '0' and ci = '0') else
'1' when (a = '0' and b = '0' and ci = '1') else
'1' when (a = '1' and b = '1' and ci = '1') else
'0' ;
co<= '1' when (a = '1' and b = '1' and ci = '0') else
'1' when (a = '0' and b = '1' and ci = '1') else
'1' when (a = '1' and b = '0' and ci = '1') else
'1' when (a = '1' and b = '1' and ci = '1') else
'0' ;
end *Add*;

As we will learn in Chapter 7, there are more elegant ways to describe this function. However, the dataflow approach is useful in that we do not need to know any more about the details of the function itself. It can be applied even if we do not have the logic expression at hand.

The alert reader may have noticed that the use of *Condition* in the above statements is different from that of the *bit* identifier type defined earlier. A *bit* is a binary quantity that has values of '0' and '1' only. However, *Condition* is properly classified as a **Boolean** type quantity that has values of "True" or "False" only. VHDL defines the values that an identifier may have by means of a **type** declaration.

We have been using *bit* as a predefined data type. If we wish to define the *bit* formally, then we use a listing of the form

type *bit* **is** ('0', '1');

This defines the *bit* as quantities that can have the values of '0' and '1' only. The logical statements above are classified as *Boolean* type quantities with different characteristic values. Boolean parameters are defined by

type *boolean* **is** (FALSE, TRUE);

Other data **type** parameters are useful in more complex situations.

Boolean conditions are evaluated using **relational operators** such as " = ." The relational operators defined in VHDL are given in Figure 5.18; these have their usual meanings, making them easy to remember. When the relational operations are used in the same expressions as logical operations (such as not and or), the not always has the highest level of precedence, followed by relational operations. The remaining logical operators and, or, nand, nor, xor, and xnor have the lowest level of precedence. Of course, we may use parentheses to insure that the VHDL statement accurately reflects the proper logic.

Operator	Meaning
=	equal to
/=	not equal to
>	greater than
<	less than
>=	greater than or equal to
<=	less than or equal to

Figure 5.18 VHDL relational operators

5.5 Binary Words

Basic logic operations are performed using individual bits. However, since a single bit can only represent two values (0 or 1), n-bit binary words with 2^n distinct values are much more useful in practical applications. As we will see in later chapters, a binary word is usually treated as a single object, not as a group of individual bits. VHDL allows us to handle words using **arrays** of bits that greatly simplify the code.

Consider the 4-bit word

$$In\_a = a_3 a_2 a_1 a_0$$

that is used as an input to the module shown in Figure 5.19. The output of the module is a single bit x. To describe this configuration, we introduce the concept of the bit_vector in which the components of In_a are specified by their subscript number as shown by the port listing

$$port(\textit{In\_a} : in \; bit\_vector \; (3 \; downto \; 0);$$
$$x : out \; \textit{bit});$$

This defines $Input\_a$ as having 4 components that are labeled by the indexing downto command as

$$In\_a(3) = a_3$$
$$In\_a(2) = a_2$$
$$In\_a(1) = a_1$$
$$In\_a(0) = a_0$$

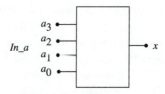

Figure 5.19 Example of a *bit_vector* input

in descending order. Once the order has been defined by this command, the positions of the individual bits cannot be changed.

Now let us apply the concept of a *bit_vector* to describing the module shown in Figure 5.20. The inputs are taken as the 4-bit words *In_a* and *In_b*, and the output is the 4-bit word

$$Out = In\_a \text{ OR } In\_b$$

where the OR operation is implied to be indexed. This means that *Out* is a 4-bit word that has bit values of

$$Out\,(3) = In\_a\,(3) \text{ OR } In\_b(3)$$

$$Out\,(2) = In\_a\,(2) \text{ OR } In\_b(2)$$

$$Out\,(1) = In\_a\,(1) \text{ OR } In\_b(1)$$

$$Out\,(0) = In\_a\,(0) \text{ OR } In\_b(0)$$

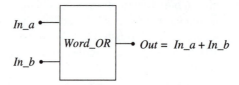

Figure 5.20 A vector module

The VHDL module *Word_OR* may be described by the following listing.

```
entity Word_OR is
        port(In_a, In_b : in bit_vector (3 downto 0);
                Out : out bit_vector (3 downto 0));
end Word_OR;
architecture Listing of Word_OR is
begin
        Out(3) <= In_a(3) or In_b(3);
        Out(2) <= In_a(2) or In_b(2);
        Out(1) <= In_a(1) or In_b(1);
        Out(0) <= In_a(0) or In_b(0);
end Listing;
```

The equivalent network with all bits shown explicitly is provided in Figure 5.21.

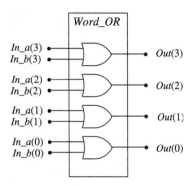

Figure 5.21 Details of the *Word_Or* Module

Individual elements of a bit_vector may be used in separate calculations. The logic network *Bit_ops* in Figure 5.22 shows two functions f_1 and f_2 that are generated from different bits of the two input words. The listing for this module can be written as

entity *Bit_ops* is
    port(*In_a, In_b* : in *bit_vector* (3 downto 0);
        f1, f2 : out *bit*);
end *Bit_ops* ;
architecture *Basic* of *Bit_ops* is
begin
    f1 <= (*In_a*(3) and *In_b*(3)) or (*In_a*(2) and *In_b*(2));
    f2<= (*In_a*(1) and *In_b*(1)) or (*In_a*(0) and *In_b*(0));
    end *Basic*;

Constructing arrays using the *bit_vector* description provides a very powerful tool in VHDL.

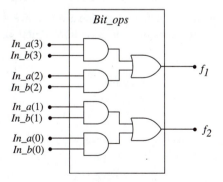

Figure 5.22 Example of another *bit_vector* module

5.6 Libraries

The concept of a library is very powerful. All VHDL simulation packages include various libraries to ease the burden on the designer. Libraries are collections of predefined quantities and procedures that are used by the VHDL compiler to interpret your code. A library is introduced into the VHDL listing using a statement of the form

library *Library_name*;

For example, one particularly useful standard library is called *ieee*; this stands for the Institute of Electrical & Electronics Engineers (IEEE), which is a professional society for electrical and electronics engineers. It is introduced by means of the command

library *ieee*;

Once a library is referenced, any definition from the library may be used in writing a VHDL description of a digital system.

What is actually contained in a library? Everything from fundamental definitions of types and basic physical units to large groups of predesigned logic blocks that can be directly instanced into your design. As an example, the library *ieee* contains items such as

- type identifiers such as bit and boolean
- time definitions for ns, ps, and others useful scaling values

Information is contained in subgroups called **packages**; the items above are found in the *ieee* library in the package standard. A particularly useful package is called IEEE 1164 which is based on the VHDL language definition. This defines types and functions that are convenient to use in writing code, and it is usually placed in the library *ieee*. Among the items defined are

- signal type such as '0' and '1'
- basic function operations for and, or, nor, and all others
- bit, vectors and bit_vectors

which collectively provide a very powerful set of definitions.

To introduce the *ieee* library and IEEE 1164 package, we initiate a VHDL description by

library *ieee*;
use *ieee.std_logic_1164.all*;

One change that arises when using the IEEE 1164 package is that we use different notation for the signal types. In particular,

- bit is replaced by std_ulogic

This allows us to change from bit, which has two values ('0' and '1'), to the more useful std_ulogic, which has nine possible values:

'0' = forced 0

'1' = forced 1

'X' = forcing unknown

'Z' = high impedance (open circuit)

'W' = a "weak" unknown

'L' = a "weak 0"

'H' = a "weak 1"

'U' = uninitialized unknown

'-' = don't care

The reader will recognize the Don't Care condition that arises in basic logic design. The remaining six values are variations due to the hardware considerations. For example, suppose that the electronics defines a perfect logic 1 as a voltage with a value of 5 *V.* A "weak 1" would be a high voltage, but not a perfect value; for example, a value of 3.8 *V* might qualify as a "weak 1" from the electronics viewpoint. In a similar manner, we note that with the IEEE 1164 package,

- bit_vector is replaced by std_ulogic_vector

which has the same range of variations for each component of the vector.

Once a library has been introduced, the VHDL description may access any of the defined quantities. The main rules and examples that we have examined up to this point remain valid. However, using a package may make the modeling more general.

Consider the following example.

```
library ieee;
use ieee.std_logic_1164.all;

entity and_gate is
        port(a, b : in std_ulogic;
                f : out std_ulogic);
end and_gate;

architecture Logic of and-gate is
begin
        f <= a and b ;
end and_ gate;
```

The main body has the same form as before, only now the library signal types *std_ulogic* are used.

If you are using a VHDL compiler package in the laboratory, then you should consult the documentation to learn more about the different libraries and what they include. You will find that a wide variety of constructs and commercial devices have been predefined and are readily available for use. Alternatively, you should realize that you are developing the skills to define modules that can be used in creating your own library.

5.7 Learning VHDL

As with any language, learning VHDL requires practice. The more you practice, the more familiar and natural it becomes. You are encouraged to write code for as many simple networks as you can; save the complicated systems for later. We will learn additional VHDL constructs as we progress in our study of digital systems design.

Try to access a VHDL compiler in your computer laboratory; the documentation provided with the program provides the details that you need to operate the compiler. There is no better way to learn VHDL than through hands-on experience, so take advantage of your situation! If you intend to pursue digital logic design as a profession, then it is highly recommended that you take an entire course in an HDL to increase your background in the subject.

5.8 Problems

[5.1] Write the VHDL **entity** and **architecture** declaration for the module *Gate_X* defined by the truth table in Figure P5.1 below.

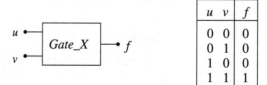

u	v	f
0	0	0
0	1	0
1	0	0
1	1	1

Figure P5.1

[5.2] Write the VHDL **entity** and **architecture** declaration for the module *Excl_2* defined by the truth table in Figure P5.2.

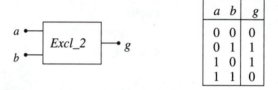

a	b	g
0	0	0
0	1	1
1	0	1
1	1	0

Figure P5.2

[5.3] Write the VHDL statement for Boolean function

$$g = a \cdot b + c \cdot d$$

using concurrent operations.

[5.4] Write the VHDL statement for Boolean function

$$output = \overline{(x \oplus y) \cdot (w + \bar{z})}$$

using concurrent operation statements.

[5.5] Write the VHDL statement for Boolean function

$$s = a \oplus b \oplus \bar{c}$$

using concurrent operation statements

[5.6] Find the simplest Boolean equation that corresponds to the VHDL statement

$$f <= (a \text{ and } b) \text{ or } (c \text{ and } (w \text{ xor } z));$$

[5.7] Translate the following VHDL statement into a Boolean expression

$$Out <= \text{not}((a \text{ or } b) \text{ and } (\text{not}(c) \text{ and } d));$$

[5.8] Construct the VHDL listing that describes the logic circuit shown in Figure P5.3 using concurrent statements.

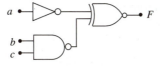

Figure P5.3

[5.9] Construct the VHDL listing that describes the logic circuit shown in Figure P 5.4 using concurrent statements.

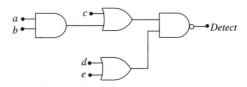

Figure P5.4

[5.10] Construct the VHDL listing that describes the logic circuit shown in Figure P5.5 using concurrent statements.

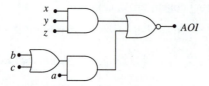

Figure P5.5

[5.11] Consider the logic cascade shown below in Figure P5.6. The "Xs" are labels for the internal nodes, and the "Gs" are used to number the gates. Write the VHDL description of this network using port maps.

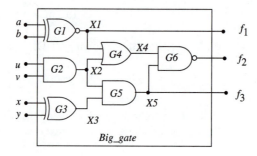

Figure P5.6

[5.12] Construct the VHDL listing that describes the network shown in Figure P5.7. Use VHDL vector statements.

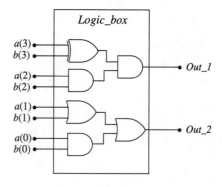

Figure P5.7

[5.13] Construct the VHDL listing that describes the network shown in Figure P5.8. Use vectors.

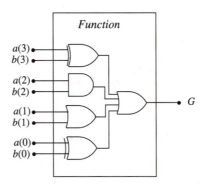

Figure P5.8

[5.14] Construct the VHDL listing that describes the network shown in Figure P5.9. Use the dataflow methodology, and create the simplest listing that you can.

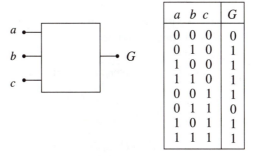

a	b	c	G
0	0	0	0
0	1	0	1
1	0	0	1
1	1	0	1
0	0	1	1
0	1	1	0
1	0	1	1
1	1	1	1

Figure P5.9

[5.15] Construct the VHDL listing that describes the network shown below in Figure P5.10. Use the dataflow methodology. The name of the unit provides a hint as to a simple approach.

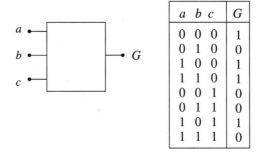

a	b	c	G
0	0	0	1
0	1	0	0
1	0	0	1
1	1	0	1
0	0	1	0
0	1	1	0
1	0	1	1
1	1	1	0

Figure P5.10

[5.16] Construct the VHDL listing that describes the network *Zero* shown below in Figure P5.11. Use the dataflow methodology. The name of the unit provides a hint as to a simple approach.

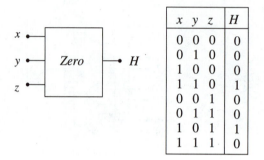

x	y	z	H
0	0	0	0
0	1	0	0
1	0	0	0
1	1	0	1
0	0	1	0
0	1	1	0
1	0	1	1
1	1	1	0

Figure P5.11

[5.17] Consider two vectors

$$A = a_3 a_2 a_1 a_0$$

$$B = b_3 b_2 b_1 b_0$$

that are used as inputs to a digital network. We want to define a circuit that has an output *compare* defined by the following statement:

$$compare = 1 \text{ if } \quad a_3 = b_3$$
$$a_2 = \bar{b}_2$$
$$a_1 = b_1$$
$$a_0 = \bar{b}_0$$

else *compare* = 0.

Write the VHDL description for this digital unit.

CHAPTER 6

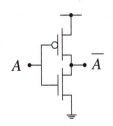

CMOS Logic Circuits

Now that we have studied both the theoretical and practical aspects of logic networks, our next step will be to see how a modern digital logic circuit is created in what is called a **CMOS technology**. This particular logic family allows us to design extremely dense digital circuits on a small piece of silicon and forms the basis for modern computer chips. This chapter will concentrate on examining how CMOS logic gates are built using the basics of digital switching logic.

6.1   CMOS Electronics

CMOS[1] is an acronym that stands for **C**omplementary **M**etal-**O**xide-**S**emiconductor, and refers to a specific type of electronic **integrated circuit** (IC). An integrated circuit is a tiny piece of silicon that may contain several million electronic components. The term IC is meant to imply that the individual circuits are integrated (combined) together into a single circuit, resulting in a very compact device. In general terms, an integrated circuit is often referred to as a **chip**, and ICs that are designed for computer applications are called **computer chips**.

Although there are several ways to make integrated circuits, CMOS is the dominant technology used for digital logic circuits. Desktop personal computers, workstations, video games, and thousands of other products rely on CMOS integrated circuits for the necessary functions. The importance of CMOS ICs becomes obvious when it is noted that all personal computers use specialized CMOS chips known as **microprocessors** to achieve their computing characteristics. Some of the reasons for the popularity of CMOS are

- Logic functions are very easy to implement using CMOS circuits.

---

[1] CMOS is pronounced as *see-moss*.

- CMOS allows for very high logic integration density. This means that the logic circuits are very small and can be built in extremely small areas.

- The technology used to make silicon CMOS chips is very well known, and the chips can be manufactured and sold for a reasonable price.

These and other features have provided the basis for making CMOS such a dominant technology.

In this book, we will use CMOS as the vehicle for studying how logic functions are implemented in an electronic network. CMOS has been chosen because it allows us to build logic circuits using simple ideas and models. And only basic electrical concepts are needed to understand the concepts, so that we do not have to get bogged down by too many details.

The important points that will evolve in this chapter are critical to understanding modern digital logic design. We will list a few of them here:

- Digital logic circuits use low and high voltages to represent logical 0 and 1 values.

- An electronic circuit can be viewed as a basic logic cell. The cell has a defined size (area), with input and output connections. The logic delay through the cell depends on the physical construction.

- A system is created by connecting cells together. The size and shape of each cell are important to the system design.

These concepts generally apply to any type of digital system, not just those created in CMOS. But they will be immediately obvious in a CMOS circuit, and the material in this chapter is designed to allow you access to many advanced topics in the current literature.

6.2 Electronic Logic Gates

Let us now turn to the problem of building an electronic circuit that behaves as a logic gate. The main idea can be understood by referring to the inverter (NOT) gate shown in Figure 6.1(a). Suppose that we want to create an electronic network that implements the NOT function. Logically, an input A gives an output of $\overline{A}$; the circuit must perform the same operation using voltages. The symbol in Figure 6.1(b) shows the transformation to include electronic parameters. In this case, we have added a power supply V_{DD}. This is used to provide electrical energy to the electronic circuit. In the same manner, the ground connection at the bottom provides a return path for the current to the power supply.

The logic variables have been replaced by voltages. The input voltage V_{in} represents the input variable A, and the output voltage V_{out} provides the result, which would be $\overline{A}$ for the inverter. The voltages are applied to wires that connect the outside world to the electronics. The logic truth table in Figure 6.2(a) is replaced by the voltage table in Figure 6.2(b) to describe the behavior of the electronic circuit. Recall the ideal definitions for 0s and 1s from Chapter 4:

$$\text{Logic } 0 \Leftrightarrow 0v$$
$$\text{Logic } 1 \Leftrightarrow V_{DD}$$

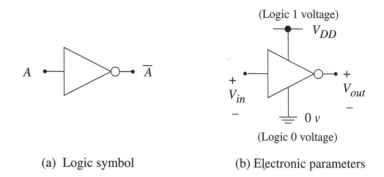

| (a) Logic symbol | (b) Electronic parameters |

Figure 6.1 Electronic characteristics of an inverter

These may be used to construct the table by noting that an input voltage of $V_{in} = 0v$ must give an output voltage of $V_{out} = V_{DD}$, while $V_{in} = V_{DD}$ yields $V_{in} = 0v$. By inspection, we see that the two truth tables are equivalent.

One important observation can be made at this point. Since the ground connection is an ideal logic 0 voltage, while the power supply V_{DD} is an ideal logic 1 voltage, we can create a NOT circuit by using a simple switch that connects either the ground to the output, or the power supply V_{DD} to the output, as shown in Figure 6.3. The state of the switch (UP or DOWN) must be controlled by the value of the input voltage V_{in} for proper operation. In CMOS, the switching action is implemented using electronic devices called **transistors**. These are discussed in the next section.

6.3 MOSFETs

The acronym *MOSFET* stands for **M**etal-**O**xide-**S**emiconductor **F**ield-**E**ffect **Tra**nsistor. MOSFETs are electronic devices that are used as switches in CMOS digital logic circuits. Although the name may be somewhat intimidating, the structure and operation of a MOSFET can be understood using basic arguments from electrostatic physics. In our viewpoint, the MOSFET will be modeled as a simple switch that is either OPEN or CLOSED; this action is very similar to the behavior of the

A	$\overline{A}$
0	1
1	0

V_{in}	V_{out}
$0v$	V_{DD}
V_{DD}	$0v$

| (a) NOT truth table | (b) Electronic equivalent |

Figure 6.2 Ideal Boolean-to-voltage translation

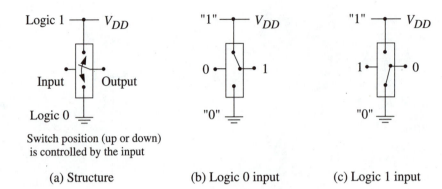

Switch position (up or down)
is controlled by the input

(a) Structure   (b) Logic 0 input   (c) Logic 1 input

Figure 6.3 Electronic view of the inverter output

switch that turns a room light on or off, except that the electronic counterpart will be controlled by logic signals!

Figure 6.4 shows the circuit symbol for one type of basic MOSFET. This particular component is called an **n-channel** MOSFET, or **nFET** for short, as it relies on negatively charged electrons for current flow. The device has three terminals that have been labeled as the **gate**, the **drain**, and the **source**. Note that the gate terminal has been designated as the "control" electrode. In essence, the voltage applied to the gate determines whether there is current flow from the drain to the source. In other words, the value of the current I shown in the symbol depends on the value of the voltage applied to the gate. The designation of the gate electrode as the control terminal is the key to understanding the operation of a transistor as a logic switch.

The electrical operation of the nFET is summarized in Figure 6.5. In electrical terms, it is the value of the voltage V_{GS} applied between the gate and the source terminals[2] that controls the operation of a FET. For our purposes, we will only consider two values of the voltage V_{GS}. In Figure 6.5(a), a gate-to-source voltage of $V_{GS} = 0v$ is applied. This results in zero current flow ($I = 0$) between the drain and source terminals, and the transistor is said to be in **cutoff**; physically, $I = 0$ is

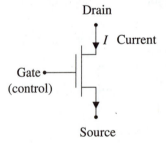

Figure 6.4
Circuit symbol for an n-channel MOSFET (nFET)

---

[2] This means that V_{GS} is the voltage at the gate minus the voltage at the source.

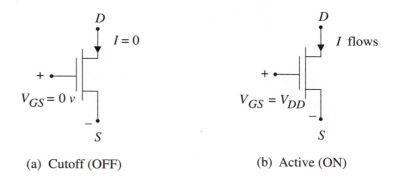

(a) Cutoff (OFF) (b) Active (ON)

Figure 6.5 Current flow through an nFET

equivalent to there being no connection between the two terminals. Alternatively, we say that the transistor is **OFF** for simplicity. If, on the other hand, the gate-to-source voltage is set at a high value $V_{GS} = V_{DD}$, the current I can flow, entering the drain and exiting at the source. When current flows between two terminals, they are electrically connected to each other. This is shown in Figure 6.5(b). In this case, the transistor is said to be **ACTIVE** or **ON**.

The ability to turn a transistor OFF and ON (i.e., control whether the device blocks or allows current flow) makes a MOSFET an ideal type of electronic switch. Although transistors can be a bit complicated to describe, we can introduce the **ideal switch model** shown in Figure 6.6 when we are only concerned with the ON/OFF characteristics of a MOSFET. In this model, the gate voltage has been replaced by a logic variable G. The value of the logic variable determines whether the switch is OPEN or CLOSED. An OPEN switch does not allow any current to flow through and corresponds to a transistor in cutoff, as in Figure 6.6(a). Conversely, a CLOSED switch allows current to flow between the drain and source terminals and describes an ACTIVE or ON transistor; this situation is shown in Figure 6.6(b).

To define the operation of the switch, note that a voltage of $V_{GS} = 0v$ turns the MOSFET OFF. Logically, this says that a gate input of $G = 0$ should have the same effect, so that the switch is OPEN for this case. Conversely, if $V_{GS} = V_{DD}$, corresponding to an ideal value of $G = 1$, then the MOSFET is ACTIVE, so the switch is in a CLOSED

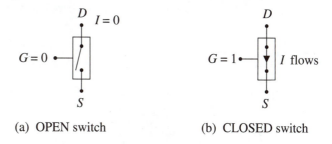

(a) OPEN switch (b) CLOSED switch

Figure 6.6 Characteristics of the nFET as a controlled switch

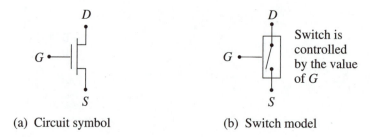

(a) Circuit symbol   (b) Switch model

Figure 6.7 Switching model of an nFET

state. The switch itself is called a **voltage-controlled**, or **logic-controlled**, device depending upon whether we are using the value of the voltage or the value of a logic variable in our description. The switch characteristics are summarized in Figure 6.7.

Although it is possible to build logic circuits using only nFETs, CMOS design relies on also using another, "complementary" MOSFET that employs positive charge for current flow. This second type of transistor is called a **p-channel MOS-FET**, or **pFET** for short, and is the electrical and logical **complement** of the nFET. What does this mean exactly? Well, all voltage polarities (+ and – sides) and current flow directions are opposite. And the control properties are opposite to those of an nFET.

Figure 6.8 shows the circuit symbol for a pFET. Note that it has the same features as those used for the nFET, except that the gate has an inversion bubble at the gate. This allows us to distinguish it from an nFET, but, more important, implies that the logical control applied to the gate is going to have the opposite effect of that discussed for an nFET. Also note that the source and drain are reversed, so that the current flows into the source and out of the drain terminal.

Since the pFET is opposite to the nFET, the operational characteristics of the device can be understood by just reversing everything discussed for the nFET. Let us first examine the switching behavior using voltages. For the case of the pFET, it is the source-to-gate voltage V_{SG} (the voltage at the source minus the voltage at the gate) that controls the behavior. If the source-to-gate voltage is set at $V_{SG} = V_{DD}$, then the transistor allows current to flow and is ACTIVE or ON. If the source-to-gate voltage is small with $V_{SG} = 0v$, then the pFET is in CUTOFF (or simply OFF), and no current flows. The operation is summarized in Figure 6.9. Comparing this

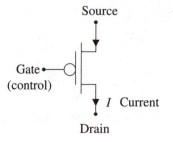

Figure 6.8
Circuit symbol for a p-channel
MOSFET (pFET)

with Figure 6.5 for the nFET, we see that the switching operation is exactly opposite. Once again, this is because the nFET and pFET are complementary electronic devices.

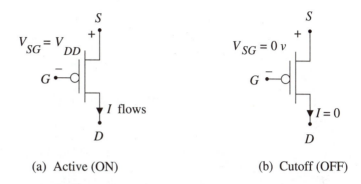

(a) Active (ON)   (b) Cutoff (OFF)

Figure 6.9 Current flow through a pFET

Alternatively, we may introduce the symbol for a pFET logic switch model in which the gate voltage is interpreted as a logic variable G, and the value of G (0 or 1) determines if the pFET switch is OPEN or CLOSED. The switch symbol is shown in Figure 6.10 and is the same as that used for the nFET except for the inversion bubble at the gate. This reminds us that the gate of a pFET has the characteristics opposite to the gate of an nFET. The operation of the pFET is straightforward. If $G = 0$, then the switch is CLOSED, while a logic value of $G = 1$ gives an OPEN switch. These are summarized in Figure 6.11.

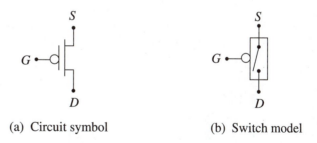

(a) Circuit symbol   (b) Switch model

Figure 6.10 pFET switch equivalent

6.4 The NOT Function in CMOS

Logic circuits are electronic networks that are designed to implement logic functions. Although the analysis of electronic circuits is beyond the scope of this book, the operation and concept of a logic gate can be understood using simple models.

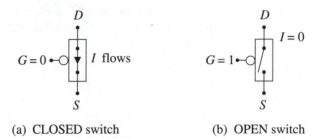

(a) CLOSED switch   (b) OPEN switch

Figure 6.11 pFET switching characteristics

Consider first the task of creating an electronic logic gate. A simplified switching model that describes the operation is shown in Figure 6.12. In this network, the output of the gate is connected to a switch whose position is controlled by an input. As shown in the drawing, the switch can be either UP or DOWN, depending on the voltage that is applied to the input terminal.

Let us use this model to construct an inverter (NOT) circuit. The operation of the switch is then summarized by the two cases shown in Figure 6.12. With a 0 at the input, the output is to be a 1, so that the switch must be connected to V_{DD} to give a logic 1 [see Figure 6.12(a)]. Conversely, applying a 1 at the input should connect the output to ground to produce a logic 0, as shown in Figure 6.12(b). It is important to note the following aspects of the operation.

- The output logic value (0 or 1) corresponds to the value of the voltage measured at the output terminal. The ground and power supply connections provide these output voltage values.

- The state of the switch (up or down) is controlled by the value of the input.

As will be seen below, these general principles can be applied to help us create many different logic gates using simple MOSFET arrangements.

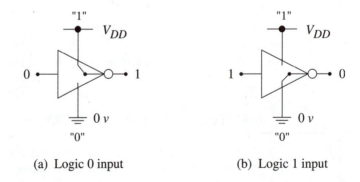

(a) Logic 0 input   (b) Logic 1 input

Figure 6.12 Electronic switch viewpoint of inverter operation

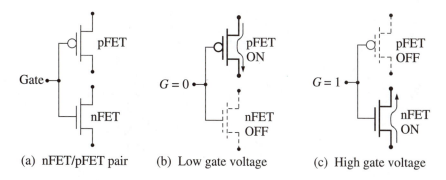

Figure 6.13 A complementary pair of FETs

6.4.1 Complementary Pairs

CMOS logic circuits use nFET and pFET transistors that are arranged as **complementary pairs**; this is the origin of the "C" in the acronym CMOS. A complementary pair consists of an nFET and a pFET that have the gate electrodes connected together to form a single terminal, as shown in Figure 6.13(a). The common gate signal G controls both transistors. However, since the transistors have opposite characteristics, this pair gives the useful characteristic that one of the transistors is ON while the other transistor is OFF. When $G = 0$, the pFET is ON and the nFET is OFF, as shown in Figure 6.13(b). If $G = 1$, then the opposite is true: the pFET is OFF and the nFET is ON; this is seen in Figure 6.13(c). The term "complementary" is used because of this behavior with regard to the gate signal G.

The operation of the complementary pair can be understood using MOSFET switch models. Figure 6.14 shows the same arrangements with the transistors replaced by logic-controlled switches. When $G = 0$ is applied, the pFET switch is CLOSED while the nFET switch is OPEN. Conversely, a value of $G = 1$ yields the opposite situation, with the pFET switch OPEN and the nFET switch CLOSED. An important characteristic of the complementary pair is that a stable (well-defined) value of G always gives the situation where one switch is open and the other switch is closed.

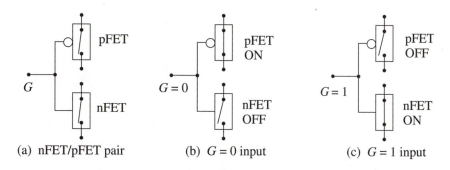

Figure 6.14 Switching behavior of a complementary MOSFET pair

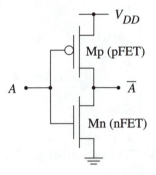

Figure 6.15
The CMOS inverter

In basic CMOS circuits, every input to a logic gate is connected to a complementary pair. A single-input gate, such as an inverter, consists of two MOSFETs (one pair), while a 2-input gate requires four transistors (two pairs), and so on. Complementary pairs are used to create a current flow path between the output and either the power supply V_{DD} or the ground connection. This gives CMOS logic circuits many of their unique characteristics. From the logic viewpoint, the use of complementary pairs provides a simple technique for designing basic and complex logic gates solely by the proper placement of the nFET and pFET transistor pairs.

6.4.2 The CMOS Inverter

A CMOS inverter gate that performs the NOT operation is constructed by connecting a complementary MOSFET pair between the power supply V_{DD} and ground, as shown in Figure 6.15, and providing a common output terminal. The input is labeled as A, and the output is $\overline{A}$.

The operation of the inverter can be understood by applying the switching properties of a complementary pair to the circuit. Consider first the case where the input is $A = 1$, as illustrated in Figure 6.16(a). The pFET Mp is OFF and acts like an open switch; this disconnects the power supply from the output. The nFET Mn, on the other hand, is ON. Since Mn is effectively a closed switch, the output is connected

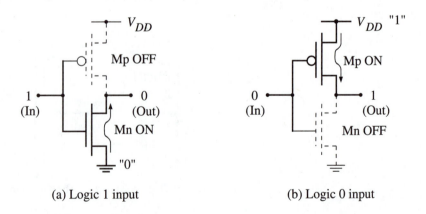

(a) Logic 1 input　　　　　　　　(b) Logic 0 input

Figure 6.16 Operation of the CMOS inverter

to the ground, giving 0 v there. This, of course, is equivalent to having a logic 0 output. Thus, we see that an input of $A = 1$ gives and output of $\overline{A} = 0$ as desired.

For the case where the input is $A = 0$ [shown in Figure 6.16(b)], the situation is exactly opposite. Now Mp is ON and provides a connection between the output and the power supply. Since Mn is OFF, the output voltage has a voltage value of V_{DD}, which is a logic 1 output. Thus, an input of $A = 0$ yields an output of $\overline{A} = 1$, and our analysis is complete.

It is worthwhile to reflect on the simplicity of the NOT gate. Using only two FETs (or, equivalently, one complementary pair), we have been able to create an electronic network that provides the logical NOT operation. The behavior is due to the characteristics of MOSFETs when used as a complementary pair. These observations provide us with motivation for delving deeper into the structure of CMOS logic, where we shall see that it is straightforward to use the NOT operation as a basis for designing more complex (and useful) logic gates.

6.5   Logic Formation Using MOSFETs

The ideas introduced to build the NOT operation can be extended to design NOR and NAND logic circuits. To create this "family" of gates, we will use the basic structure shown in Figure 6.17 for the case where there are three input variables A, B, and C. Complementary pairs will be used to construct the logic gates, so that each variable is connected to both an nFET and a pFET. This is represented in the drawing by extending each input to separate switch boxes. For three inputs, the upper box (close to V_{DD}) labeled "pFET array" consists of three p-channel MOSFETs, while the lower switch box (close to ground) contains three n-channel MOSFETs. The boxes labeled FET arrays act as "giant switches" between the top and bottom. The transistors inside of a box determine whether the array switch is OPEN (with no connection between the top and bottom) or CLOSED (which gives the connection). Note once again that the power supply connection labeled V_{DD} is an ideal logic 1 voltage, while the ground terminal gives us an ideal logic 0 voltage.

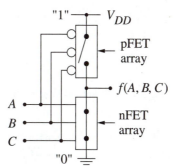

Figure 6.17
General structure of a 3-input CMOS logic gate

To create a logic gate using this circuit, we will require that one of the following statements is true for every combination of input values:

- The pFET array switch is CLOSED (from the top to the bottom) and the nFET array switch is OPEN,

or

- The pFET array switch is OPEN and the nFET array switch is CLOSED.

The output $f(A,B,C)$ is then assured of attaining only two values: a logic 1 voltage of V_{DD} if the pFET switch is closed, or a logic 0 voltage of 0 v if the nFET switch is closed. This is similar to the operation of the inverter circuit, except that each inverter transistor has been replaced by an array of FETs and the array acts as a single "composite" switch.

Now that we have introduced the general structure of a CMOS logic gate, let us proceed to study how the nFETs and pFETs must be arranged to synthesize a desired logic function. Recall that a MOSFET has three main terminals: the gate, the source, and the drain. The logic state G (or, equivalently, the gate voltage) on the gate terminal controls the current flow between the drain and source terminals, such that nFETs and pFETs have opposite characteristics. Suppose that we want to connect the drain and source terminals of two FETs together to create a switching circuit.[3] There are only two ways that the two transistors can be connected, in **series** or in **parallel**.

A series connection of two nFETs is shown in Figure 6.18(a). As seen from the drawing, the series connection can be described as an "end-to-end" arrangement which is usually drawn with an implied connection. Electrically, series-connected devices are identified as having the same current flowing through them.

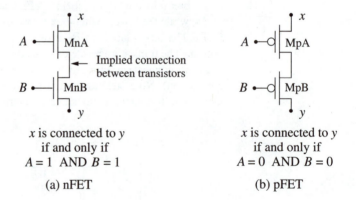

x is connected to y
if and only if
$A = 1$ AND $B = 1$

(a) nFET

x is connected to y
if and only if
$A = 0$ AND $B = 0$

(b) pFET

Figure 6.18 Series-connected MOSFETs

The switching properties of series-connected FETs depend on their polarity (n-type or p-type). For the two nFETs as shown in Figure 6.18(a), both transistors

---

[3] Note that for our logic circuit, the gate terminal is already defined to be connected to an input.

must have logic 1s applied to their gates in order to turn on both transistors and create a connection between the top (x) and bottom (y). This can be stated by writing

If both $A = 1$ AND $B = 1$,

then x is connected to y.

If either $A = 0$ or $B = 0$ (or both), then at least one of the FETs is OFF and acts like an open switch; terminals x and y are then electrically separate from one another.

Series-connected pFETs are shown in Figure 6.18(b). Since a gate signal of 0 is required to turn on a pFET, we must have logic 0s applied to both MpA and MpB to obtain a connection between the top and bottom. In other words,

If both $A = 0$ AND $B = 0$,

then x is connected to y.

If either input is 1, then x and y are not connected.

Parallel-connected transistors can be analyzed in the same way. As shown in Figure 6.19, MOSFETs are said to be in parallel if both the drain terminals are common, and the source terminals are common. Consider first the nFET pair shown in Figure 6.19(a). If $A = 1$, then x and y are connected through transistor MnA. Similarly, if $B = 1$, then x and y are connected through MOSFET MnB. If both A and B are 1s, then both transistors act like closed switches. Thus, we may write

If $A = 1$ OR $B = 1$ OR both,

then x is connected to y

to describe the combination. If we place two pFETs in parallel, as in Figure 6.19(b), then the statement becomes

If $A = 0$ OR $B = 0$ OR both,

then x is connected to y

since pFETs require a logic 0 at the gate to act as a closed switch.

Examining the switching properties of series and parallel FETs thus illustrates that the AND and OR functions are possible in both nFETs and pFETs. By themselves, none of the transistor groups discussed above yield a complete logic function since we have only looked at how terminal x can be connected to terminal y.

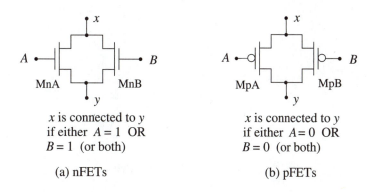

x is connected to y
if either $A = 1$ OR
$B = 1$ (or both)

(a) nFETs

x is connected to y
if either $A = 0$ OR
$B = 0$ (or both)

(b) pFETs

Figure 6.19 Parallel-connected MOSFETs

For example, suppose that we have two series-connected nFETs as in Figure 6.18(a) with $A = 1$ and $B = 0$. In this case, we cannot write any relationship between x and y. However, by using groups of nFETs and pFETs in a single gate as shown in Figure 6.17, we will learn how to overcome this problem.

We are now in a position to synthesize a set of CMOS logic circuits using complementary MOSFET pairs. To effect this goal, let us note the following observations:

- Series-connected nFETs can be used to create the AND function.

- Parallel-connected nFETs can be used to create the OR function.

Series and parallel groups are thus duals of one another. To write equivalent statements for pFETs, we must consider the fact that logic 0 inputs are needed to close a pFET switch. Then, recalling the DeMorgan relations

$$\overline{A} \cdot \overline{B} = \overline{A + B}$$

$$\overline{A} + \overline{B} = \overline{A \cdot B}$$

(6.1)

allows us to state that

- Series-connected pFETs can be used to create the NOR function.

- Parallel-connected pFETs can be used to create the NAND function.

In CMOS, the NOR and NAND gates, along with the NOT operation, are used as the basic gates.

6.5.1 The NOR Gate

To construct a CMOS NOR2 logic gate, we will use two complementary pairs with inputs A and B as shown in Figure 6.20(a). The two nFETs, MnA and MnB, are in parallel, while the pFETs MpA and MpB are in series.

The operation of the circuit can be understood by the table provided in Figure 6.20(b). Let us view the transistors as switches. If either $A = 1$, or $B = 1$, or both, then at least one of the nFET switches MnA or MnB is closed, connecting the

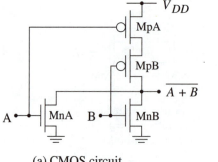

A B	MnA	MnB	MpA	MpB	Out
0 0	OFF	OFF	ON	ON	1
0 1	OFF	ON	ON	OFF	0
1 0	ON	OFF	OFF	ON	0
1 1	ON	ON	OFF	OFF	0

(a) CMOS circuit   (b) Operation summary

Figure 6.20 The NOR2 gate in CMOS

output $\overline{A + B}$ to ground. Since at least one of the pFETs will be off, this gives an output value of 0. The only time that the output is connected to the power supply V_{DD} (for a logic 1 output) is when $A = 0$ and $B = 0$, so that both MpA and MpB are closed; in this case, both nFETs are open. The four input possibilities are summarized by the circuits presented in Figure 6.21(a)–(d). These are drawn to indicate whether a MOSFET is ON (conducting as a closed switch) or OFF (an open switch). To verify these cases, just remember that a 0 to the gate of an nFET turns it OFF, while a 1 turns the transistor ON; a pFET is just the opposite. Once the state of every transistor has been determined, the resulting switching diagram will show whether the output node is connected to the power supply V_{DD} or ground, hence giving the output value.

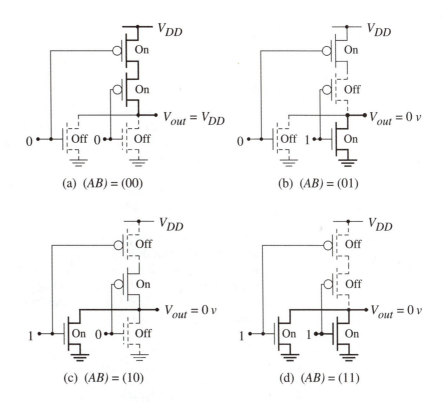

(a) $(AB) = (00)$ (b) $(AB) = (01)$

(c) $(AB) = (10)$ (d) $(AB) = (11)$

Figure 6.21 Input combinations for the CMOS NOR2 gate

The structure of the NOR2 gate can be easily extended to create NOR gates with three inputs. Calling the input variables A, B, and C, we construct the NOR3 gate by using three complementary pairs of MOSFETs, with the three nFETs in parallel and the three pFETs in series. This results in the circuit shown in Figure 6.22. This methodology can be applied to NOR gates with four (or more) inputs. However, the switching times of CMOS NOR gates with more than four inputs get quite long and are usually avoided in high-performance designs.

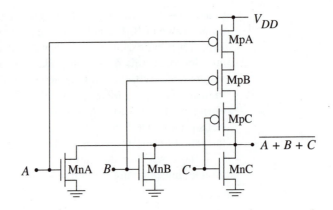

Figure 6.22 A NOR3 gate in CMOS

6.5.2 The NAND Gate

A CMOS NAND2 gate is constructed by again using two complementary pairs, only now the two nFETs are placed in series and the two pFETs are parallel. This results in the circuit shown in Figure 6.23(a). The operation of the circuit is summarized in the table of Figure 6.23(b).

Let us search for the case(s) where the output $\overline{A \cdot B}$ is connected to ground and evaluates to a logic 0. In order for this to occur, we must have $A = 1$ and $B = 1$ to insure that both MnA and MnB act like closed switches. If either input is 0, then at least one of the nFETs is off, while the corresponding pFET in the pair is on. This connects the output to the power supply V_{DD}, giving a logic 1 output. The operation table thus verifies that this circuit does indeed produce the NAND function. Every input combination is shown in Figure 6.24(a)–(d) to aid in verifying the behavior of the gate. The important procedure in finding the output of the gate is first to determine the state (ON or OFF) of every transistor and then trace the output to either ground or the power supply V_{DD}.

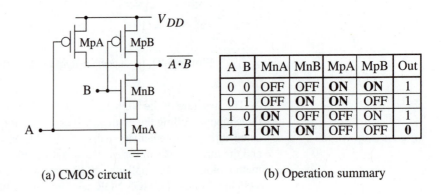

A B	MnA	MnB	MpA	MpB	Out
0 0	OFF	OFF	**ON**	**ON**	1
0 1	OFF	**ON**	**ON**	OFF	1
1 0	**ON**	OFF	OFF	**ON**	1
1 1	**ON**	**ON**	OFF	OFF	**0**

(a) CMOS circuit   (b) Operation summary

Figure 6.23 The CMOS NAND2 gate

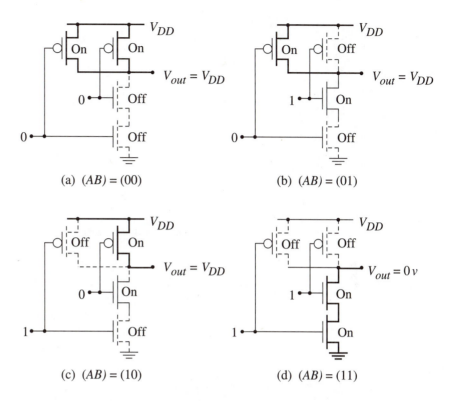

(a) $(AB) = (00)$

(b) $(AB) = (01)$

(c) $(AB) = (10)$

(d) $(AB) = (11)$

Figure 6.24 Input combinations for the CMOS NAND2 gate

The NAND2 structure can be extended to create a NAND3 gate by placing three nFETs in series and three pFETs in parallel, then connecting the transistors to the inputs as complementary pairs. As for the case of the NOR gate, it is rare to find a NAND gate with more than four inputs since the switching times are long.

6.5.3 The CMOS-Logic Connection

CMOS integrated circuits provide an excellent technology base for logic design. In these circuits, it is important to remember that Boolean logic variables such as A and B can only assume binary values of "0" and "1"; the CMOS network provides the connection between these logic bits and the electronic circuit implementation. Once this is recognized, all of the logic design techniques we have studied may be applied directly to CMOS.

To understand this concept, consider the NOR2 gate shown in Figure 6.25. This gate has three input combinations that result in an output of $f = 0$: $(AB) = (10)$, (01), and (11). In terms of the general CMOS gate shown earlier in Figure 6.17, an output of 0 means that

- The nFET logic block is acting as a closed switch, and
- The pFET logic block is acting as an open switch.

If we want to design the nFET logic block, then we need to connect two nFETs together to insure that all three input combinations that produce $f = 0$ result in a

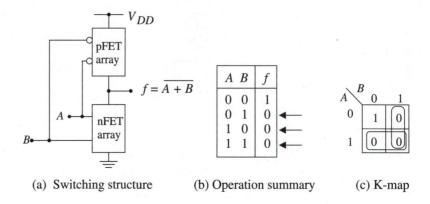

(a) Switching structure   (b) Operation summary   (c) K-map

Figure 6.25 Truth table analysis of a NOR2 gate

closed switch from the top to the bottom of the 2-transistor group. In terms of the Karnaugh map in Figure 6.25(c), the 0 entries thus provide the information on the nFET array. Grouping these entries as shown gives the complement of the output as

$$\bar{f} = A + B \tag{6.2}$$

From our earlier studies in this section, we know the OR operation implies that the two nFETs must be in parallel, which demonstrates the link between CMOS circuits and conventional logic. Since the nFETs connect a value of 0 (ground) to the output, we must complement the expression to arrive at

$$f = \overline{A + B} \tag{6.3}$$

as expected. Once the structure of the nFET array is known, the pFET connections are "opposite" in the sense that since the nFETs are in parallel, the pFETs must be in series.

The NAND2 gate may be analyzed in the same manner. The link between the Karnaugh map and the logic array is shown in Figure 6.26. In this gate, there is only one input combination that has a 0 output. The map thus yields the single term

$$\bar{f} = A \cdot B \tag{6.4}$$

as verified by inspection. The AND operation implies that the nFETs must be connected in series by applying our earlier arguments. Similarly, we see that the pFETs must be wired in parallel.

In practice, the K-map can be used to determine the wiring connections for one of the transistor arrays; either the nFET or the pFET maybe used. If the nFET array is created first, the "0" entries are analyzed, while the pFET array will be structured by the "1" entries. The other array is then created using series-parallel arguments discussed below. Although it is tempting to treat CMOS logic as an independent subject, a good designer will view the two as being equivalent through the circuit-to-logic connection discussed here.

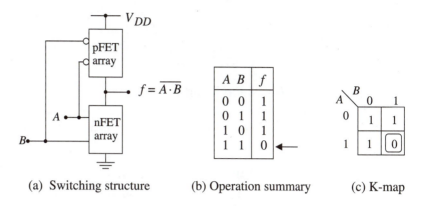

(a) Switching structure   (b) Operation summary   (c) K-map

Figure 6.26 Truth table analysis of the NAND2 gate

6.6   Complex Logic Gates in CMOS

Once we understand how to build NOT, NAND, and NOR gates, it is theoretically possible to construct the circuit for any logic function using the techniques introduced in Chapter 3. This would result in a group of cascaded CMOS logic gates that provide the necessary operations. While this approach is straightforward, CMOS often allows us to create logic gates that are much simpler and use fewer transistors than what would be obtained using only the basic gates. This is, in fact, one reason why CMOS has evolved to such a prominent status in modern technology.

Let us first introduce some terminology. In this section, we will study how to build a **complex** logic gate in CMOS. A complex logic gate is one that implements a function that can provide the basic NOT, AND, and OR operations but integrates them into a single circuit. The logic gates will have the same characteristics as the basic gates, namely,

- Complementary pairs of MOSFETs are used, so that every input is connected to both an nFET and a pFET.

- Complex logic gates will be constructed using the general circuit shown in Figure 6.17.

In our approach, we will use **series-parallel** combinations of nFETs and pFETs. This means that nFETs and pFETs are connected in opposite ways. For example, when two nFETs are in parallel, then the corresponding pFETs (with the same inputs) are in series, and vice versa. This is identical to the design used for NOR and NAND circuits. Finally, we will find that CMOS is ideally suited for creating gates that have logic equations that exhibit

- **AOI**: AND-OR-INVERT

or

- **OAI**: OR-AND-INVERT

forms. We note that an AOI logic equation is equivalent to a complemented SOP form, while an OAI equation is equivalent to a complemented POS structure. Because of this correlation, it is a straightforward matter to construct complex CMOS logic gates for the structured logic forms examined in Chapter 2.

6.6.1 3-Input Logic Gates

The first class of logic gates we will examine have three inputs that will be denoted as A, B, and C. Various functions can be obtained by changing the manner in which the FETs are connected.

AND-OR-Invert Example

Consider the logic gate shown in Figure 6.27. If we examine the A and B inputs, the NAND2 gate structure can be seen by noting that the A and B nFETs are in series, while the A and B pFETs are in parallel. The difference between this circuit and the NAND2 gates is transistors associated with the complementary pair that have C as an input.

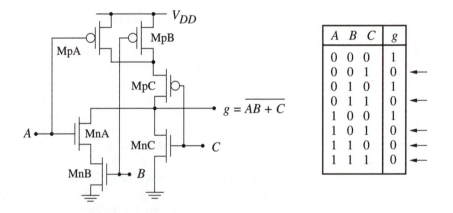

A	B	C	g
0	0	0	1
0	0	1	0
0	1	0	1
0	1	1	0
1	0	0	1
1	0	1	0
1	1	0	0
1	1	1	0

Figure 6.27 AND-OR-Invert logic gate example

This gate implements the complex logic function

$$g(A, B, C) = \overline{A \cdot B + C} \tag{6.5}$$

which has a basic AOI structure. To verify this statement, let us examine the nFET transistor array to determine what input combinations will yield an output value of $g = 0$. This is possible either if

$$A \cdot B = 1$$

since this provides a conduction path between the output and ground through the left side of the nFET network, OR if

$$C = 1$$

as this connects the output to ground through the right side nFET (that has an input of C). Figure 6.28(a) and (b) shows these two cases using the circuit to help clarify the discussion. These statements may be summarized by writing that

$$g = 0 \text{ if } A \cdot B + C = 1$$

which is equivalent to writing that

$$g = \overline{A \cdot B + C} \qquad\qquad \textbf{(6.6)}$$

as stated.

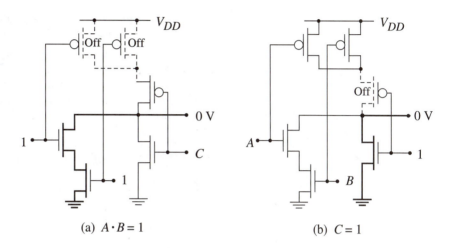

(a) $A \cdot B = 1$ (b) $C = 1$

Figure 6.28 Input combinations that yield a logic 0 output

This function could have been obtained directly from the function table by extracting SOP terms. Since SOP is noninverting, let us search for the terms where $g = 0$ as indicated in the table. Writing each term and combining gives

$$\begin{aligned}
\bar{g} &= \overline{A} \cdot \overline{B} \cdot C + \overline{A} \cdot B \cdot C + A \cdot \overline{B} \cdot C + A \cdot B \cdot \overline{C} + A \cdot B \cdot C \\
&= A \cdot B \cdot \overline{C} + C \qquad\qquad\qquad\qquad\qquad\qquad \textbf{(6.7)} \\
&= A \cdot B + C
\end{aligned}$$

so that taking the complement to obtain g yields the same result as using the simpler qualitative switching analysis.

Now let us examine the nature of the transistor arrays. The nFETs that have inputs A and B are connected in series, so that the pFETs with the same inputs are connected in parallel. This is the same as the NAND2 gate. To see how the additional transistors have been added, note that the C-input nFET has been placed in *parallel* with the **group** of A and B nFETs; this achieves the OR operation when analyzing $\bar{g}$. The pFET that uses C as an input has been placed in *series* with the group of A and B pFETs. Thus, the series-parallel structure between the nFETs and the pFETs has been expanded to be applied to groups of transistors, not just individual devices. This series-parallel wiring insures that, for a given input combination, the output is connected either to the power supply V_{DD} *OR* to the ground node, which always gives a well-defined voltage. This approach eliminates the possibility of the output node being connected to both the power supply and ground at the same time, or the situation where the output is not connected to the power supply or ground even though the inputs are well defined.

OAI Example

Next consider the logic gate shown in Figure 6.29. This implements the function

$$f = \overline{(A + B) \cdot C} \qquad (6.8)$$

which is in OR-AND-Invert form. The circuit uses the NOR2 gate as a basis, as can be seen by examining the A and B inputs; for these two variables, the nFETs are in parallel, while the pFETs are in series. The transistors associated with the additional input C have been added in series-parallel fashion. This is verified by noting that the C-input nFET has been placed in *series* with the **group** consisting of MnA and MnB, while the pFET that is driven by C has been placed in *parallel* with the **group** of pFETs made up of MpA and MpB.

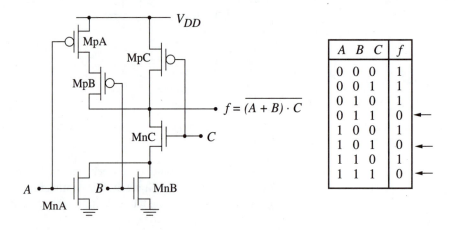

A	B	C	f
0	0	0	1
0	0	1	1
0	1	0	1
0	1	1	0	←
1	0	0	1
1	0	1	0	←
1	1	0	1
1	1	1	0	←

Figure 6.29 CMOS OR-AND-Invert gate

Let us verify the output function by examining the nFET array and finding all possibilities that give $f = 0$. First, we see that $C = 1$ is required to place MnC into a conducting state. Once this has been established, then either $A = 1$ OR $B = 1$ must be applied to the respective nFETs to complete a path to ground. This behavior may be summarized by saying that

$$f = 0 \text{ if } (A + B) \cdot C = 1$$

so inverting to obtain the $f = 1$ conditions gives the function

$$f = \overline{(A + B) \cdot C} \qquad (6.9)$$

as originally stated. The behavior of the switching network for the two cases $A \cdot C = 1$ and $B \cdot C = 1$ is shown in Figure 6.30(a) and (b). Once again, the series-parallel structure automatically insures a well-defined output, making it very useful in practical engineering.

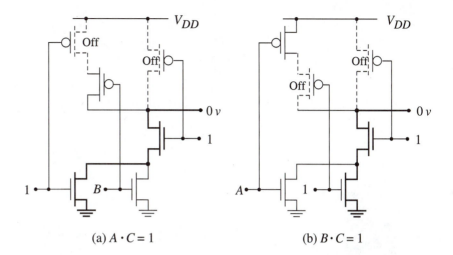

(a) $A \cdot C = 1$ (b) $B \cdot C = 1$

Figure 6.30 OAI input combinations that give a logic 0 output

Of course, the same function can be obtained using standard reduction techniques. For this circuit, let us construct a 3-variable Karnaugh map for the $f = 0$ input combinations shown in the function table. Grouping as shown in Figure 6.31 gives directly the reduced expression

$$\bar{f} = B \cdot C + A \cdot C \qquad (6.10)$$

so that taking the complement yields the same result.

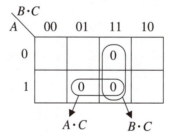

Figure 6.31
Karnaugh map analysis of the OAI function

6.6.2 A General 4-Input Gate

The series-parallel gate structuring can be extended to a larger number of inputs. Figure 6.32 shows a general gate with four inputs that gives the AOI logic function

$$h = \overline{A \cdot B + C \cdot D} \qquad (6.11)$$

as can be verified by examining the behavior of the switching network. The output has a value of $h = 0$ either if $A \cdot B = 1$, providing a conduction path to ground

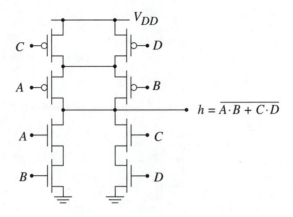

Figure 6.32 4-input AOI logic gate in CMOS

through the left side of the nFET array, or if $C \cdot D = 1$, which gives a conducting path on the right side of the circuit. These arguments are equivalent to stating that

$$\bar{h} = A \cdot B + C \cdot D \tag{6.12}$$

which can be complemented to give the function h. Note that the pFET array is connected using the proper series-parallel arrangements:

- nFETs with A and B are in series, so pFETs with A and B are in parallel.

- nFETs with C and D are in series, so pFETs with C and D are in parallel.

- The nFETs groups (A, B) and (C, D) are in parallel, so that pFETs groups (A, B) and (C, D) are in series.

These arrangements insure proper behavior.

One application of this circuit is to build XOR and XNOR circuits. For example, let us define the values $C = \bar{A}$ and $D = \bar{B}$. Then,

$$
\begin{aligned}
h_1 &= \overline{A \cdot B + \bar{A} \cdot \bar{B}} \\
&= \overline{(A \cdot B)} \cdot \overline{(\bar{A} \cdot \bar{B})} \\
&= (\bar{A} + \bar{B}) \cdot (A + B) \\
&= \bar{A} \cdot A + \bar{A} \cdot B + A \cdot \bar{B} + \bar{B} \cdot B \\
&= \bar{A} \cdot B + A \cdot \bar{B} \\
&= A \oplus B
\end{aligned}
\tag{6.13}
$$

gives the exclusive-OR function. Conversely, if we $B = \bar{C}$ and $D = \bar{A}$, then

$$
\begin{aligned}
h_2 &= \overline{A \cdot \bar{C} + C \cdot \bar{A}} \\
&= \overline{A \oplus C}
\end{aligned}
\tag{6.14}
$$

as can be verified by an algebraic reduction; this shows that the gate can be used to create the equivalence (XNOR) function. These special cases are shown in Figure 6.33.

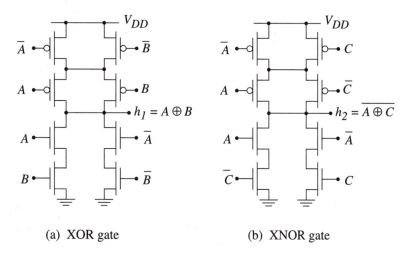

(a) XOR gate   (b) XNOR gate

Figure 6.33 CMOS XOR and XNOR logic gates

A 4-input OAI logic gate is shown in Figure 6.34. This circuit provides the general function

$$p = \overline{(A + B) \cdot (C + D)} \tag{6.15}$$

as can be verified by the techniques presented above. An output of $p = 0$ requires that either $A = 1$ OR $B = 1$ to provide a conducting path from the output node through the first "layer" of nFETs with inputs A and B. The remaining path to ground is achieved if either $C = 1$ OR $D = 1$. Combining these statements allows us to write

$$\bar{p} = (A + B) \cdot (C + D) \tag{6.16}$$

for the complement $\bar{p}$, so applying the NOT operation gives the correct function. It is left as an exercise to show that the FETs are properly connected using the series-parallel rules.

6.6.3 Logic Cascades

Cascaded logic networks are easily realized in CMOS by just using the output of one gate as the input to the next gate. Although we can build digital systems using the basic NOT, NAND, and NOR gates, it is usually more efficient to merge the logic into AOI and OAI logic functions first. This minimizes the amount of connections by performing several primitive operations in a single circuit.

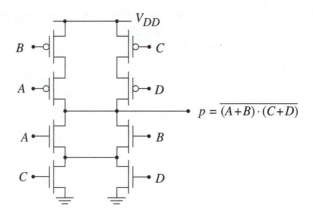

$$p = \overline{(A+B)\cdot(C+D)}$$

Figure 6.34 A 4-input OAI logic gate circuit

As an example, consider the logic diagram shown in Figure 6.35(a). This is based on two complex logic gates; the first has an OAI structure, while the second is based on the AOI technique. The output is given by

$$g = \overline{\overline{f\cdot y}+x} \qquad\qquad (6.17)$$

where we have defined the output of the OAI gate (not shown explicitly in the diagram) as

$$f = \overline{(a+b)\cdot c} \qquad\qquad (6.18)$$

The CMOS cascade for the logic function g is shown in Figure 6.35(b). This clearly demonstrates the ability to perform several basic logic operations using a small number of gates and transistors. The challenging work is to achieve a form for the logic function that allows for this type of CMOS gate structuring.

6.7 MOSFET Logic Formalism

The series-parallel logic gates discussed thus far are only one of several possible approaches to creating logic circuits in CMOS. The flexibility of CMOS as a logic design platform has allowed it to develop into the primary technology for high-density logic networks.

One important characteristic that makes CMOS so powerful for creating logic circuits is the fact that it uses MOSFETs, which are the simplest and smallest switches that can be made in silicon technology. The details of design at the silicon level are introduced in the next chapter. For the present discussion, we will concentrate on modeling FETs as logic-controlled devices that have a natural fit into standard Boolean algebra.

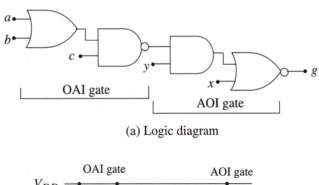

(a) Logic diagram

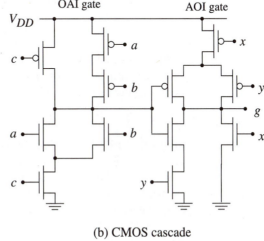

(b) CMOS cascade

Figure 6.35 Example of a logic cascade in CMOS

6.7.1 FET Logic Descriptions

Consider the nFET shown in Figure 6.36(a). The switching characteristics are obtained by viewing A as the input and B as the output. Control is established through C, which is applied to the gate. Figure 6.36(b) shows the characteristics of the nFET when $C = 0$, corresponding to a low voltage. Since the nFET is in cutoff, there is no relationship between A and B, and it is not possible to write a formal

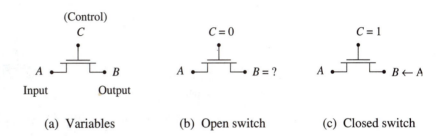

Figure 6.36 Logic model for an nFET

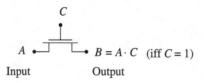

$$B = A \cdot C \quad (\text{iff } C = 1)$$

Figure 6.37 nFET logic equation summary

logic expression. If we change the gate variable to a value of $C = 1$, then the nFET turns on as in Figure 6.36(c), and we may write

If $C = 1$ then

$$B \leftarrow A$$

where the second line is read "A is transferred to B."

This is sufficient for a formal description, but is not in a form that can be used in Boolean algebra. A more suitable form is

$$B = A \cdot C \quad (\text{iff } C = 1) \tag{6.19}$$

so that the nFET by itself performs the AND operation between the input and the variable that is applied to the gate. It is important to note that the condition "iff $C = 1$" means that the equation is not valid if $C = 0$. Although this may seem like a limit on when the device may be used, a CMOS circuit will also contain pFETs with complementary features that will take care of the problem.

Next, let us study the pFET shown in Figure 6.38(a). As with the nFET, A and B are the input and output variables, respectively, and C is used as the control bit. For this device, we may write the operation formally as

If $C = 0$ then

$$B \leftarrow A$$

since the pFET acts as a closed switch when the gate has $C = 0$ applied to it. A value of $C = 1$ produces an open switch, and there is no relationship between A and B. Combining these two conditions gives a Boolean expression of

$$B = A \cdot \overline{C} \quad (\text{iff } C = 0) \tag{6.20}$$

where we have expressed the condition $C = 0$ by using $\overline{C}$ in the AND expression. This is summarized by the logic model shown in Figure 6.38(b).

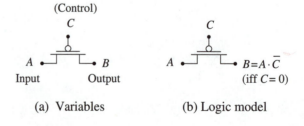

(a) Variables   (b) Logic model

Figure 6.38 pFET logic model

6.7.2 Voltage Transmission Characteristics

These simple arguments illustrate that both nFETs and pFETs can intrinsically provide the AND operation between the input variable and the control bit applied to the gate. Why then do we need both types of transistors in CMOS? It is because of the voltage-to-binary interface that makes the relationship between circuit voltages and Boolean 0 and 1 values. Consider the nFET voltage pass characteristics shown in Figure 6.39(a), where we apply to the gate a high voltage that is equal to the power supply value V_{DD}; this is the largest voltage in the circuit, so it provides good conduction. The pass characteristics are illustrated in Figure 6.39(b). If the input voltage V_A is allowed to range from 0 v to V_{DD}, we find that the output voltage V_B is limited to the range 0 v to a value $V_1 < V_{DD}$. The maximum voltage at the output is given by

$$V_1 = V_{DD} - V_{Tn} \tag{6.21}$$

where V_{Tn} is called the **threshold voltage** of the nFET. The threshold voltage is discussed in the next chapter. In essence, it is an electrical characteristic of the transistor with a typical value of around 0.5 v. Regardless of the numerical value, we usually summarize the transfer by saying

- An nFET passes a strong logic 0, but a weak logic 1.

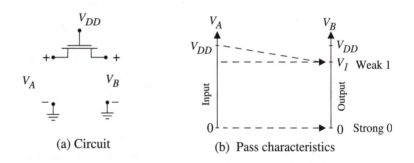

(a) Circuit (b) Pass characteristics

Figure 6.39 Voltage transmission through an nFET

This problem in the voltage-to-logic translation implies that we should avoid situations where an nFET is used to pass a logic 1 voltage level unless the electronic circuitry is modified to compensate for the lower value.

A pFET has voltage passing characteristics that are exactly opposite of the nFET, as expected since they are complementary devices. To turn on a pFET, we apply 0 v at the gate; this is equivalent to connecting the gate to ground as in Figure 6.40(a). The transfer characteristics in Figure 6.40(b) show that the pFET can pass a high voltage V_{DD}, but the lowest output voltage is $V_B = V_0 > 0$ v, where

$$V_0 = |V_{Tp}| \tag{6.22}$$

with V_{Tp} the threshold voltage of the pFET; by convention, V_{Tp} is a negative number and has a typical value of $|V_{Tp}| \approx 0.5\ v$. The pFET transmission characteristics may be summarized by saying

- A pFET passes a strong logic 1, but a weak logic 0.

This is, of course, exactly opposite to the nFET.

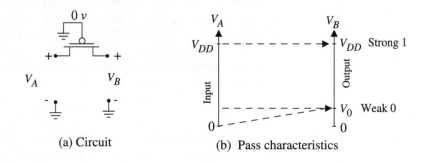

(a) Circuit   (b) Pass characteristics

Figure 6.40 Voltage transmission through a pFET

6.7.3 The Complementary Principle

The discussions above provide the circuit design philosophy for CMOS logic gates that are based on pairs of transistors. A complementary pair consisting of an nFET and a pFET uses devices that have opposite switching characteristics and also have opposite voltage transfer characteristics. Applying the results above shows that

- nFETs should be used to pass low voltages (0 v), and
- pFETs should be used to pass high voltages (V_{DD}).

This can be seen in all of the logic gates presented in this chapter. Every gate uses nFETs to connect the output to ground and uses pFETs to connect the output to the power supply.

Logic formation can also be understood using this analysis. Consider the inverter in Figure 6.41. To analyze the logic characteristics, note that the power supply (V_{DD}) is a logic 1, while ground (0 v) is a logic 0. The output function f may be viewed as consisting of two terms, one from the pFET and one from the nFET. The pFET logic is described by $\overline{A} \cdot 1$, while the nFET contribution is $A \cdot 0$. Since A can only assume one value (0 or 1), the output can be expressed as an OR between the two possibilities by writing

$$f = \overline{A} \cdot 1 + A \cdot 0 \tag{6.23}$$

Physically, the first term contributes when the pFET is ON, while the second term provides the output when the nFET is ON. Obviously, this reduces to

$$f = \overline{A} \tag{6.24}$$

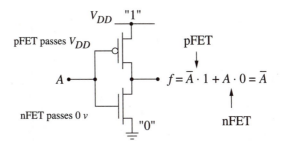

Figure 6.41 Logic analysis of the inverter circuit

as expected. However, note that the complementary structure provides for strong voltage levels at both 0 and 1 levels and eliminates the undefined cases in the FET logic equations by insuring that when one transistor is OFF, then the output is defined by the other transistor.

This type of logic analysis can be extended to any CMOS gate. Figure 6.42 shows the logic formation in a NAND2 gate. The nFET chain connects the output to ground, which is a perfect logic 0. The nFETs thus contribute a term in the form $A \cdot B \cdot 0$, which evaluates to 0 but is needed to insure that the output has a well-defined connection to the ground. The pFETs respectively contribute terms $\overline{A} \cdot 1$ and $\overline{B} \cdot 1$. As shown in the drawing, the NAND function is obtained by ORing these three possibilities via

$$
\begin{aligned}
f &= \overline{A} \cdot 1 + \overline{B} \cdot 1 + A \cdot B \cdot 0 \\
&= \overline{A \cdot B}
\end{aligned}
\tag{6.25}
$$

where we have used a DeMorgan reduction. As our final example, consider the NOR gate in Figure 6.43. The logic is formed by the expression

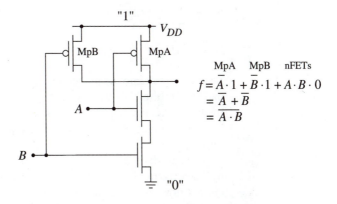

$$
\begin{aligned}
f &= \overline{A} \cdot 1 + \overline{B} \cdot 1 + A \cdot B \cdot 0 \\
&= \overline{A} + \overline{B} \\
&= \overline{A \cdot B}
\end{aligned}
$$

Figure 6.42 Logic analysis of a NAND2 gate

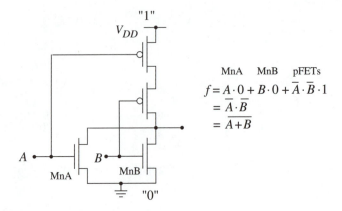

Figure 6.43 Logic analysis of a NOR2 gate

$$f = A \cdot 0 + B \cdot 0 + \overline{A} \cdot \overline{B} \cdot 1$$
$$= \overline{A + B} \qquad\qquad (6.26)$$

which is easily verified by ORing each possibility. These examples illustrate the fact that MOSFET logic equations have a direct connection to the structure of CMOS logic gates.

6.7.4 Current Switching

As was discussed above, nFETs can pass a strong logic 0 voltage, but only a weak logic 1 voltage. If, however, we choose the current I as the logic variable such that

- $I = 0$ is a logic 0, and
- $I \neq 0$ is a logic 1

then the reduced logic 1 voltage is not significant. In practice, networks that represent logic variables by current flow are very difficult to implement. This is because current tends to split every time it hits a junction where three or more wires intersect. Applying the conservation of charge to the circuit in Figure 6.44 gives that

$$I = I_1 + I_2 \qquad\qquad (6.27)$$

i.e., the current into the junction is equal to the total current flowing out of it. This is called **Kirchhoff's Current Law** in electrical circuit analysis. The problem is that the two currents I_1 and I_2 are generally unequal and can be unpredictable. Regardless, it is interesting to examine one type of logic network that uses current and **light** as logic variables.

The interface between electric current flow and light is through the use of **optoelectronic devices**. These are electronic devices that either emit light when current flows through them or absorb light, resulting in a current flow. The former are called **light emitters** and usually refers to a **light-emitting diode (LED)** or a

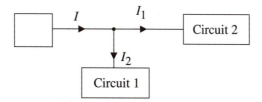

Figure 6.44 Current splitting at a node

semiconductor **laser**. Devices that absorb light are called **light detectors** and are used in fiber-optic transmission systems and as solar cells.

Let us examine how a light-emitting diode can be used in a simple logic network. Figure 6.45 shows the important characteristics of an LED. In Figure 6.45(a), the switch is open, and no current flows through the LED, which is represented by the symbol on the right side of the circuit. When $I = 0$, no light is emitted by the device. This defines the light variable L as having a value $L = 0$ when there is no emission. If we close the switch, as in Figure 6.45(b), the circuit is closed and current flows. The LED turns on and emits light, which defines the state of $L = 1$.

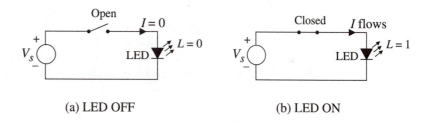

(a) LED OFF   (b) LED ON

Figure 6.45 Characteristics of a light-emitting diode (LED)

The obvious connection between the switch and transistors is to use nFETs as logic-controlled switches. The AND function is created by placing FETs in series as in Figure 6.46(a). In this circuit, we may write

$$L = A \cdot B \tag{6.28}$$

since we must have both $A = 1$ and $B = 1$ to achieve current flow and turn on the LED. The OR function is constructed in a similar fashion. Paralleling two nFETs yields the circuit of Figure 6.46(b), which gives

$$L = A + B \tag{6.29}$$

This is obvious, since having either $A = 1$ or $B = 1$ allows current flow, thus giving light from the LED.

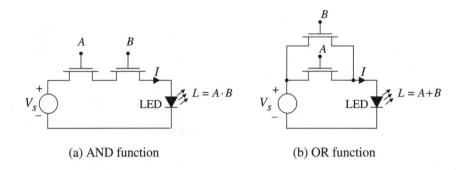

(a) AND function   (b) OR function

Figure 6.46 Basic logic functions using currents and light

These simple examples are not meant to illustrate a powerful logic implementation technique. Instead, the significance is in the relationship between logic voltages (that control the FETs) and the optical output L of the LED.

6.7.5 Fiber-Optic Transmission Networks

A fiber-optic data link is a system that is designed to transmit a serial data stream using the concept of the light variable L discussed above. A generic system is shown in Figure 6.47. Electronic data are sent to a **transmitter** unit, which accepts the voltage pulses and drives an LED or a laser. The output from the light source is focused into the **optical fiber**. Once the light enters the fiber, it remains inside and follows the path of the fiber itself. At the **receiver** end, the light exits the fiber and strikes an optoelectronic photodetector. This device gives a current pulse when light is incident on it.

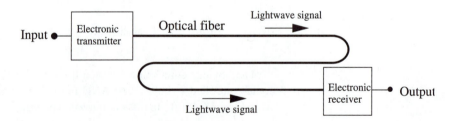

Figure 6.47 A fiber-optic transmission system

The real heart of a fiber-optic network is the means by which the light pulses are guided from the transmitter to the receiver. The fiber itself is a thin strand of glass that consists of a central region called the **core**, which is surrounded by an outer coating called the **cladding**. The optical characteristics of the two regions are designed in a manner that allows light to be trapped and guided within the core This is shown in Figure 6.48. A cross-section of a typical optical fiber is shown in Figure 6.49. The outer diameter of the cladding is typically about $125 \, \mu m = 0.125 \, mm$; this is usually coated with a protective plastic jacket that has a diameter of about

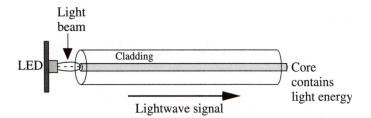

Figure 6.48 Light guiding in an optical fiber

250 μm = 0.25 *mm*. The inner core region is very small in comparison, with a typical diameter of about 9 μm. Optical fibers are used extensively in communication systems because they have the following characteristics:

• Fibers are immune to electromagnetic interference (EMI) and crosstalk.

• Optical fibers are lightweight.

• Fibers are surprisingly strong.

• Optical fibers are able to carry very large amounts of data in a serial format.

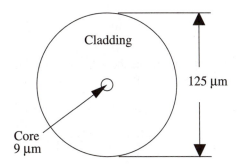

Figure 6.49 Fiber cross-section

A detailed discussion of optical fibers is beyond the scope of the treatment here. The important point to remember is that the lightwave signal L is just a binary-formatted variable like that found in any logic network.

Fiber-optic transmission systems are widely used in state-of-the-art digital communication systems. This includes systems such as telephone links, computer networks, and cable television distribution systems. Although the use of lightwaves and optical fibers makes the applications look unique, they all operate on the basic principles of digital logic introduced here.

6.8 Problems

[6.1] Determine the conducting state (ON or OFF) for each of the n-channel MOS-FETs shown below.

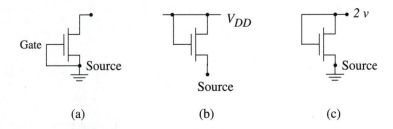

(a) (b) (c)

[6.2] Determine the conducting state (ON or OFF) for each of the p-channel MOS-FETs shown below.

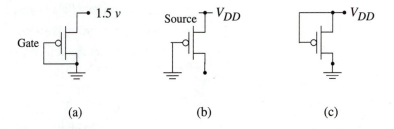

(a) (b) (c)

[6.3] Determine the value(s) of the inputs that must be applied to turn the n-channel MOSFET ON.

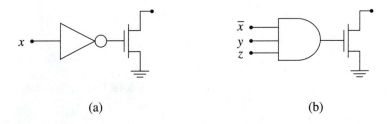

(a) (b)

[6.4] Determine the values of the inputs that must be applied to turn the pFET ON.

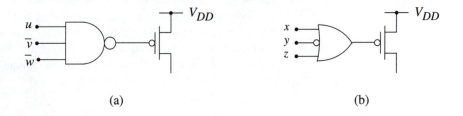

(a) (b)

[6.5] Each MOSFET shown below has an input of a 0 or a 1. For each circuit, determine whether or not the nodes x and y are electrically connected.

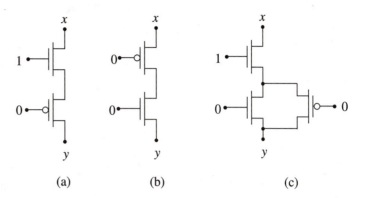

(a)   (b)   (c)

[6.6] For each circuit shown, determine the values of the input variables that will insure that nodes x and y are electrically connected.

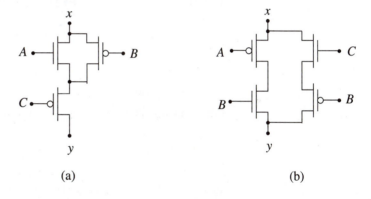

(a)   (b)

[6.7] Determine the input combinations that will yield a circuit in which nodes x and y are electrically connected.

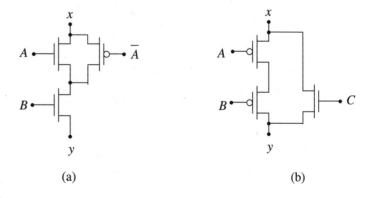

(a)   (b)

[6.8]   Draw the CMOS logic gate for a 4-input NOR gate (NOR4).

[6.9]   Draw the CMOS logic gate for a 3-input NAND gate (NAND3).

[6.10]   Draw the CMOS logic cascade that produces the AND function $X \cdot Y$.

[6.11]   Design a CMOS cascade that has an output of

$$g = A \cdot \bar{B}$$

[6.12]   Design a CMOS logic circuit that produces the output function

$$f = \bar{A} + B$$

You may use a simple logic cascade.

[6.13]   Design the CMOS logic cascade for the OR function $X + Y$.

[6.14]   Construct a single CMOS logic gate that provides the logic function

$$F(A, B) = \bar{A} \cdot \bar{B}$$

and uses A and B directly as the inputs (not their complements). Hint: Use a DeMorgan theorem that allows you to write F in a form that can easily be built in CMOS.

[6.15]   Construct a single CMOS logic gate that gives an output of the form

$$F = \bar{x} + \bar{y} \tag{6.30}$$

with x and y as inputs.

[6.16]   Design the CMOS logic gate for the function

$$g = \overline{A \cdot B + A \cdot \bar{B}}$$

using complementary structure. The objective is to design the simplest circuit, so analyze the function for possible reductions before constructing the circuit.

[6.17]   The CMOS logic gate below shows only the nFETs. (a) Determine the function $f(x, y, z)$, and then (b) construct the truth table for the gate.

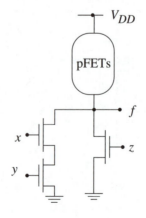

[6.18] Design the CMOS logic gate that will implement the function

$$f = \overline{A \cdot B + A \cdot C + B \cdot C} \qquad\qquad (6.31)$$

using complementary structure. Use the smallest number of transistors that you can. Note that each variable occurs twice in the function, so that it may be possible to share MOSFETs.

[6.19] Design a CMOS logic circuit for the function

$$h = x \cdot y + z$$

by using an AOI gate cascaded into an inverter. Can this function be implemented in CMOS by only using one gate (instead of a 2-gate cascade?).

[6.20] The CMOS logic gate below shows only the nFETs. Determine the function $g(x, y, z)$ and construct the truth table for the gate.

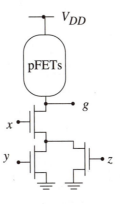

[6.21] The CMOS logic gate below shows only the nFETs. Determine the function $F(x, y, z, u, v)$ and construct the truth table for the gate.

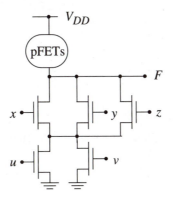

[6.22] Design the pFET network for the CMOS logic gate in Problem [6.17].

[6.23] Design the pFET network for the CMOS logic gate in Problem [6.20].

[6.24]   Design the pFET network for the CMOS logic gate in Problem [6.19].

[6.25]   The CMOS logic gates below show only the pFETs. Determine the function that each gate is designed to provide.

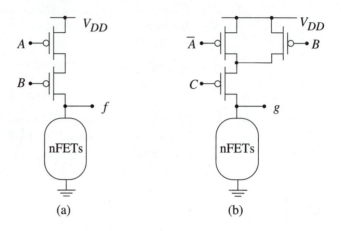

(a)   (b)

[6.26]   Design the nFET circuits for the logic gates in Problem [6.25].

[6.27]   The CMOS logic gates below show only the pFETs. Determine the function that each gate is designed to provide.

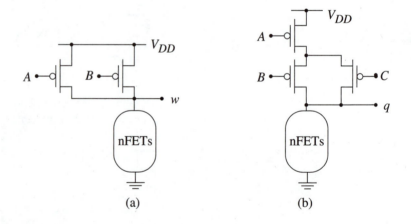

(a)   (b)

[6.28]   Design the nFET circuits for the logic gates in Problem [6.27].

[6.29]   Suppose that you want to implement the following function using CMOS logic circuits:

$$F = a \cdot b + c \cdot (d + e)$$

Determine the number of separate logic circuits you need for an efficient implementation. Then design the circuits.

[6.30] Consider the circuit below. Fill in the nFETs with the appropriate inputs such that the gate implements the following functions:

(a) $f = \overline{x + y + z}$

(b) $f = \overline{x \cdot y + z}$

(c) $f = \overline{(x + y) \cdot z}$

[6.31] Consider the circuit below. Fill in the pFETs with the appropriate inputs such that the gate implements the following functions:

(a) $g = \overline{(x \cdot y) \cdot z}$

(b) $g = \overline{(x + y) \cdot z}$

(c) $g = \overline{(x + y) \cdot z + w}$

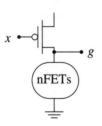

[6.32] Design an nFET-LED logic network like those shown in Figure 6.46 for the function

$$L = (a \cdot b) \cdot (c + d)$$

[6.33] Design an nFET-LED logic network like those shown in Figure 6.46 for the function

$$L = (a \cdot b) + (c \cdot d)$$

[6.34] Consider the basic LED-logic circuit shown below. Design the logic network for each logic function listed. Use one FET per input variable, but you may use both nFETs and pFETs in your design.

(a) $L = x \cdot \bar{y} + w$

(b) $L = (x + \bar{z}) \cdot (y + \bar{w})$

(c) $L = x + (\bar{y} + z \cdot w)$

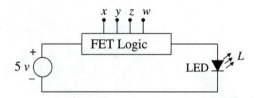

[6.35] Choose a "popular" magazine that deals with computing and see if you can find real-world examples of fiber-optic systems. List the characteristics of any systems you find, such as usage, speed, etc.

CHAPTER 7

Silicon Chips and VLSI

VLSI is an acronym that stands for *Very Large Scale Integration*. This engineering field deals with the never-ending quest to create electronic circuits that are physically small yet contain massive amounts of computing power. In this chapter, we will investigate how digital circuits are created in tiny pieces of silicon and learn the true foundations of modern electronics.

7.1 What Is VLSI Engineering?

VLSI engineering is concerned with the integration of circuits and systems onto small pieces of silicon. A typical VLSI engineer has a working knowledge of silicon devices, CMOS circuits, logic design, and systems architecture and is usually a specialist in one or more of these areas. Up to this point, we have examined logic design and CMOS circuits. This chapter will take us to the bottom of the hierarchy, to the electrons that give rise to current flow.

The field of **microelectronics** has made spectacular advances in recent years. Overall, the objective has been to reduce the size of electronic devices and circuits to provide the technology needed to build large information systems on tiny chips. The VLSI engineer uses this technology for creating powerful "systems on silicon." Perhaps the most visible examples of this type of design are the current generations of microprocessor chips that provide unprecedented computational power.

A typical VLSI circuit may contain several million transistors. Complex systems of this type cannot be designed using pencil and paper. Instead, many computer programs have been written to help the engineer design, debug, simulate, and test the integrated circuits. In the jargon of chip design, the programs are called **tools**, since they help solve problems. An integrated collection of tools is called a **tool set**. Modern tool sets are extremely sophisticated programs that consist of several linked modules; each module is designed to perform a specific task, such as logic testing or VHDL simulation. Most of the commercial tool sets run on powerful workstations. These are usually Unix-based machines with upwards of 512 MB system memory, very large hard disks, and a powerful network and server. With the increasing power of desktop PCs, however, it is possible to find excellent tool sets that run on more modest hardware.

To understand the role of the VLSI engineer, it is first necessary to start with the basics—all the way down to electrons! To this end, we will examine the electrical characteristics of silicon and then advance to the point where MOSFETs can be discussed. Circuit integration can then be presented in the context of designing logic gates.

7.1.1   Inside a Computer Chip

Modern computing hardware is based on the ability to create and manufacture tiny pieces of silicon that act as electronic logic networks. In the language of the mass media, these are generically referred to as *computer chips*. However, in the world of engineering, the silicon marvels are more properly named **integrated circuits**, as they contain many subcircuits that are integrated into a single equivalent circuit.

What exactly is an integrated circuit? Well, that question has a somewhat complicated answer that will take the remainder of the chapter to answer. For now, just think of an integrated circuit as the physical realization of a logic network. In other words, we will find that a chip consists of electronic logic gates that are connected together to perform various logic functions.

An integrated circuit in the form that you find on a computer board is shown in Figure 7.1. As discussed in Chapter 5, only the features of the package are obvious from the outside. This includes the cover and several metal pins that are used to connect the device electrically to other electronic elements. The actual silicon chip that provides the electronic switching functions is inside the package as shown and is called the **die**.

Figure 7.1 A silicon integrated circuit

To understand the concept of a digital integrated circuit, recall that a digital system can be studied from many different viewpoints. The chip constitutes the most basic physical viewpoint where logical functions are synthesized using transistor switches and voltage levels. The operation of the switches (FETs) can be understood by examining how current flow through silicon can be controlled using logic voltages.

7.1.2 A Silicon Primer

First, let us look at a chip from the physical level. This device is created on the top of a thin wafer of silicon using very advanced techniques from physics and chemistry. An integrated circuit consists of several layers of materials, such as silicon, glass, and metal, that have been chosen for their electrical properties. For example, metals such as aluminum are **conductors** that allow electricity to flow very easily. Glass, on the other hand, is an **insulator**, meaning that it cannot pass electric current. Silicon (Si) is a very special type of material called a **semiconductor**; this term is derived from the phrase "partial conductor" and is used because silicon allows only small levels of current to flow.

Current flow through a material is the result of moving charges. In pure crystal silicon (also called **intrinsic** silicon), only a small number of electrons are available to conduct current. This is portrayed in Figure 7.2(a). At low temperatures, most of the electrons are trapped in orbit about atomic nuclei and cannot move through the crystal to give current flow. A few electrons can gain energy from thermal (heat) processes and then break away from their bonds as shown. Since an electron carries a negative charge of $-q$, it can contribute to current flow in the sample.[1] When it breaks away, it leaves an empty bond that acts like a positively charged particle. We call these particles **holes** and assign them a charge of $+q$ since they have characteristics opposite to those of electrons. It is useful to count the number of electrons and holes in the sample. We define n as the number of free electrons per cm^3, and p as the number of holes per cm^3. Since a free electron and a hole are created at the same time, we have that

$$n = p \qquad\qquad (7.1)$$

The two types of charges are thus balanced in number.

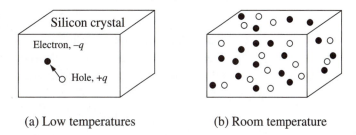

(a) Low temperatures (b) Room temperature

Figure 7.2 Electrons and holes in silicon

---

[1] Remember that $q = 1.602 \times 10^{-19}$ coulombs is the fundamental charge unit.

If we elevate the temperature of the sample to about $27°C$ (which is called *room temperature* in thermal physics), then we find that more electrons are released for conduction as shown in Figure 7.2(b). However, the current levels are still very low, and $n = p$ is still valid. To increase the number of charges available for current flow, small amounts of impurity atoms are added to the otherwise pure silicon crystal. The impurities are called **dopants**, and the process of adding the impurity atoms is called **doping**.

There are two types of dopants that can be added to silicon. If we use arsenic (As) or phosphorus (P) as a dopant, we find that the resulting crystal has many more negatively charged electrons available for conduction and

$$n > p \tag{7.2}$$

This is called an **n-type semiconductor** region due to the enhanced electron population. An n-type semiconductor can be visualized by the drawing in Figure 7.3(a). If, on the other hand, we add boron (B) as a dopant, we obtain a **p-type** region where there is a larger number of holes that are available for conduction. This is described by writing

$$p > n \tag{7.3}$$

as shown in Figure 7.3(b). The ability to create n-type and p-type regions on silicon forms the basis for integrated microelectronics.

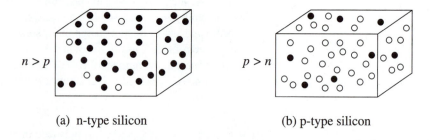

(a) n-type silicon   (b) p-type silicon

Figure 7.3 Charges in doped silicon samples

Some Quantitative Details[2]

Electrical conduction in silicon depends upon the number of free charge carriers n and p that are in the sample. The larger the number of free charges, the better the sample will conduct electricity.

Device analysis can be based on the characteristics of an intrinsic sample in which there are only silicon atoms. If we examine a silicon crystal, then the number of atoms per cubic centimeter is found to be

$$N_{Si} = 5 \times 10^{22} \ cm^{-3} \tag{7.4}$$

---

[2] This subsection may be omitted in a first reading without a loss of continuity.

where the unit cm^{-3} is used for simplicity and means "per cubic centimeter." The number of free electrons n and free holes p per cubic centimeter depends upon the temperature of the material. Choosing a temperature of T = 27°C = 300 K gives the **intrinsic carrier density** n_i as

$$n_i = 1.5 \times 10^{10} \ cm^{-3} \tag{7.5}$$

Since releasing a free electron automatically creates a free hole, we must have

$$n = p = n_i \tag{7.6}$$

in the intrinsic sample. This gives us the **mass-action law,** which states that

$$np = n_i^2 \tag{7.7}$$

It is worthwhile to note that only a small fraction of electrons in the crystal are actually released for conduction. This is why silicon is classified as a semiconductor: it only conducts small amounts of current because only a relatively small number of charges are available.

To create n-type silicon, we add phosphorus or arsenic atoms as impurities. These "donate" an extra free electron to the sample, so that we call these dopants **donor** atoms. The number of donors per cubic centimeter is denoted by N_d. Typically, N_d ranges from about 10^{15} to $10^{20} \ cm^{-3}$, depending upon the desired characteristics. Every donor atom that can be placed into a crystal site that is normally occupied by a silicon atom releases a free electron that can be used for conduction. Thus, the number of electrons per cm^3 n_n in an n-type sample can be estimated by

$$n_n \approx N_d \tag{7.8}$$

The n-subscript is used to remind us that we are talking about an n-type material. The number of holes per cm^3 in the n-type material, denoted as p_n, is given by

$$p_n \approx \frac{n_i^2}{N_d} \tag{7.9}$$

by rearranging the mass-action law and substituting the equation for n_n.

A p-type semiconductor is treated in a similar manner. In this case, the boron atoms create holes that can "accept" electrons, so that p-type dopants are called **acceptors**. An acceptor concentration of N_a atoms per cubic centimeter gives a hole concentration of

$$p_p \approx N_a \tag{7.10}$$

where p_p means the number of holes per cm^3 in a p-type sample. The electron density n_p in the p-type material is calculated from

$$n_p \approx \frac{n_i^2}{N_a} \tag{7.11}$$

This is obtained using the mass-action law.

Example 7-1

Suppose that we have a sample of n-type silicon that has phosphorus atoms added with a density of $N_d = 10^{19}\ cm^{-3}$. The concentration of free electrons is

$$n_n \approx N_d = 10^{19}\ cm^{-3} \tag{7.12}$$

while the density of free positively charged holes is

$$p_n \approx \frac{n_i^2}{N_d}$$

$$= \frac{(1.5 \times 10^{10})^2}{10^{19}} \tag{7.13}$$

$$= 22.5\ cm^{-3}$$

Obviously, $n_n \gg p_n$, so the term n-type is well chosen.

7.1.3 The pn Junction

If an n-type region of the silicon is interfaced to a p-type region, then we form a very special structure called a **pn junction**. This combination, shown in Figure 7.4(a), creates an electronic component called a **diode**. A diode has two terminals, one called the **anode** (the p-type side) and the other called the **cathode** (the n-type side). The circuit symbol for a diode is shown in Figure 7.4(b). Diodes exhibit a property called **rectification**; in ordinary language, a rectifier only allows current to flow in one direction. The diode symbol reflects this characteristic, indicating that current can only flow from the anode to the cathode. This requires that the voltage polarities be applied "+" to the p-type anode, and "–" to the n-type cathode (making them easy to remember). If we reverse the polarities and apply "–" to the anode and "+" to the cathode, then the diode acts like an open circuit and blocks the current flow.

The current flow characteristics of a pn junction are summarized by the current versus voltage, or *I-V*, plot shown in Figure 7.5. In this curve, the applied

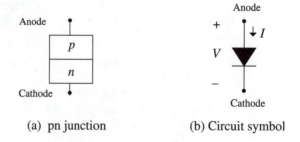

(a) pn junction   (b) Circuit symbol

Figure 7.4 The pn junction diode

voltage V is defined to be **positive** when "+" is applied to the p-side and "–" is applied to the side. This places a **forward bias** on the diode, and current flows through the device. **Reverse bias,** defined as a negative voltage $V < 0$, means that the polarities are reversed: "–" is applied to the p-side and "+" is applied to the n-side. This blocks most of the current flow, although a small amount leaks through the device.

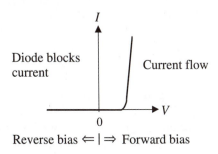

Figure 7.5 Current as a function of voltage in a pn junction diode

7.1.4 Silicon Devices

Diodes are created by "stacking" two distinct types of materials, p-type and n-type, in a manner that creates an electronic device. In CMOS circuits, diodes occur as a natural consequence of the manner in which the chips are created, but they are not used by themselves very often. Instead, we use MOSFETs as the switching devices. These are obtained by first building a capacitor in silicon and then using reverse bias pn junctions to force the current to flow through a specific region of the silicon.

To build an integrated capacitor on silicon, we will start with the basic capacitor structure shown in Figure 7.6(a) and design a silicon structure that has the same characteristics. We recall that a capacitor consists of two conducting plates that are

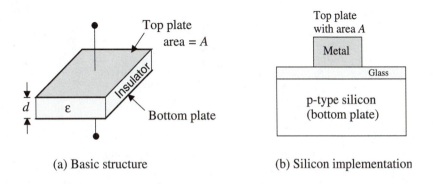

(a) Basic structure

(b) Silicon implementation

Figure 7.6 An integrated MOS capacitor

separated from each other by an insulating layer. For the device shown, the capacitance is given by

$$C = \frac{\varepsilon A}{d} \qquad (7.14)$$

where A is the area of the plates. To build a MOSFET capacitor, we use a p-type silicon region as the bottom plate, and create an insulating glass layer on top of it. The glass material is silicon dioxide (SiO_2) which is quartz glass, and is commonly referred to as the "oxide" layer. The top plate is obtained by adding a layer of metal (or other conducting material) on top of the glass, and then "patterning" it to have an area A. The silicon implementation of the capacitor structure is shown in Figure 7.6(b). This provides the basis for creating a MOSFET on silicon.

The important features of the integrated capacitor are that it consists of several layers of materials (p-type silicon, glass, and metal) and that it is necessary to pattern the metal layer to obtain a well-defined structure. This simple example actually gives a simple definition of an integrated circuit as a set of patterned material layers. Of course, it is a little more complicated than it sounds. Modern integrated circuit technology is based upon the ability to create layers of materials on a sample of silicon and then create patterns on each layer as needed. Devices such as MOSFETs are then classified as three-dimensional structures created in this manner. Before we dive headfirst into MOSFETs, however, let us examine some of the details of transferring a desired pattern to a material layer.

7.2 Lithography and Patterning

Silicon integrated circuits are fabricated on large circular wafers of silicon as illustrated in Figure 7.7(a). Typically, the wafer is about 8 to 10 inches in diameter, and many individual circuits are created on the wafer at the same time. This is shown in Figure 7.7(b) where each square is an integrated circuit.

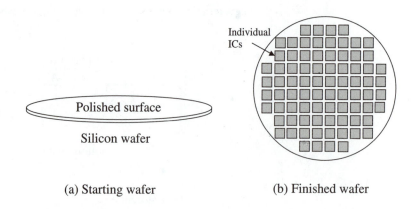

(a) Starting wafer　　　　　　　　　(b) Finished wafer

Figure 7.7 A silicon wafer

An integrated circuit is made up of many different layers of materials, with each layer having its own distinct pattern and electrical characteristics. The process of creating the pattern on the layer is called **optical lithography**. The technique is similar to taking a photograph and developing film, but it is highly advanced and allows us to create patterns that are less than 0.5 **micrometer** (μm) wide.[3] In everyday usage, a micrometer is often called 1 **micron** for simplicity. The ability to create circuits this small has allowed us to achieve VLSI-level circuits.

Let us examine the principal stages of the patterning process by means of an example. Suppose that we want to create a metal line on the surface of the silicon that can be used to "wire" two circuits together. The starting point is taken to be a wafer that has a glass layer (silicon dioxide) on top as shown in Figure 7.8(a). The first step in the sequence is to deposit a uniform layer of the metal material over the surface of the wafer. This results in the structure shown in Figure 7.8(b). To produce

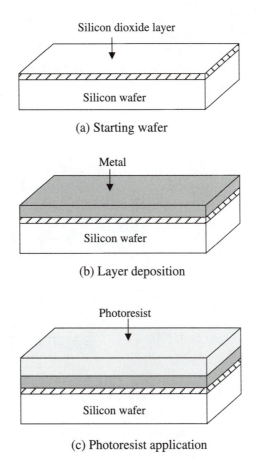

(a) Starting wafer

(b) Layer deposition

(c) Photoresist application

Figure 7.8 First steps in the photolithographic process

---

[3] One micrometer is equal to 10^{-6} meters.

a patterned layer, we start the lithographic steps by applying a thin layer of a material called **photoresist**. This is a light-sensitive organic polymer (i.e., plastic) that acts very similarly to photographic film: light alters its chemical characteristics. Figure 7.8(c) shows the sample after the photoresist has been applied.

The next step is called **exposure**, and it is where the pattern is transferred to the surface of the wafer. The pattern itself is first designed using a computer drawing program and then transferred to a special plate called a **reticle**. The reticle is a plate of glass that has been coated with a reflective metal layer (such as chromium) that has the same pattern as that drawn on the computer. The reticle is made larger than the actual silicon pattern size to allow higher precision.

The pattern transfer process is shown in Figure 7.9. The back side of the reticle is illuminated with UV light. The shadow of the reticle is focused through an optical imaging system onto the surface of the photoresist. This causes some regions to be exposed to the light, while other areas are shielded. After the exposure step, the photoresist is developed much like photographic film. After rinsing, the sections of the photoresist that were shielded from the light remain in a hardened state, while those that were exposed to the light are rinsed away. The photoresist now has the same pattern as the reticle, as portrayed in Figure 7.10(a).

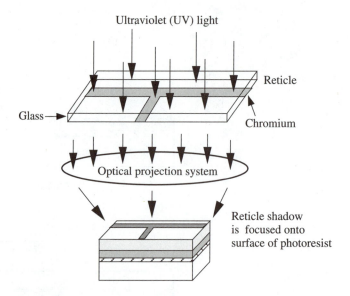

Figure 7.9 Exposure of the photoresist in the printing process

To complete the patterning process, the wafer is subjected to an **etching** step in which chemicals in an inert gas (such as Ar) are used to react with the metal and remove it from the wafer. Regions that are protected by the photoresist are not etched away, leaving the metal with the same pattern as the photoresist. The surface now appears as in Figure 7.10(b). The final step in this sequence removes the photoresist, which leaves the desired result, the patterned metal layer shown in Figure 7.10(c). Thus, the sequence has taken the pattern from a computer program and

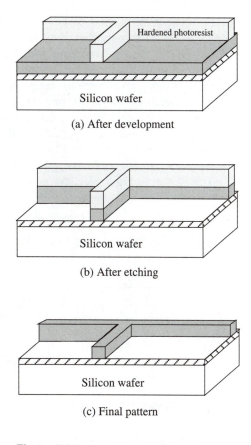

(a) After development

(b) After etching

(c) Final pattern

Figure 7.10 Etching of the material layer

transferred it to the metal layer on the silicon. The lithographic printing sequence must be performed for every patterned layer on the wafer.

7.2.1 The Importance of Physical Layout

This discussion serves to highlight two major concepts about VLSI circuits. The first one is due to the fact that the lithographic process produces features that have measurable sizes:

- A physical structure on an integrated circuit occupies a finite (non-zero) area.

In other words, we can make statements such as "Two transistors occupy more area than one transistor" and "A metal line 'consumes' area on the chip." This tells us that the larger the electronic network, the more area we must allocate for it. The second important point is that

- Adjacent material patterns on an integrated circuit must be separated from each other by a specified minimum distance.

This is due to the fact that because of various physical effects, the lithographic process cannot reliably separate two line patterns to an arbitrarily small distance. For example, the "shadow" of the masking reticle is not a perfectly sharp boundary but requires a small distance to change from light to dark. The spacing cannot be eliminated and serves no purpose in the operation of the circuit.

Considerations that arise from the limitation of how fine of a pattern can be printed onto the chip surface lead to a set of geometrical specifications that are called **design rules**. A design rule set tells the engineer the limitations of the lithographic and fabrication process by listing the smallest value for critical dimensions for every layer used on the integrated circuit. An example of a design rule specification is shown in Figure 7.11 for patterns created on a metal layer. There are two important dimensions shown:

- The width w of a metal line.

- The line-to-line spacing d between adjacent metal lines.

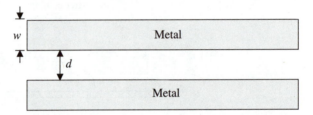

Figure 7.11 Example of design rule specifications

A design rule set specifies the minimum values for both w and d that can be created in the manufacturing line. If the design rules are not obeyed, then there is a high probability that the resulting physical structure will not behave as expected. Most high-density integrated circuit processes require that every dimension be checked before an attempt is made to fabricate the chip.

Layout design is an important aspect of VLSI. As we will see in this chapter, the layout and fabrication of silicon integrated circuits are closely related to the logic and system characteristics.

7.3 MOSFETs

In the previous chapter, our studies were directed towards the switching properties of MOSFETs when they are used to create CMOS logic gates. We will now examine the physical construction and manufacture of these switching elements as they are used to build physically tangible electronic networks.

A MOSFET is a three-dimensional structure. It is therefore simplest to examine the device from two different perspectives that separately illustrate the **side view** and the **top view**. Consider first the side view shown in Figure 7.12; this is useful for understanding the electronic operation of the device. Figure 7.12(a) shows the symbol for an nFET, and the physical realization of the transistor is shown in Figure

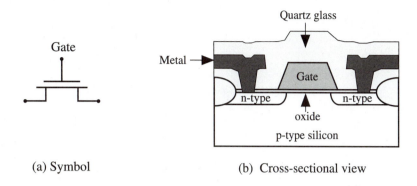

(a) Symbol   (b) Cross-sectional view

Figure 7.12 An integrated MOSFET

7.12(b).The side view (or cross-sectional) drawing represents what one would see if the chip were cut and analyzed.

At the silicon level, the transistor is created on the top of a p-type silicon wafer that is called the **substrate**; the substrate provides mechanical strength and is a few millimeters thick. The gate is created using **polycrystal silicon**, which is just silicon that is not a single crystal but many small regions of "crystallites"; we will usually refer to this material as **polysilicon,** or **poly** for short. The left and right (drain and source) terminals of the transistor are created by patterning regions of n-type silicon as shown and then adding a layer of metal to provide electrical contacts.

The drawing also shows the presence of quartz glass, chemically known as silicon dioxide (SiO_2). It is used because it is easy to "grow" on silicon via the reaction

$$Si + O_2 \rightarrow SiO_2$$

and it acts as an excellent insulator. We often simplify the terminology and refer to this material as **oxide**. The most important aspect of the oxide is that it electrically separates (insulates) the poly gate from the p-type silicon. In other words, there is no path for current flow between the gate and the substrate region of the transistor. This forms an electrical structure known as a **MOS capacitor** that is used to control current flow through the transistor. The acronym "MOS" stands for Metal-Oxide-Semiconductor and represents this layering in older technologies where aluminum metal was used for the gate. Even though we no longer use metal as the main gate material, the "M" remains in the acronym.

A close-up of the MOS capacitor is shown in Figure 7.13(a). This shows that the oxide acts as the insulating layer between the conductive polysilicon and the p-type semiconductor. The charge on the semiconductor is controlled by the voltage applied to the gate, as in any capacitor. The unique aspect of this structure is that the semiconductor is p-type, which indicates a large number of positively charged holes. However, if we apply a positive voltage V_G on the gate, then we can create a layer of negatively charged electrons underneath the oxide insulator as shown in Figure 7.13(b). This provides the necessary action to create a MOSFET.

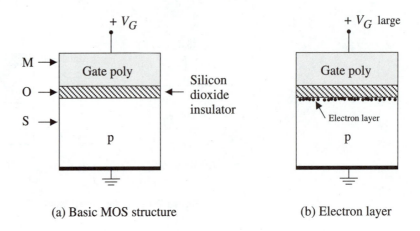

(a) Basic MOS structure

(b) Electron layer

Figure 7.13 Behavior of the MOS capacitor

7.3.1 MOSFET Switching

The operation of a MOSFET can be understood by studying the behavior of the central MOS gate capacitor. This section of the transistor has been enlarged and redrawn in Figure 7.14. The gate acts as the top plate of the capacitor, while the p-type substrate acts as the bottom plate. The oxide insulator between the two blocks the direct current flow. The n-type drain and source regions can be visualized as containing many electrons as shown. To achieve current from one side to the other, there must be a conduction path established between the two sides. This action allows us to model the MOSFET as a voltage-controlled switch where the value of the gate-source voltage V_{GS} determines the current flow through the device.

Applying a small gate-source voltage gives the situation shown in Figure 7.14(a). A voltage of V_{GS} less than about 0.5 v is not sufficient to alter the charge distribution in the semiconductor. The p-type region blocks any motion of electrons between the two n-type regions, and no current can flow between them. In this case,

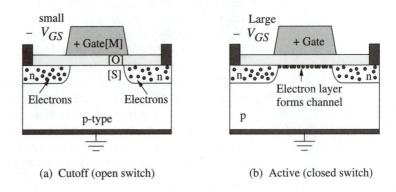

(a) Cutoff (open switch)

(b) Active (closed switch)

Figure 7.14 MOSFET operation

the transistor is in **cutoff** and acts like an open switch between the left and right sides. If we increase the value of V_{GS}, the gate becomes positive enough to attract a negative electron layer underneath the oxide layer as shown in Figure 7.14(b). The electron layer is called the **channel** since it provides an electrical conduction path between the two n-type regions. Current can flow through the device, defining what is called **active operation**. In electrical terms, this means that the two sides are in good electrical contact; this is a CLOSED switch in the switching model of a MOS-FET that was introduced in Chapter 6.

A top view of the nFET is shown in Figure 7.15(b). This is called a **layout** drawing, since it shows the patterns of the materials. There are three layers shown: the gate (which is polycrystal silicon), the n-type regions, and the metal connections. An electrical contact between the metal and n-type silicon is denoted by a box with an "X" in it and represents a **contact cut** in the silicon dioxide, which is a hole in the oxide that allows a connecting material to pass through it. Note that the presence of the oxide itself is implied in the entire drawing.

There are two important dimensions shown in the transistor. L is the distance that current must travel to get from one n-type region to the other. Since this looks like a "current flow channel," L is called the **channel length** of the MOSFET. Usually, L is the smallest dimension allowed for the gate material. In modern CMOS, L is less than about 0.35 micron. The other dimension shown is W, which is the **channel width** of the transistor. The amount of current that can flow through the transistor is determined by the **aspect ratio (W/L)** of the device. Both the current flow level and the switching time depend on the value of (W/L). An analysis of the MOSFET shows that the maximum current that can flow through an nFET is given by

$$I_n = \frac{k'_n}{2}\left(\frac{W}{L}\right)_n (V_{DD} - V_{Tn})^2 \tag{7.15}$$

where k'_n is a constant for the device, and V_{Tn} is called the **threshold voltage** of the nFET. Physically, V_{Tn} is the value of V_{GS} that just starts current flowing through the transistor. Note that we will occasionally place an n-subscript on the nFET aspect ratio and write $(W/L)_n$ in order not to confuse it with a pFET.

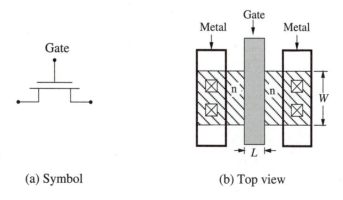

(a) Symbol   (b) Top view

Figure 7.15 Top (layout) view of an n-channel MOSFET

If we use an nMOSFET with a small channel width W_n, the current is forced through a narrow region, which limits its value. Thus, a small value of $(W/L)_n$ gives small currents, while a large value of $(W/L)_n$ allows larger currents to flow through the transistor. In terms of circuit switching times, a larger current allows us to charge capacitors faster. To model the dependence of current flow on the aspect ratio, let us rewrite the current in the form

$$I_n = I_{n1}\left(\frac{W}{L}\right)_n \tag{7.16}$$

where

$$I_{n1} = \frac{k'_n}{2}(V_{DD} - V_{Tn})^2 \tag{7.17}$$

is the current through the transistor when the aspect ratio has a value of $(W/L)_n = 1$. As we will see, fast logic gates require large aspect ratio transistors, since this gives higher current flow values. Note, however, that the area of the transistor is proportional to W_n. Combining these statements allows us to conclude that

• Fast switching circuits use large transistors, requiring more area.

This represents the **area-versus-speed** tradeoff in VLSI. It is important because of the companion observation that

• High integration requires small transistors.

In other words, we cannot simultaneously achieve high-density circuits without losing some speed.

The problem is illustrated graphically using the two FETs shown in Figure 7.16. In Figure 7.16(a), the transistor has a small aspect ratio with $(W/L) = 2$; a larger transistor with $(W/L) = 6$ is shown in Figure 7.16(b). Obviously, we can fit more small FETs into a given area than large ones. However, the large transistor has a current that is 3 times larger than that in the smaller device.

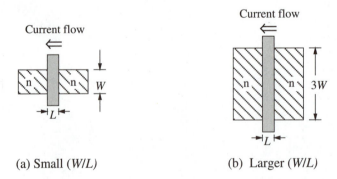

(a) Small (W/L) (b) Larger (W/L)

Figure 7.16 Small and large MOSFET geometries

Example 7-2

Consider a CMOS process in which $k'_n = 100 \times 10^{-6} \, A/V^2$, $V_{DD} = 5 \, v$, and $V_{Tn} = 0.7 \, v$. We first calculate

$$I_{n1} = \frac{(100 \times 10^{-6})}{2}(5 - 0.7)^2 \tag{7.18}$$
$$= 0.9245 \, mA$$

where 1 *mA* (milliampere) is 10^{-3} amperes. This is the maximum current through a MOSFET with an aspect ratio of $(W/L)_n = 1$. For an nFET with $(W/L)_n = 4$, the maximum current is

$$I_n = 4(0.9245) = 3.698 \, mA \tag{7.19}$$

while a transistor with an aspect ratio of $(W/L)_n = 12$ can support a maximum current flow of

$$I_n = 12(0.9245) = 11.094 \, mA \tag{7.20}$$

Since the logic signal is physically transmitted by means of the current, large transistors are required for fast switching.

RC Model

MOSFETs are relatively complex devices, and the equations used to relate the current I to the voltages are nonlinear. It is possible, however, to create a set of simple equivalent circuit models using resistors, capacitors, and switches that can be used to approximate the behavior of a MOSFET in a digital VLSI network.

Consider first the situation where we "look" into the gate terminal of a MOSFET. Since this is the top plate of the MOS capacitor (as implied by the circuit symbol itself), we can model the gate input as being the **gate capacitance** C_{Gn} shown in Figure 7.17. The value of the capacitance is proportional to the area of the gate, and so we will write that

$$C_{Gn} = C_{ox}W_nL \tag{7.21}$$

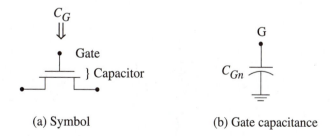

(a) Symbol   (b) Gate capacitance

Figure 7.17 Gate capacitance of a MOSFET

where C_{ox} is called the oxide capacitance per unit area; the value of C_{ox} depends upon the thickness of the oxide and is a known parameter at this level of analysis.

Example 7-3

Consider a CMOS process where $C_{ox} \approx 1.8 \times 10^{-7}$ F/cm^2 is specified. To calculate the gate capacitance of an nFET that has $W_n = 14\mu m$ and $L = 1$ μm, we note that

$$1 \; \mu m = 10^{-6} \; m = 10^{-4} cm \tag{7.22}$$

so that

$$\begin{aligned} C_{Gn} &= (1.8 \times 10^{-7})(14 \times 10^{-4})(1 \times 10^{-4}) \\ &= 25.2 \; fF \end{aligned} \tag{7.23}$$

where 1 fF (femtofarad) is equal to 10^{-15} F.

Next, suppose that we want to model the current flow between the drain and source terminals. The structure of the transistor shown in Figure 7.18(a) introduces unwanted (or, **parasitic**) resistance and capacitance that must be included in our calculations. Since every current flow path has resistance, we will introduce the nFET resistance R_n as representing the internal transistor resistance. Since the maximum current through the device is proportional to the aspect ratio $(W/L)_n$, we will compute the value of R_n using

$$R_n = \frac{r_n}{\left(\dfrac{W}{L}\right)_n} \tag{7.24}$$

where r_n is the resistance for an nFET that has $(W/L)_n = 1$. This shows that the resistance decreases as the aspect ratio $(W/L)_n$ increases, which is the behavior that we would expect. The nFET resistance is shown in Figure 7.18(c) along the voltage-controlled switch. Two capacitors, C_S and C_D, have also been included to represent the parasitic capacitance at the terminals; this originates from both the MOS structure and the **pn** junction region formed with the substrate. In general, these are estimated using

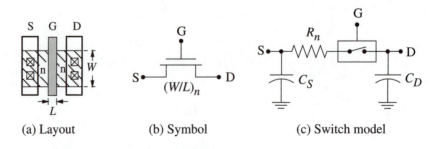

| (a) Layout | (b) Symbol | (c) Switch model |

Figure 7.18 Electrical modeling of a MOSFET

$$C_n = c_n \left(\frac{W}{L}\right)_n \qquad (7.25)$$

where c_n is a constant determined by the physics. Note that capacitance increases with the aspect ratio, which is exactly opposite to the resistance. In particular,

$$R_n C_n = r_n c_n \qquad (7.26)$$

shows that the RC product is a constant for all nFETs.[4]

It is important to note again that both the resistance and the capacitances of a MOSFET are classified as parasitic values in that they are inherent to the structure of the device but are not introduced on purpose. As we will see, both slow the switching speed of an integrated circuit.

Example 7-4

Suppose that the process parameters specify that $r_n = 2325 \ \Omega$ and $c_n = 2.0fF$ for a MOSFET with $(W/L)_n = 1$. Then a MOSFET with an aspect ratio of $(W/L)_n = 14$ is characterized by

$$R_n = \frac{2325}{14} \approx 166 \ \Omega$$
$$C_n = 14(2) \approx 28 \ fF \qquad (7.27)$$

by direct application of the formulas.

7.3.2 pFETs

A pFET is identical to an nFET, except that its polarities are exactly opposite. This means that the ends are p-type regions (instead of n-type), which reverses the sense of the voltages. An added complication is the presence of an **n-well** around the device. This is just an n-type region that is introduced into the p-type substrate for the purpose of building pFETs. For the discussion here, it suffices to state that the presence of the n-well complicates the placement of transistors, since all pFETs must be surrounded by n-well regions. Figure 7.19 shows the layout geometry for a pFET. The important dimensions are still given by the channel width W, the channel length L, and the aspect ratio $(W/L)_p$.

The maximum current that flows through a pFET is given by

$$I_p = \frac{k'_p}{2}\left(\frac{W}{L}\right)_p (V_{DD} - |V_{Tp}|)^2 \qquad (7.28)$$

---

[4] Although this is only an approximation, it remains reasonably accurate even if we perform a more rigorous analysis of the device.

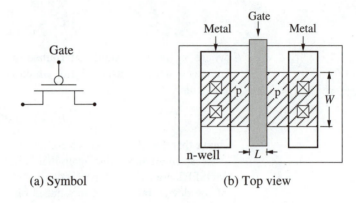

(a) Symbol   (b) Top view

Figure 7.19 Layout geometry for a p-channel MOSFET (pFET)

where k'_p is a constant for the p-channel transistor, and V_{Tp} is the pFET threshold voltage.[5] Note that the pFET current is also proportional to the aspect ratio $(W/L)_p$. As in the case of the nFET, we will write the current in the form

$$I_p = I_{p1}\left(\frac{W}{L}\right)_p \tag{7.29}$$

where

$$I_{p1} = \frac{k'_p}{2}(V_{DD} - |V_{Tp}|)^2 \tag{7.30}$$

is the current through a pFET with an aspect ratio of $(W/L)_p = 1$.

The primary differences between the nFET and the pFET are as follows:

- The pFET conducts current using positively charged holes.

- The pFET voltages and currents are reversed from those of the nFET.

These differences explain why the pFET has switching characteristics that are opposite to those of nFETs. One additional difference is that

- The value of k'_p is smaller than the value of k'_n due to the differences between electron and hole motion in silicon.

In general, $k'_n \approx (2.5)k'_p$, which implies that the pFET doesn't conduct as much current as the same size nFET. This is very important in advanced CMOS circuit design, where speed is critical.

---

[5] The pFET threshold voltage is usually quoted as a negative number. However, only the magnitude is important in our formulas.

Example 7-5

Suppose that our CMOS process has $k'_p = 40 \times 10^{-6}$ A/V^2, $V_{DD} = 5$ v, and $|V_{Tp}| = 0.8$ v. The maximum current is based on the value

$$I_{p1} = \frac{(40 \times 10^{-6})}{2}(5 - 0.8)^2 \tag{7.31}$$
$$= 0.3528 \ mA$$

which is the current for a pFET with an aspect ratio of $(W/L)_p = 1$. Let us consider a pFET with $(W/L)_p = 12$; the maximum current through the transistor is

$$I_p = 12(0.3528) = 4.2336 \ mA \tag{7.32}$$

The same size nFET in the process can conduct about 2.5 times more current.

The RC model for a pFET is constructed in the same manner as for the nFET and is summarized in Figure 7.20. The gate capacitance C_{Gp} is given by

$$C_{Gp} = C_{ox}W_pL \tag{7.33}$$

which is identical to the nFET gate capacitance equation, except that we use the pFET values for W_p and L.

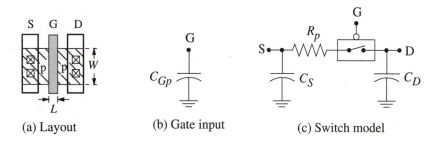

(a) Layout   (b) Gate input   (c) Switch model

Figure 7.20 Electrical model for a pFET

Example 7-6

Usually, the value of C_{ox} is the same for both nFETs and pFETs. Assuming the value of $C_{ox} \approx 1.8 \times 10^{-7}$ F/cm^2 for a pFET that has $W_p = 14$ μm and $L = 1$ μm gives

$$C_{Gp} = (1.8 \times 10^{-7})(14 \times 10^{-4})(1 \times 10^{-4}) \tag{7.34}$$
$$= 25.2 \ fF$$

which is the same as for an nFET of the same size.

The pFET resistance is computed using

$$R_p = \frac{r_p}{\left(\dfrac{W}{L}\right)_p} \tag{7.35}$$

where r_p is the resistance of a pFET with $(W/L)_p = 1$. One important point here is that pFETs do not conduct current as well as nFETs, so that $r_p > r_n$. Finally, the drain and source capacitors are estimated by

$$C_p = c_p \left(\frac{W}{L}\right)_p \tag{7.36}$$

with c_p a constant that is obtained from an analysis of the semiconductor structure. These equations clearly demonstrate the similarities between nFETs and pFETs.

Example 7-7

Suppose that the process parameters specify that $r_p = 5800\ \Omega$ and $c_p = 2.8\ fF$ for a pFET with $(W/L)_p = 1$. These are different than for an nFET on the same chip because the doping densities are different for the two types of transistors. With these values, a pFET that has an aspect ratio of $(W/L)_p = 14$ is characterized by

$$R_p = \frac{5800}{14} \approx 414\ \Omega$$
$$C_p = 14(2.8) \approx 39.2\ fF \tag{7.37}$$

As with the nFET, the product

$$R_p C_p = r_p c_p \tag{7.38}$$

is a constant for the process.

7.3.3 MOSFET Design Rules

To create MOSFETs on an integrated circuit, we must use the photolithographic sequence to define the patterns needed on each layer. The most important dimensions are the channel width W and the channel length L. The value of W is given by the size of the n$^+$ (or p$^+$) region, while the channel length L is specified by the size of the polysilicon gate. While these are sufficient to build a simple FET, we must also provide for additional wiring and contacts from a metal edge to the n$^+$ or p$^+$ region. The minimum size of the transistor is limited by the fabrication process through the design rules. It is therefore useful to study this problem in more detail.

The basic design rules are those that tell us the minimum widths and minimum spacings on each layer. Examples of these are shown in Figure 7.21 for polysilicon and metal line patterns. The numbers specified in the drawings are representative of those found in some real-world processes, but it must be remembered that the actual values depend upon the details of the manufacturing line and must always be checked. With this caution noted, we see that the minimum width of a polysilicon

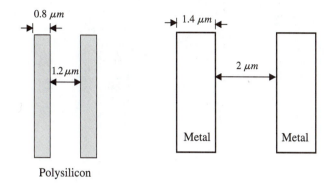

Figure 7.21 Design rule examples

line is specified as 0.8 μm, while the minimum spacing between two adjacent lines is 1.2 μm. The metal line has a minimum width specification of 1.4 μm and a minimum spacing of 2.0 μm. Metal lines are usually wider because they must carry higher current flow levels, and also because there is a stronger tendency to crack since the metal is deposited on top of other materials that leave the surface terrain with "peaks" and "valleys" that it must go over.

MOSFET design rules include these rules but also require additional specifications. These are shown in Figure 7.22 and have the following meaning.

- d_o is the polysilicon **gate overhang** distance that the poly must extend beyond the edge of the n^+ region. This ensures proper formation of the transistor.

- d_{p-c} is the **poly-to-contact** spacing between the edge of the polysilicon gate and the edge of the n^+ contact (that is outlined in black). This is needed to insure that the contact does not interfere with the gate material.

- d_{m-c} is the **metal-to-contact** spacing between the edge of the metal connection gate and the edge of the n^+ contact.

- s_c is the **contact size.** These are usually specified to help ensure a good connection.

- d_c is the **contact-to-contact** spacing.

Note that the channel length L of the MOSFET is set by the minimum width specification of the polysilicon. The size of the channel width W is determined by the circuit design and is almost always larger than the smallest size possible.

7.3.4 The Incredible Shrinking Transistor

Advances in lithographic patterning have resulted in MOSFET sizes that were thought impossible only a few years ago. Current production lines produce transistors with channel lengths as small as 0.18 μm, and values down to around 0.1 μm are expected for the next generation of manufacturing plants. Although this may sound like a natural evolution of the field, it is very difficult and expensive to create

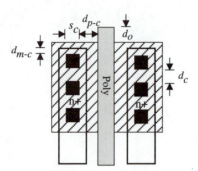

Figure 7.22
MOSFET design rules

a CMOS fabrication facility that can produce tiny transistors with these dimensions. It is therefore worthwhile to examine the general philosophy that drives the industry to shrink the transistor.

Let us start with the MOSFET shown in Figure 7.23(a) and consider this to be the current size of a typical device. The important device parameter is the aspect ratio (W/L) of the transistor. The smallest value of L is limited by the lithographic resolution of the pattern transfer equipment. Once this is specified, then the channel width W is chosen according to the desired current flow level. The area of the gate region is given by $A = WL$, and can be used as a measure of the overall size of the transistors.

Now suppose that we improve the lithography and fabrication process in a manner that allows us to reduce the value of the channel length to $L' < L$. This can be described by introducing the **scaling factor** $\alpha > 1$ such that

$$L' = \frac{L}{\alpha} \tag{7.39}$$

This is illustrated by the layout in Figure 7.23(b). For example, if $\alpha = 2$, then the new device has a channel length that is one-half the value of the original transistor.

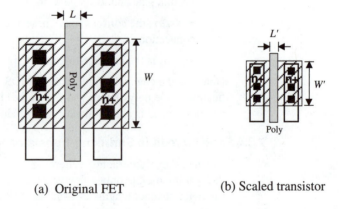

(a) Original FET

(b) Scaled transistor

Figure 7.23 MOSFET scaling

Since the important quantity is the width-to-length ratio (W/L), then we should also shrink the channel width to the new value

$$W' = \frac{W}{\alpha} \qquad (7.40)$$

as this gives

$$\frac{W}{L} = \frac{W'}{L'} \qquad (7.41)$$

i.e., the two transistor have the same aspect ratio. Note, however, that the gate area of the new transistor is

$$A' = W'L' \qquad (7.42)$$

Applying the scaling equations gives

$$A' = \left(\frac{W}{\alpha}\right)\left(\frac{L}{\alpha}\right)$$
$$= \frac{A}{\alpha^2} \qquad (7.43)$$

This shows that the area has been reduced by a factor of α^2.

Scaling theory illustrates the important driving force behind microelectronics. Decreasing the size of the basic transistor allows us to place more of them in a given area. This increases the **integration density** of the design and allows us to create very complex digital systems in very tiny areas.

Example 7-8

Suppose that we start with a transistor with $L = 1$ μm and $W = 20$ μm, and scale the size using $\alpha = 2$. The scaled device has sizes

$$L' = \frac{1}{2} = 0.5 \;\mu m$$
$$W' = \frac{20}{2} = 10 \;\mu m \qquad (7.44)$$

and a gate area of

$$A' = (0.5)(10) = 5 \;\mu m^2 \qquad (7.45)$$

Since the original area was

$$A = (1)(20) = 20 \;\mu m^2 \qquad (7.46)$$

this shows that the new device only occupies 25% of the area.

Moore's Law

Scaling allows us to increase the circuit density by shrinking the size of the transistors. Advances in modern silicon technology have resulted in devices that are only a few tenths of a micron on a side. This in turn allows much more sophisticated logic to be compressed into a tiny chip. Microprocessors now exhibit computing power that was restricted to large machines only a few years ago. The future continues to look bright as the technology improves.

How far can this trend go? In 1965, Gordon Moore, one of the co-founders of Intel Corporation, predicted that the number of transistors on an integrated circuit would double with every succeeding generation, which is about every 2 years. This statement has become known as **Moore's Law**, and it has been surprisingly accurate, as shown by the plot in Figure 7.24 obtained using average trends since 1970. What can stop this trend? From the technological viewpoint, it is thought that certain **quantum mechanical** effects will limit the use of MOSFETs when channel lengths fall below perhaps a tenth of a micron. At the other end of the spectrum, the cost of a manufacturing plant is now well over one billion dollars and rising. This makes it more difficult to make a new plant cost-effective. Regardless of these perceived barriers, we anticipate that integration densities will continue to increase well into the foreseeable future.

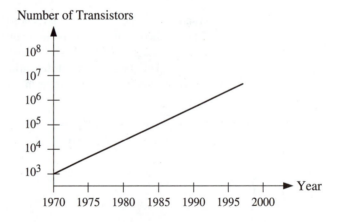

Figure 7.24 Moore's Law

7.4 Basic Circuit Layout

CMOS logic circuits are created using complementary pairs of nFETs and pFETs that are fabricated on a silicon wafer. Both the sizes of the transistors and the surface area required for interconnect wiring determine the overall size of a logic circuit when it is built as a chip. In this section we will examine some of the basic problems involved in actually constructing gates in CMOS integrated circuits.

We will simplify our discussion by viewing FETs as basic "building blocks" that are used to construct logic gates physically. Figure 7.25 introduces the concept of

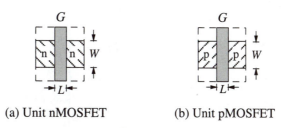

(a) Unit nMOSFET (b) Unit pMOSFET

Figure 7.25 Unit MOSFET building blocks

"unit" nFET and pFET structures for this purpose. To create a CMOS circuit, we will "wire" MOSFETs together using interconnect lines made out of metal; gate polysilicon can also be used for short lines. This is similar to the way in which we constructed logic diagrams using logic symbols in Chapter 3, only we are at a lower level in the hierarchy.

7.4.1 The CMOS Inverter

Let us first examine the transistor layout for a simple inverter. The basic requirement is to have an nFET and a pFET connected together to provide the NOT function. There are several ways to accomplish the layout and still have proper electronic operation.

An example layout is shown in Figure 7.26. The basic circuit in Figure 7.26(a) has been translated to the silicon pattern in Figure 7.26(b) by using a 1-to-1 correspondence between each device and wire. The input variable A enters the circuit at

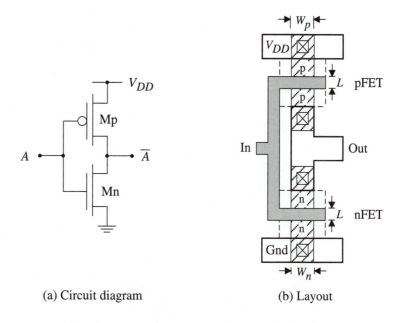

(a) Circuit diagram (b) Layout

Figure 7.26 Layout of a simple inverter circuit

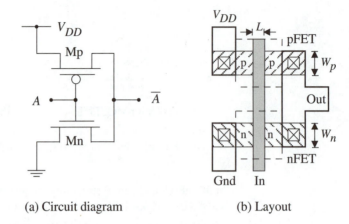

(a) Circuit diagram (b) Layout

Figure 7.27 Alternative layout for an inverter

the "In" point, which is at the polysilicon gate level. Note that the common gate consists of a single pattern of poly that extends to both transistors. The output $\overline{A}$ is taken from the right side, which is a metal interconnect line.

An alternative approach to layout is shown in Figure 7.27. In this example, the inverter circuit has been redrawn in Figure 7.27(a) to emphasize the use of horizontal-oriented transistors. Although the circuit is identical in operation, the layout shown in Figure 7.27(b) looks quite different. The polysilicon gate input consists of a single vertical poly line that connects the two transistors together. The output is once again a vertical metal line. The reader can easily verify that the transistors are electrically connected in the proper manner by tracing the lines.

These examples illustrate how different layout patterns can be obtained. Although both provide the NOT operation, they may exhibit different switching speeds. And, of course, their overall shapes are different.

7.4.2 Electrical Modeling

The switching performance of a CMOS gate is a result of the currents and voltages flowing through the electronic network. Since the values of resistance and capacitance in a transistor depend upon the aspect ratio (W/L), the layout geometry is critical to the speed of a gate. This dependence can be understood by using our transistor models to create an electrical equivalent circuit for the logic gate.

General Switching Model

The electrical performance of the inverter is analyzed by adding the output capacitor C_{out} to the circuit as shown in Figure 7.28(a). As discussed below, the value of the output capacitance is due to contributions from both the MOSFETs and the next stage in the logic cascade. In Figure 7.28(b), we have replaced the MOSFETs by equivalent resistances R_n for the nFET and R_p for the pFET. Note that the switches used in this circuit are complementary in nature. This means that when the pFET switch is CLOSED, the nFET switch is OPEN, and vice versa.

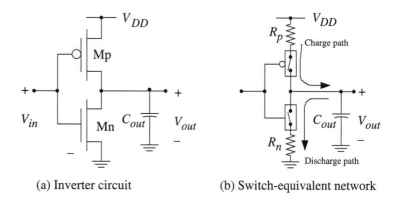

(a) Inverter circuit   (b) Switch-equivalent network

Figure 7.28 Electrical model for the inverter response

To calculate the transient response times, let us first consider the case where the pFET is ON and the nFET is OFF. In this case, current flows through R_p and charges the capacitor C_{out} as described by the RC time constant

$$\tau_p = R_p C_{out} \tag{7.47}$$

Let us relate this to the low-to-high time t_{LH} by assuming that the output voltage is described by

$$V_{out}(t) = V_{DD}[1 - e^{-(t/\tau_p)}] \tag{7.48}$$

For V_{out} to reach a particular value $V_{out}(t_x) = V_X$ requires a time of

$$t_x = \tau_p \ln\left[\frac{1}{1 - \dfrac{V_X}{V_{DD}}}\right] \tag{7.49}$$

which is obtained by rearranging the original equation. Now then, recall that the low-to-high time is defined as the time needed for the output to rise from the 10% voltage $0.1V_{DD}$ to the 90% voltage $0.9V_{DD}$. This means that we can write

$$t_{LH} = t_{90\%} - t_{10\%} \tag{7.50}$$

and substitute the values of $t_{90\%}$ and $t_{10\%}$ by using the appropriate voltages. In other words,

$$t_{LH} = \tau_p \ln\left[\frac{1}{1 - \dfrac{0.9V_{DD}}{V_{DD}}}\right] - \tau_p \ln\left[\frac{1}{1 - \dfrac{0.1V_{DD}}{V_{DD}}}\right] \tag{7.51}$$

$$= \tau_p\left(\ln\left[\frac{1}{0.1}\right] - \ln\left[\frac{1}{0.9}\right]\right)$$

$$= \tau_p \ln[9]$$

where we have used the identity

$$\ln(a) - \ln(b) = \ln\left(\frac{a}{b}\right) \tag{7.52}$$

in arriving at the last line. Noting that $\ln(9) \approx 2.2$, we have finally that

$$t_{LH} \approx 2.2\tau_p = 2.2R_pC_{out} \tag{7.53}$$

This shows that the rise time depends upon the value of the capacitance C_{out} and the resistance of the pFET through R_p.

The high-to-low time t_{HL} is calculated in a similar manner. This is a result of C_{out} discharging through the nFET as described by the time constant

$$\tau_n = R_nC_{out} \tag{7.54}$$

Assuming that the output voltage falls exponentially as

$$V_{out}(t) = V_{DD}e^{-(t/\tau_n)} \tag{7.55}$$

gives the fall time as

$$t_{HL} \approx 2.2\tau_n = 2.2R_nC_{out} \tag{7.56}$$

By definition, this is the time needed for the output voltage to fall from $0.9V_{DD}$ to $0.1V_{DD}$. Although the derivations for the switching times require a little algebra, the results are very straightforward to apply.

Example 7-9

Suppose that we have an inverter with a total output capacitance of $C_{out} = 150\ fF$, and transistors that are characterized by resistances of $R_p = 414\ \Omega$ and $R_n = 166\ \Omega$. The switching times are given by

$$t_{LH} = 2.2(414)(150 \times 10^{-15}) \approx 0.137\ ns$$

$$t_{HL} = 2.2(166)(150 \times 10^{-15}) \approx 0.055\ ns \tag{7.57}$$

where we recall that $1\ ns$ is defined as 10^{-9} second. The important dependence to keep in mind is that both t_{LH} and t_{HL} are proportional to C_{out}. Small output capacitance allows faster switching. Alternatively, both times can be reduced by decreasing the FET resistances by using larger transistors.

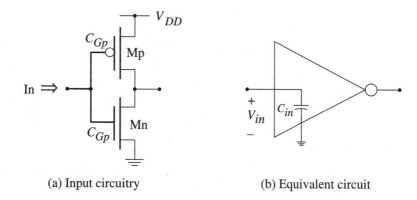

(a) Input circuitry   (b) Equivalent circuit

Figure 7.29 Inverter input equivalent circuit

Capacitance Analysis

The value of the output capacitance is important to the switching behavior, so it is worth studying how it enters into the circuit operation.

Let us start by noting that the input to an inverter circuit is connected to the gate of both an nFET and a pFET. From our modeling discussion in Section 7.3.1, the gate of a MOSFET looks like a capacitor, so we can construct the input circuit shown in Figure 7.29. The input capacitor C_{in} is calculated using

$$C_{in} = C_{Gn} + C_{Gp} \tag{7.58}$$

where the value of each depends upon the gate area as in equations (7.18) and (7.30). If large transistors are used, then C_{in} is large.

The output circuit can be modeled in a similar manner. Figure 7.30 illustrates how the we may replace the transistors in the circuit (Figure 7.30a) with their equivalent RC switch model to arrive at the general network shown in Figure 7.30(b). The capacitors C_{Dn} and C_{Dp} represent the parasitic MOSFET capacitances

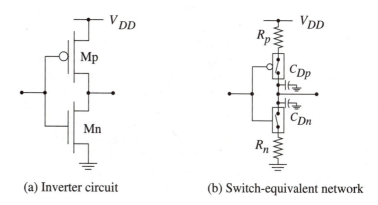

(a) Inverter circuit   (b) Switch-equivalent network

Figure 7.30 Inverter equivalent circuit based on MOSFET models

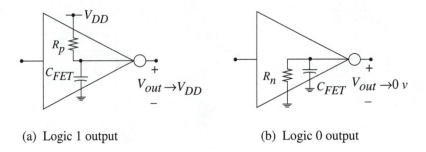

(a) Logic 1 output   (b) Logic 0 output

Figure 7.31 Charge and discharge equivalent circuits

seen at the drains of the nFET and pFET, respectively. These allow us to construct the simplified switching model in Figure 7.31 to calculate the important parameters. Although both the nFET and pFET resistors are shown, the complementary behavior of the two transistor switches says that only one resistor will be important during any given switching event. The equivalent circuit in Figure 7.31(a) is used if the output is undergoing a transition from $V_{out} = 0$ v towards V_{DD} with the pFET conducting current from the power supply. The important resistor is R_p, which is set by the size of the pFET. The parasitic capacitor C_{FET} given by

$$C_{FET} = C_{Dn} + C_{Dp} \tag{7.59}$$

has been included to represent the internal MOSFET capacitances. If the nFET is conducting such that the output voltage is falling from $V_{out} = V_{DD}$ towards 0 v, then the equivalent circuit in Figure 7.31(b) is valid. Since the discharge of the capacitor is through the nFET, the important resistance is R_n as shown. Again note the inclusion of the parasitic capacitor C_{FET} in our model.

Origin of Time Delay Contributions

Let us now examine the relationship of the various capacitors to the overall performance of the circuit. As always, the important parameter is the time constant

$$\tau = RC_{out} \tag{7.60}$$

which characterizes the response time of a simple circuit that consists of a resistance R and a capacitance C_{out}. For this analysis, C_{out} will be broken into individual contributions that illustrate the origin of the delays. Since small values of τ are desirable, it is important to understand which contributions to C_{out} can be reduced.

Consider again the equivalent circuits shown in Figure 7.31. These may be used to estimate the response of an isolated inverter, i.e., an inverter whose output is not connected to anything. An output transition from a logic 0 to a logic 1 state is characterized by using the pFET resistance R_p in Figure 7.31(a). For this case, the important time constant is

$$\tau_p = R_p C_{FET} \tag{7.61}$$

Similarly, the time constant for an output transition from a logic 1 to a logic 0 state is obtained from Figure 7.31(b) as

$$\tau_n = R_n C_{FET} \tag{7.62}$$

These may be used as first-order estimates for the propagation times

$$t_{PLH} \approx \tau_p \quad \text{(Low-to-high)}$$
$$t_{PHL} \approx \tau_n \quad \text{(High-to-low)} \tag{7.63}$$

so that

$$t_p = \frac{1}{2}(t_{PLH} + t_{PHL})$$
$$= \frac{1}{2}(R_p + R_n)C_{FET} \tag{7.64}$$

is the propagation delay time for an isolated inverter (one that is not connected to any other gate).

A more practical case is where the inverter is driving a "load" such as another logic gate. Figure 7.32(a) shows the situation where we cascade the inverter into another inverter. We can now see that the total output capacitance for an inverter cascaded into another inverter is given by

$$C_{out} = C_{FET} + C_{in} \tag{7.65}$$

Since the input to the load is characterized by a capacitance C_{in}, the time constants for both the low-to-high and the high-to-low transitions are modified. This means that the switching response of the circuit is limited by both the external load (contained in C_{in}) and the internal parasitic capacitances as calculated from C_{FET}. Figure 7.32(b) shows the equivalent circuit for the case when the output voltage V_{out} of the first gate INV makes a transition from 0 v towards V_{DD}. Physically, this occurs because the capacitance is being charged through the pFET resistor R_p. However, both the internal capacitance C_{FET} of the gate and the input capacitance C_{in} of the load are charged. This increases the time constant to

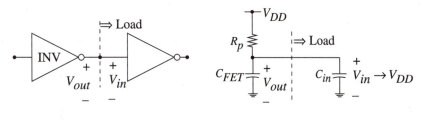

(a) Logic cascade   (b) INV output to logic 1

Figure 7.32 Transient response for cascaded inverters

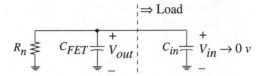

Figure 7.33 Equivalent circuit for the high-to-low time constant

$$\tau_p = R_p(C_{FET} + C_{in}) \tag{7.66}$$

which is larger than the time constant for an isolated gate. If V_{out} changes from V_{DD} towards ground as in Figure 7.33, then the time constant is

$$\tau_n = R_n(C_{FET} + C_{in}) \tag{7.67}$$

since C_{in} must be included. Using these for the individual components of the propagation delay gives

$$\begin{aligned} t_{PLH} &\approx R_p(C_{FET} + C_{in}) \\ t_{PHL} &\approx R_n(C_{FET} + C_{in}) \end{aligned} \tag{7.68}$$

so that the total propagation delay for the gate is

$$t_p = \frac{1}{2}(R_p + R_n)(C_{FET} + C_{in}) \tag{7.69}$$

This illustrates the important fact that increasing the load slows down the switching response of a logic gate (i.e., it takes a longer time for the gate to change). Moreover, increasing the fan-out of the gate makes the situation worse. To understand this comment, consider the case where a fan-out of 3 is used. As illustrated in Figure 7.34, each gate adds a capacitance of C_{in} to the load so that the propagation delay is

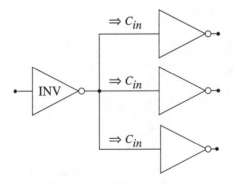

Figure 7.34 Increased load capacitance due to a larger fan-out

$$t_p\big|_{FO=3} = \frac{1}{2}(R_p + R_n)(C_{FET} + 3C_{in}) \tag{7.70}$$

In general, a fan-out of N gives

$$t_p\big|_{FO=N} = \frac{1}{2}(R_p + R_n)(C_{FET} + NC_{in}) \tag{7.71}$$

so that the logic designer must be very careful to insure that the fan-out does not slow down the logic gate too much.

High-Speed Design

The VLSI designer is often faced with the problem of increasing the speed of a logic circuit. In basic terms, this means that the circuit must be designed to reduce the delay times. Examining the delay equations above shows that one way to reduce the time constants is to make the resistors smaller. Physically, this means that we use larger values of the aspect ratio (W/L), since this allows a larger current, which in turn reduces the resistance. This is expressed by the resistor dependence

$$R = \frac{r}{\left(\dfrac{W}{L}\right)} \tag{7.72}$$

where r is the resistance for an aspect ratio of (W/L) = 1.

Figure 7.35 shows inverter circuits that have been designed with larger transistors than in the original design of Figure 7.27. In Figure 7.35(a), the aspect ratios have been doubled. This reduces the resistances to effective values of ($R_n/2$) and ($R_p/2$) so that the resistor sum is reduced to

$$\left(\frac{R_n}{2} + \frac{R_p}{2}\right) = \frac{1}{2}(R_n + R_p) \tag{7.73}$$

If the internal capacitance C_{FET} is small compared to the load capacitance term NC_{in} in equation (7.68), then the new propagation delay t'_p is given by

$$t_p' = \frac{1}{2}t_p \tag{7.74}$$

This means that the switching speed has itself been doubled. Tripling the aspect ratio of the transistors as shown in Figure 7.35(b) reduces the resistor term to

$$\frac{1}{3}(R_n + R_p) \tag{7.75}$$

so that the propagation time is reduced to a new value of

$$t_p'' = \frac{1}{3}t_p \tag{7.76}$$

which then triples the switching speed over the original, "small-FET" design shown in Figure 7.27.

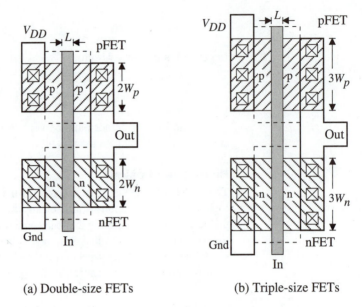

(a) Double-size FETs　　　(b) Triple-size FETs

Figure 7.35 Larger transistors used to speed up the switching

These examples illustrate the limiting feature of high-speed, high-density VLSI design. Fast circuits require large transistors, but using large transistors reduces the number that can be placed on a chip. In other words, there is a tradeoff between **area** and **speed**: you cannot achieve fast circuits without using large transistors. In the jargon of VLSI engineers, the area on an integrated circuit is referred to as **real estate.** Using large transistors increases the real estate "costs" in that we have less area left for the other circuits. Since our goal is to integrate literally tens of thousands of gates on a single high-speed chip, the tradeoffs are critical. This is an excellent example of real-world engineering. Most problems have more than one solution; the difficult part of engineering is deciding which approach will yield a better end result.

7.5　MOSFET Arrays and AOI Gates

Complex logic gates are extremely useful in CMOS logic design as they allow us to combine several primitive logic operations into a single circuit. Since both the pFET and nFET logic arrays use series- and parallel-connected transistors, it is important to see the layout strategies.

7.5.1　Wiring Strategies

Layout design is classified as an art, with many approaches used in practice. We will concentrate on the basics for now and leave the fine points to be studied in a later course.

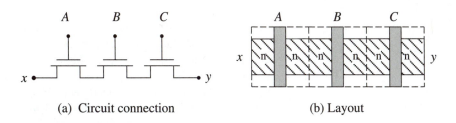

(a) Circuit connection   (b) Layout

Figure 7.36 Series-connected MOSFETs

Series-connected MOSFETs

When two transistors have the same current, they are said to be in series. In terms of MOSFETs, this means that the drain/source terminals of adjacent MOSFETs are connected. Figure 7.36(a) illustrates the case with three series-connected nMOS-FETs. The end points are labeled x and y such that the value of the current flowing into one end is the same as the current flowing out of the other.

The layout of the three series-connected MOSFETs is shown in Figure 7.36(b). In the drawing, the dashed-line boxes define the individual transistors. Since each side of a MOSFET is n-type silicon, adjacent transistors can be connected electrically by just "touching" the n-type regions; the two n-type regions then merge to form one. This illustrates that series-connected MOSFETs are very easy to lay out on the chip.

Parallel-connected MOSFETs

Two transistors that have both their drain and source terminals connected as shown in Figure 7.37(a) are said to be in parallel with each other. The layout of parallel devices can be accomplished by making a one-to-one correspondence with the circuit diagram. In Figure 7.37(b), the dashed-line boxes show the unit transistors;

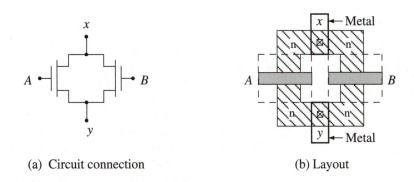

(a) Circuit connection   (b) Layout

Figure 7.37 Parallel-connected MOSFETs

additional n-type sections have been added to connect the top and bottom ends of the transistors. Metal interconnect lines complete the connections.

This example illustrates that parallel-connected MOSFETs generally take up more area than series-connected MOSFETs. However, the situation can be improved somewhat by remembering that the length of a conducting wire can be changed without affecting the electrical performance too much, so long as we don't make it too long. Figure 7.38(a) shows an alternative circuit diagram for two parallel nMOSFETs. This looks similar to a series group, but we have taken the connections to be in the middle (x) and at the ends (y). Electrically, this is equivalent to the circuit shown in Figure 7.37(a), as can be verified by tracing the wire connections. The layout drawing in Figure 7.38(b) shows this connection scheme with the dashed-line boxes again indicating unit transistors. The x terminal has been obtained by adding contacts and a metal line from the central n-type region. To accommodate the metal lines for the y terminal, the n-type regions on the left and right sides have been extended, while the routing is below the FETs.

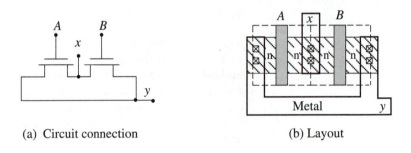

(a) Circuit connection   (b) Layout

Figure 7.38 Alternate wiring scheme for parallel MOSFETs

VLSI layout designers spend a lot of time varying the MOSFET placement strategies and wiring patterns in an effort to create a minimum-size layout that exhibits fast switching response. As can be imagined, sophisticated CAD tools are crucial for achieving good designs.

7.5.2 NAND and NOR Layout

Let us now apply the series and parallel MOSFET layouts to the basic gates. A NAND2 gate is shown in Figure 7.39(a), with a possible layout provided in Figure 7.39(b). The layout strategy is the same as that used for the inverter example of Figure 7.27. Input lines A and B at the polysilicon level have been chosen to run vertically, while n- and p-type FET regions are oriented in a horizontal direction. Metal is used for wire in the series nFET arrangement and the parallel pFETs, and also for V_{DD} and ground connections.

A NOR2 gate is shown in Figure 7.40(a) with its layout in Figure 7.40(b). Since the same layout philosophy has been used, there are many similarities between this example and the NAND2 gate. This is due to the fact that both circuits are created using two series-connected FETs and two parallel-connected FETs. The difference between the NAND2 and NOR2 gates is just the type (polarity) of transistor used for each array. If you examine the patterns of the metal layer and the contacts in the

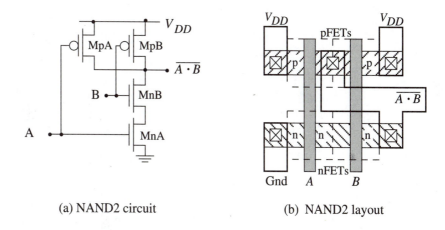

(a) NAND2 circuit

(b) NAND2 layout

Figure 7.39 Layout for a NAND2 gate

two drawings, you will notice that they are identical in shape and placement, except that one is "flipped" around an imaginary horizontal line that cuts the center of the cell. This observation is important because the NAND and NOR operations are duals of each other.

7.5.3 Complex Logic Gates

The layout strategy above can be applied to an arbitrary gate. Consider the AOI logic circuit shown in Figure 7.41(a). This provides the generic function

$$f = \overline{A \cdot B + C \cdot D} \tag{7.77}$$

A possible layout is provided in Figure 7.41(b). The nFET array is straightforward to see because it consists of two groups of two series-connected nFETs. The pFET wiring is more complex. The power supply connection to V_{DD} is taken between the

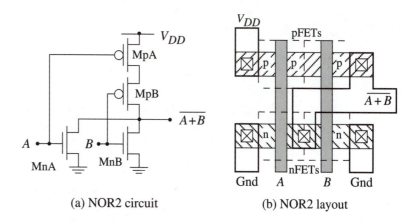

(a) NOR2 circuit

(b) NOR2 layout

Figure 7.40 Layout scheme for a NOR2 gate

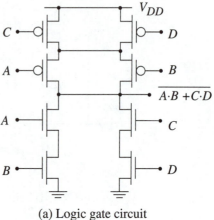

(a) Logic gate circuit

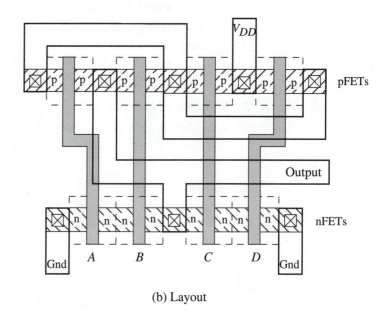

(b) Layout

Figure 7.41   Layout for a 4-input AOI logic gate

C and D pFETs; the other side of these transistors is connected to the A and B pFETs, as can be seen by tracing the metal wiring. Finally, the output is taken from between the A and B pFETs to complete the circuit.

7.5.4   General Observations

These examples illustrate some very important concepts that may be applied to the entire field of logic integration in a VLSI environment. These are as follows.

- Transistors require area on the chip. Small transistors are slow, while large transistors are faster.

•   An *n*-input logic gate uses 2*n* FETs. Increasing the fan-in of a logic gate requires additional area for the transistors.

•   Additional area must be allocated for wiring the transistors together as specified by the circuit diagram. Every interconnect wire has a minimum width and must be spaced from its neighbor according to the design rules.

The significance of the last observation can be appreciated by examining the layout examples above. Let us define A_{gate} as the total area of a gate, and A_{FET} as the area that is due to transistors. Visual inspection of all of the layouts shows that

$$\frac{A_{FET}}{A_{gate}} < 0.2 \qquad\qquad (7.78)$$

is a reasonable upper-limit guess for the ratio of the FET area to the total area of the logic circuit. This shows that the area needed for interconnect and wiring is much greater than the area used by the transistors themselves. In other words, the limiting factor in high-density VLSI is the complexity of the interconnect wiring, not the number of transistors. This makes even more sense when we note that the dimensions of a transistor typically result in an area less than a few square microns (where $1 \, \mu m^2 = 10^{-8} \, cm^2$), while metal lines have minimum widths of about 1 to 2 μm and can have lengths of several microns.

Variations to the layout design of a gate are often used to reduce the overall area. However, the problem is based on fundamental concepts that are due to physical constraints in the manufacturing process. This remains one of the big challenges of modern VLSI design.

7.6 Cells, Libraries, and Hierarchical Design

Up to this point, we have been viewing logic gates at the primitive silicon level where we must deal with the items such as the shape of the poly gates and the translation of circuits to silicon. Let us now elevate our discussion to the next hierarchical level so that we can progress to system design.

VLSI layout is performed on high-performance computers using various types of computer-aided design (**CAD**) and computer-aided engineering (**CAE**) tools. This allows us to create very powerful system design software that are based on the fact that all digital networks can be built using primitive logic cells. The meaning of "primitive" varies with the application; the most obvious case is that where we use the basic logic operations NOT, AND, and OR to create more complex logic functions. In the general sense, however, a primitive cell can just be defined as one that is at a lower level in the hierarchy than we are working at. The important concept here is that we may design a complex system by using simpler building blocks. This is analogous to viewing all matter as being made up of atoms and using atoms as the basic units.

7.6.1 Creation of a Cell Library

High-density VLSI networks are created by noting that any system can be broken down into simpler functional blocks that appear many times in the design. If we

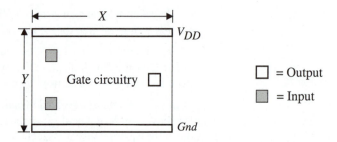

Figure 7.42 Logic gate cell

design a block once, then there is no need to do it again: we can just use a copy of the original. The originals are kept in a file called the **library,** which is available for use during the system design process.The components in a cell library are chosen for their usefulness in logic design, but they must also account for the electrical switching characteristics.

Let us examine the basic idea behind creating a cell library. Every cell provides a particular logic operation that appears several times in the system design. Each cell is created in silicon using the layout rules. Since every logic gate requires a connection to the power supply V_{DD} and ground, a general view of the finished cell will appear as shown in Figure 7.42. The vertical spacing Y is the height of the cell and is chosen to be a constant for every cell in the library; this will make it easier to arrange the cells on the chip. The cell width X varies with the complexity of the circuitry. We also note the locations of inputs and outputs on the layouts. These are called **ports** and provide us access to the interior circuits of the cell.

The specific cells that are included in the library should provide all of the basic logic gates needed to design the system. As a base set, this includes the fundamental logic operations NOT, NAND, and NOR as shown in Figure 7.43. Note that every cell has the same height, but the widths vary with the function. The input and output ports are labeled with symbols for ease of use. The cell library serves as the "central repository" for a VLSI design. The cells provide the basic building blocks for complex designs and can be copied as needed. While the NOT, NAND, and NOR gates are sufficient to create any logic network, it is convenient to provide

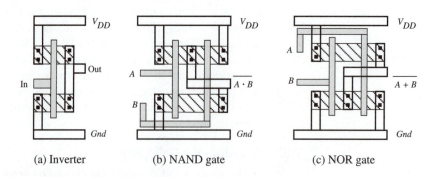

(a) Inverter (b) NAND gate (c) NOR gate

Figure 7.43 Examples of basic library cells

additional cells to help the logic designer. A simple extension of the basic set is shown in Figure 7.44. This includes the NOT-NOT cascade in Figure 7.44(a), which may be used to produce both the complement and a buffered output; the NAND2C gate (NAND2 with Complement) in Figure 7.44(b), which cascades the output of a NAND2 gate into an inverter to yield the AND operation also; and the NOR2C in Figure 7.44(c), which gives both the OR and the NOR functions.

Cell libraries often contain several hundred different entries to aid in the logic design process. It is common to provide different cells for the same logic operation, with each cell characterized by a different shape or electrical driving characteristic. This type of cell library forms the basis for building application-specific integrated circuits (**ASICs**) in which the logic or system design can use the cell library to

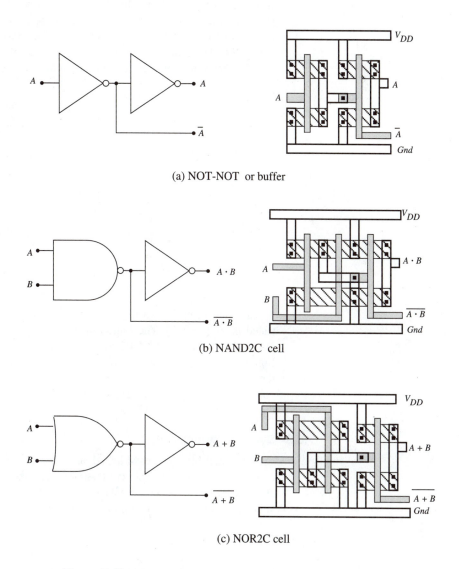

(a) NOT-NOT or buffer

(b) NAND2C cell

(c) NOR2C cell

Figure 7.44 Examples of useful library cells

pick-and-choose the design that is needed for a particular application. ASIC design is an extremely powerful, yet simple, approach to integrating digital logic networks onto a single CMOS chip.[6]

7.6.2 Cell Placement

Logic networks are created by cascading logic gates. Each logic gate corresponds to a cell in the library, so that we create a VLSI logic chain by choosing each logic gate, placing it in the chain, and wiring the inputs and outputs as required. The concept is shown in Figure 7.45. Each cell is viewed as a "tile" that has a particular width; an individual tile is shown in Figure 7.45(a). Placing tiles next to each other allows them to be connected by interconnect wiring as implied by Figure 7.45(b). In this drawing, the inputs to Gate 1 are denoted by a and b, while the output is shown as $y(a, b)$. This is used as one of the inputs into Gate 2, which has an output labeled as $f(x, y)$. Note that choosing every tile to have the same V_{DD}-Gnd spacing allows the tiles to fit together in a row. This is the reason why the value of Y is chosen to be the same for every cell in the library.

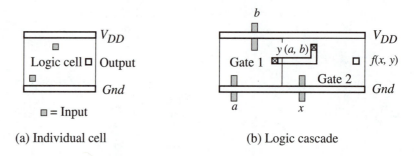

(a) Individual cell   (b) Logic cascade

Figure 7.45 Cell placement on the chip

An example of how a logic cascade is transformed into a gate layout is illustrated in Figure 7.46. The logic diagram in Figure 7.46(a) shows that we need a NAND2, a NOT, and a NOR2 gate. The cell-based equivalent network is shown in Figure 7.46(b), where the appropriate cells have been placed in a single row, and the ports have been wired according to the logic diagram. It is obvious that the more complex the logic cascade, the larger the required area, and the more complicated the interconnect wiring will be.

7.6.3 System Hierarchies

Once the cell library is established, an integrated VLSI chip is designed using the concept of hierarchies. This is illustrated schematically in Figure 7.47. MOSFETs are considered the primitive devices and are created using patterned layers of polysilicon, doped semiconductor. These are combined into logic gates, which is the

---

[6] ASIC is pronounced by saying the letter "A" followed by the word "*sic.*"

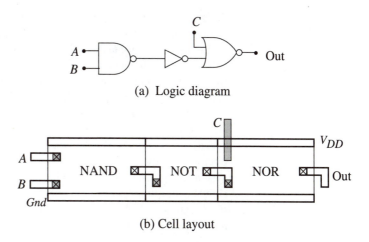

(a) Logic diagram

(b) Cell layout

Figure 7.46 Example of cell placement and wiring

next level in the hierarchy. Complex digital components are then created by combining logic gates together, resulting in the final chip design.

It is useful to compare the silicon view of hierarchies with those used in VHDL descriptions. If we want to describe a complex system in VHDL, the first step is to introduce **entity** and **architecture** declarations for every module. The architectural statements are created using primitives such as concurrent operations. Each module can be instanced in the system flow, so that they form the basic library. This is analogous to creating logic gates from MOSFETs in VLSI and using them as the cell library. Once VHDL modules are defined, they can be used as **components** to create more complex logic functions. Complex chip designs are implemented in the same manner using the cell libraries. The lesson to be learned here is that the hierarchical

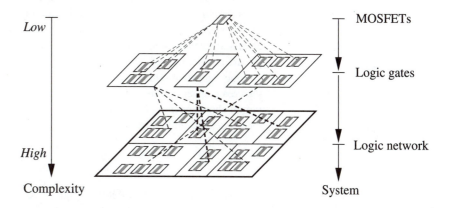

Figure 7.47 Silicon hierarchies in the layout design process

design process can be implemented in either viewpoint, and there often exists a one-to-one analogy between the seemingly distinct approaches.

Let us now return to the concept of designing a complex chip using a cell library. When we instance a cell from the library to the main chip, every characteristic of the cell is identical to the original design stored in the library. This means that we cannot change the instanced cell in our design unless we go back and modify the library entry. Usually, however, we want to leave the library intact for future usage. If we need a cell with different characteristics (such as dimensions or switching speed), there are two main options. One is to perform a custom design in our main circuit. The other is the extreme case where we **flatten** the design from a high-level system using instanced cells to the primitive silicon level. The flattening process is shown conceptually in Figure 7.48. In a flattened design, all references to cells in a library are lost. Only the primitive geometries of the polysilicon, the metal layers, and so on remain.

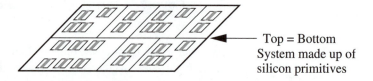

Top = Bottom
System made up of
silicon primitives

Figure 7.48 A flattened design consists only of primitives

7.7 Floorplans and Interconnect Wiring

A VLSI chip is designed section-by-section by dividing the logic up into convenient logic blocks with each performing a distinct group of functions. The complete system is then created by wiring the blocks together to give the desired result. In VLSI systems design, a **floorplan** refers the allocation of area and position to each unit in the overall system. An example is shown in Figure 7.49. Since each unit will have a characteristic size and shape, floorplanning can be viewed as being similar to designing a picture puzzle, with each unit being a piece that must fit into the system.

One important consideration that arises in the manufacturing line is the fact that we create several identical chips on one large silicon wafer, but not all of the individual chips are functional after the fabrication process is completed. The **yield** Y of the process is defined by

$$Y = \frac{\text{\# of Good Chips}}{\text{Total Number}} \times 100 \text{ \%} \tag{7.79}$$

Obviously, we would like yields as high as possible since that maximizes our profit. Although many factors influence the value of Y, the total surface area A of each chip is particularly important. With regard to the general floorplan shown above, we find that it is desirable to make the chip area

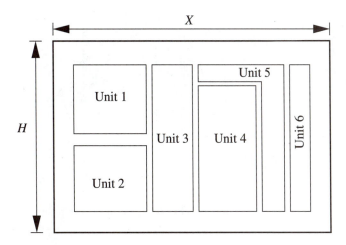

Figure 7.49 A floorplan shows the location of each unit

$$A = X \times H \qquad\qquad (7.80)$$

as small as possible.

To understand this statement, we need to study a little more about the physical characteristics of the silicon wafer used as the starting point in the CMOS VLSI process. Although we strive to create perfect silicon crystals, every silicon wafer has defects, as illustrated in Figure 7.50. These are due to many reasons, such as missing atoms and small imperfections. The number of defects is given by the defect density D, which tells the number of defects per unit area of the surface (with strict units of $1/cm^2$). Although D is very low (usually around $D = 1$), the presence of even a single defect on a chip renders the circuit nonfunctional.

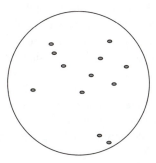

Figure 7.50
Silicon wafer showing locations of defects

Let us now examine how the area of the chip and the defect density combine to affect the yield. The drawing in Figure 7.51(a) portrays the situation where we have a design with a small area. When a die (individual chip) overlaps with a defect, then it is nonfunctional; these have been shaded in the drawing. Although a few of the die are bad, most are not affected by the defects. In Figure 7.51(b), the die area has

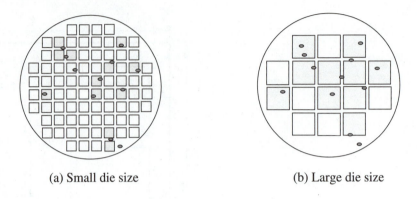

(a) Small die size   (b) Large die size

Figure 7.51 Relationship between die size and yield

been increased by a factor of 4 by doubling the side lengths. We see that the larger area greatly increases the probability of overlapping with a defect. A simple model for the yield is given by the formula

$$Y = e^{-\sqrt{AD}} \times 100\% \tag{7.81}$$

showing how a larger area results in a reduced yield. The yield problem is compounded by the fact that the area is not the only important factor. Many other physical and chemical problems enter the picture, such as dust particles destroying the patterns, breaks in the metal lines, problems in doping, and a host of other factors. In general, a low-yield design cannot be released as a product because it would have to be priced very high to allow even a small profit margin.

One of the more challenging aspects of VLSI design revolves around this discussion. We would like to provide as much logic as possible on a chip that has the smallest dimensions possible. Ideally, we would like to have an extremely complex system on a very tiny die, but this contradicts the lessons that we have learned in this chapter: logic operations are performed by circuits that require area on the silicon chip. When this is combined with the speed-versus-area tradeoff that states that high-speed circuits require large transistors, the problem becomes even more complex. Special care is needed to design circuits that are small but are capable of fast switching.[7]

Although it may appear that the MOSFETs are the culprits in increasing the size of a chip, they are completely innocent in this regard. Transistors are very tiny and require only a few tenths of a micron on each side. The real problem revolves around the **interconnect wiring** that is needed to put a complicated system together.

---

[7] This is one of the many aspects of VLSI that makes it a lucrative career choice!

7.7.1 Interconnects

VLSI design requires that we connect the different logic blocks together to form the complete system. At the silicon level, the "wiring" problem refers to using patterned layers of conductors to provide the electrical paths among the cells in the design. This is accomplished by using thin line-like patterns of metal that are collectively called **interconnects**; in most modern CMOS processes, polysilicon can also be used in this regard.

There are two major wiring considerations that directly affect the area (and hence the yield) of a chip. First, the design rules specify that patterns on a material layer are subject to minimum-width and minimum-spacing requirements. Every interconnect line consumes area and must be spaced from neighboring lines according to the design rule limitations. The second major problem is the fact that the wiring itself can be very complicated in a VLSI design and generally requires the use of several conducting layers.

An example of the interconnect area problem can be seen by the layout in Figure 7.52. The smallness of the MOSFET compared to the area needed for the metal interconnect wiring (Line 1, Line 2, and Line 3) is obvious. As was mentioned above, this is due to the minimum width and spacing values (w and S in the diagram) required for patterns on the same layer. The drawing also shows that we may use the polysilicon layer to cross under the metal lines without creating a short circuit. Additional material layers that are dedicated to interconnect lines help alleviate many of the problems, and modern CMOS processes typically have five or more dedicated layers for this purpose.

Once we recognize the main characteristics of the interconnect lines, we may tackle the problem of the wiring itself. In overview, the task is straightforward: connect the inputs and outputs of every unit to form the network specified by the system design. Every unit has ports that give the entry and exit points on the silicon.

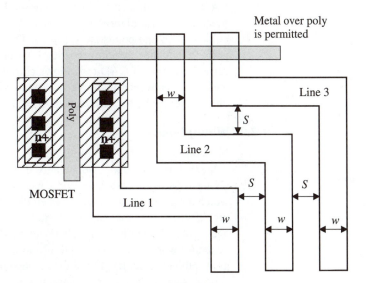

Figure 7.52 Design rules as applied to interconnect wiring

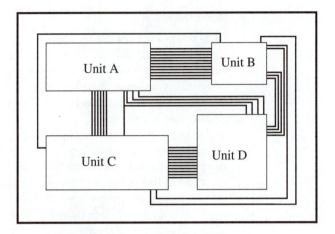

Figure 7.53 General wiring and interconnect problem

Once these units are arranged, the system schematic tells us what the necessary wiring is. This is shown schematically in Figure 7.53 for one level of interconnect that is used to wire four units together. Even with this simple drawing, it is easy to see how the interconnect problem can be difficult. For example, no line can cross another line on the same layer without causing an electrical connection between the two. When we create a large chip that requires thousands of data transmission paths, we must turn to creative layout solutions and multiple interconnect levels to accomplish the task.

In a VLSI design environment, wiring the units together can be very tedious. In addition, the large number of interconnects can increase the probability of errors. Since even a single incorrect connection will make the chip nonfunctional, various CAD tools have been written to address this problem. These are generally called **place and route** algorithms. To create a connection between two sections on the chip, the two end points are specified. The algorithm then routes the line from one point to the other, avoiding existing lines on the layer and obeying all of the design rules. This type of CAD tool allows chip designers to make more effective use of their time. However, there are occasions when the place and route solution introduces other problems, and lines must be drawn individually.

7.7.2 Wiring Delays

Interconnect routing is complicated by the fact that the lines introduce a signal delay that depends upon the length of the wiring pattern. This is because all materials have parasitic resistance and capacitance that act to slow down the speed at which a signal is transmitted between the two end points. This is illustrated in Figure 7.54(a), where the output voltage $V_{out,1}(t)$ of Inverter 1 is used as the input voltage $V_{in,2}(t)$ of Inverter 2 through a wire interconnect of length l. Although we do not generally worry about the length of the interconnection when drawing circuit diagrams, the physical length l of the finished product can be an important factor. This is because both the total resistance and the total capacitance of a wire increase with length. Figure 7.54(b) shows a simple model of the wire consisting of several RC-

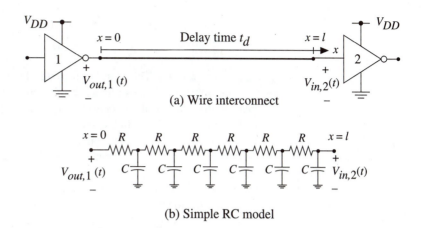

(a) Wire interconnect

(b) Simple RC model

Figure 7.54 Electrical RC model for interconnect

type elements from $x = 0$ to $x = l$. Each RC segment slows down the signal as it moves from Inverter 1 to Inverter 2.

Let us denote the time delay from $x = 0$ to $x = l$ by t_d. An analysis of this problem shows that the time delay is approximated by

$$t_d = \kappa l^2 \tag{7.82}$$

where κ is a constant. This dependence, plotted in Figure 7.55, exhibits an important characteristic: the time delay increases as the square of the length. To see the significance of this dependence, suppose that we have a line with a length l_o that exhibits a delay of

$$t_{do} = \kappa l_o^2 \tag{7.83}$$

If we double the length of the line to $2l_o$, the delay increases to

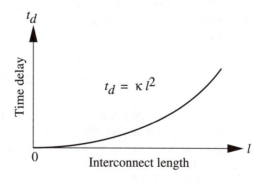

Figure 7.55 Interconnect delay as a function of length

$$t_{d1} = \kappa(2l_o)^2$$
$$= 4t_{do}$$

(7.84)

i.e., by a factor of 4. This clearly illustrates the problems with long interconnects.

The problem becomes more complicated at high current levels, where we find that another electrical parameter called the **inductance** L of the wire must be included in the analysis. Inductance is due to the magnetic field interactions that are caused by the current in the wires, and it is also a parasitic effect. This leads to many new effects in signal transmission that make problems even more difficult to deal with. These are particularly important when designing the electronic wiring networks on printed-circuit boards that hold the chip packages and insure that they are connected together correctly. This problem was discussed earlier in Chapter 4.

7.8 Problems

[7.1] In a sample of pure silicon, the number of electronics and holes are equal with a value of $n = p = 1.5 \times 10^{10}$ cm^{-3} at room temperature. Each silicon atom has 14 electrons in it. Based on this information, what is the ratio of the number of free electrons to the total number of electrons in a cubic centimeter of material? [Hint: You must use the silicon atomic density N_{Si} given in the text.]

[7.2] A sample of n-type silicon has an arsenic doping density of $N_d = 10^{17} cm^{-3}$. Calculate the density of free electrons n_n and free holes p_n in the material.

[7.3] Silicon is doped with boron at a doping density of $N_a = 2 \times 10^{16}$ cm^{-3}. Calculate the density of free holes p_p and free electrons n_p in the material.

[7.4] A sample of p-type silicon has a boron doping density of $N_a = 10^{15}$ cm^{-3}. Calculate the density of free holes p_p and free electrons n_p in the material.

[7.5] A sample of n-type silicon is doped with phosphorus with a donor density of $N_d = 10^{18}$ cm^{-3}. Calculate the density of free electrons n_n and free holes p_n for this case.

[7.6] An n-type sample of silicon has a free electron concentration of $n_n = 4 \times 10^{17}$ cm^{-3}. Find the donor doping density and the hole concentration.

[7.7] The oxide capacitance per unit area C_{ox} of a MOSFET is computed from the equation

$$C_{ox} = \frac{3.45 \times 10^{-13}}{x_{ox}}$$

which gives values in units of F/cm^2. In this expression, x_{ox} is the thickness of the silicon dioxide in the MOS structure and must be in units of centimeters. Compute the value of C_{ox} for each oxide thickness listed below. Note: 1 Å = 1 angstrom, where 1 Å = $10^{-8} cm$.

(a) $x_{ox} = 400$ Å
(b) $x_{ox} = 200$ Å
(c) $x_{ox} = 120$ Å

[7.8] As mentioned in the text, the acronym "MOS" is still used even though the "M" for metal is obsolete. Based on what you have learned in this chapter, can you suggest two other acronyms that might be more appropriate? [Even though it appears the "MOS" is here to stay!]

[7.9] The oxide gate capacitance of a particular CMOS process is $C_{ox} = 1.8 \times 10^{-7}$ F/cm^2. Find the value of C_G for FETs with the following gate dimensions. All values are in units of microns (1 micron = 10^{-4} cm). Place your answer in units of fF.

   (a) $W = 14, L = 1$
   (b) $W = 10, L = 1$
   (c) $W = 2.5, L = 1$
   (d) $W = 4, L = 1.2$
   (e) $W = 4, L = 0.8$

[7.10] Consider an nFET that has $k'_n = 0.0002$ A/V^2 and $V_{Tn} = +0.7$ v. Find the value of I_{n1} if $V_{DD} = 5$ v.

[7.11] Consider an nFET that has $k'_n = 0.0003$ A/V^2 and $V_{Tn} = +0.7$ v. Find the value of I_{n1} if $V_{DD} = 3.3$ v.

[7.12] A pFET has $k'_p = 0.00006$ A/V^2 and $|V_{Tp}| = +0.8$ v. Find the value of I_{p1} if $V_{DD} = 5$ v.

[7.13] A pFET has $k'_p = 0.000047$ A/V^2 and $|V_{Tp}| = +0.82$ v. Find the value of I_{p1} if $V_{DD} = 3.3$ v.

[7.14] An nFET with an aspect ratio of $(W/L) = 3$ allows a maximum current of $I_n = 0.23$ mA. Find the aspect ratio needed to obtain a maximum current of $I_n = 1.2$ mA.

[7.15] An nFET with an aspect ratio of $(W/L) = 10$ gives a maximum current of $I_n = 8.2$ mA. Find the aspect ratio needed to obtain a maximum current of $I_n = 0.7$ mA.

[7.16] A CMOS process creates nFETs that have parasitic parameters $r_n = 1800 \, \Omega$ and $c_n = 1.7$ fF. Find R_n and C_n for transistors with the following gate dimensions; all values are in units of microns.

   (a) $W = 8, L = 1$
   (b) $W = 4, L = 1$
   (c) $W = 2, L = 1$

[7.17] An nFET is made with an aspect ratio of $(W/L) = 1$, which yields a maximum current of $I_n = 0.5$ mA. Find the aspect ratio that increases the maximum current to $I_n = 4.7$ mA. Then calculate R_n and C_n for both transistor sizes, using $r_n = 1600 \, \Omega$ and $c_n = 2.1$ fF and then find the product $R_n C_n$ for the device.

[7.18] A pFET with an aspect ratio of $(W/L) = 6$ allows a maximum current of $I_p = 0.14$ mA. Find the aspect ratio needed to obtain a maximum current of $I_p = 2$ mA.

[7.19] A particular pFET has an aspect ratio of $(W/L) = 10$. The maximum current is $I_p = 0.24$ mA. Find the aspect ratio needed to obtain a maximum current of $I_p = 6$ mA.

[7.20] A CMOS process creates pFETs that have parasitic parameters $r_p = 2200 \, \Omega$ and $c_p = 1.4$ fF. Find R_p and C_p for transistors with the following gate dimensions; all values are in units of microns.

(a) $W = 8, L = 1$

(b) $W = 4, L = 1$

(c) $W = 2, L = 1$

[7.21] A FET is originally designed with gate dimensions of $W = 12$ and $L = 2$. It is then fabricated in a new process that can be described by a scaling factor of $\alpha = 2.3$. Find the new dimensions of the FET and the percent reduction in gate area when the transistor is made in the new process.

[7.22] Construct the FET circuit diagram for the nFET layouts below. Use the numbers to label the nodes in your drawings.

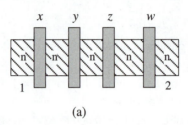

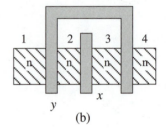

(a)   (b)

[7.23] Construct the FET circuit diagram for the nFET layouts below. Use the numbers to label the nodes in your drawings.

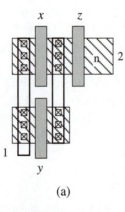

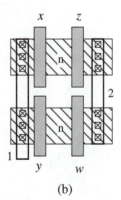

(a)   (b)

[7.24] A CMOS inverter circuit is characterized by $R_n = 800\ \Omega$, $R_p = 1000\ \Omega$, $V_{DD} = 5\ v$. The output capacitance is measured to be $C_{out} = 200\ fF$.

(a) Calculate t_{LH}.

(b) Calculate t_{HL}.

(c) Suppose that the aspect ratios of the FETs were doubled. Calculate the new switching times if the new output capacitance is estimated to be $C_{out} = 240\ fF$. [Hint: Look at what happens to R_n and R_p.]

[7.25] A CMOS inverter circuit is characterized by $R_n = 440\ \Omega$, $R_p = 440\ \Omega$, $V_{DD} = 5\ v$. The output capacitance is measured to be $C_{out} = 167\ fF$.

(a) Calculate t_{LH}.

(b) Calculate t_{HL}.

[7.26] A CMOS inverter circuit is characterized by $R_n = 600 \ \Omega$, $R_p = 590 \ \Omega$, $V_{DD} = 5 \ v$. The output capacitance is measured to be $C_{out} = 200 \ fF$ when the FO is 1; it is known that the load accounts for 62 fF in the value of C_{out}.

(a) Calculate the values of t_{LH} and t_{HL} for the circuit with FO = 1
(b) Calculate the values of t_{LH} and t_{HL} with FO increased to three identical loads

[7.27] An integrated circuit factory initiates the processing of 5000 wafers per week. Each wafer has 82 possible integrated circuits on it. The estimated yield is 41%.

(a) How many functional chips does the factory produce in one week?
(b) What would be the weekly output if the yield were increased to 45%?

[7.28] Consider a chip that has dimensions of 5000 μm × 4000 μm.

(a) Estimate the fabrication yield Y, assuming a defect density of $D = 0.5 \ cm^{-2}$.
(b) What would the projected yield be if the design were shrunk to a size of 3900 μm × 2900 μm?

[7.29] A square chip has side lengths of 6000 μm × 6000 μm.

(a) Estimate the fabrication yield Y, assuming a defect density of $D = 0.6 \ cm^{-2}$.
(b) What would be the percent increase in estimated yield if it were possible to create a better silicon wafer with $D = 0.5 \ cm^{-2}$?

[7.30] A chip has an interconnect line that is 200 microns long and exhibits a signal delay of 0.8 ps. Find the signal delays for the following lengths of the same type of interconnect line.

(a) 150 microns
(b) 350 microns
(c) 700 microns

Note: Layout design of an integrated circuit is the subject of *Physical Design of CMOS Integrated Circuits Using L-Edit* (PWS Publishers, 1995) by this author. This book was written specifically to aid persons in learning layout, and includes the L-Edit layout editor that runs on DOS-based personal computers. The first part of the book is a tutorial on circuit layout that presents the basics with hands-on examples using the supplied L-Edit program. The second part of the book is a condensed version of the L-Edit reference manual. This provides hands-on experience that complements the treatment here.

CHAPTER 8

Logic Components

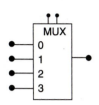

Modern digital networks can be quite complicated, consisting of millions of logic gates. To design a large system, we use the hierarchical approach where the network is broken down into smaller logic components that perform useful functions. The components themselves may be classified as basic elements for system building even though they consist of smaller components or basic logic gates. In this chapter, we will examine several useful logic components.

8.1 Concept of a Digital Component

Up to this point, we have concentrated on studying the basic logic functions NOT, AND, and OR and seeing how they can be combined to create more complex logic networks. In designing a large digital system such as a computer, we may be required to use hundreds or thousands of individual logic gates to create the desired network. Obviously, it becomes very difficult, if not impossible, to keep track of each gate. To overcome this problem, we use the hierarchical approach where a large ("macro") function is defined using a large block that we will call a **digital component**. Alternative names are **element**, **unit**, and **module**; these terms will be taken as being equivalent. They describe a logic network that consists of more than one or two gates and performs some useful function. In general, a digital component has **input** and **output** terminals. Some will also have a separate set of **control bit** lines that are not part of the data flow but are used to control the operation of the component itself.

A generic example of a digital component is shown in Figure 8.1. The inputs are a, b, c, and d, and the output is f; s_1 and s_2 are control bits. If a digital unit has control bits, they will be used to select among various possibilities for the output function f. In the present example, there are two control bits, so that there are $2^2 = 4$ possible values. This can be represented by writing the output as

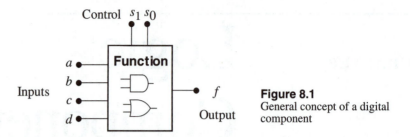

Figure 8.1
General concept of a digital component

$$f = f_0(a, b, c, d) \quad \text{if } s_1 s_0 = 00$$
$$f = f_1(a, b, c, d) \quad \text{if } s_1 s_0 = 01$$
$$f = f_2(a, b, c, d) \quad \text{if } s_1 s_0 = 10 \tag{8.1}$$
$$f = f_3(a, b, c, d) \quad \text{if } s_1 s_0 = 11$$

This shows how the value of the control bit word $s_1 s_0$ is used to select the function performed by the digital network.

The most significant characteristic of our use of a digital component is that only the overall behavior is important. While we know that the interior of the "box" consists of primitive logic gates, we will generally ignore the details. Instead, we will concentrate on how the digital component is useful in building larger networks and systems. This viewpoint is identical to that used to introduce logic gates in Chapter 2, where we defined the operation and symbol, without ever asking what was inside. Adhering to the simple rule "don't worry about the interior of a digital component" will allow you to progress to a very high level in understanding systems design. In the real world, you may use a prebuilt component and never look inside, or you may be the engineer who actually designs the unit from primitive gates. Regardless, it is important to remember the hierarchical viewpoint where we have introduced a digital component as a building block for large systems design.

8.2 An Equality Detector

Suppose that we have two 4-bit words

$$a = a_3 a_2 a_1 a_0$$
$$b = b_3 b_2 b_1 b_0 \tag{8.2}$$

that we wish to compare. Our objective is to design a logic unit that detects if the two words are equal on a bit-by-bit basis. Let us introduce an output that we will call f that will tell us whether the two words are equal or not. This can be accomplished by defining the logic network to behave according to the following description:

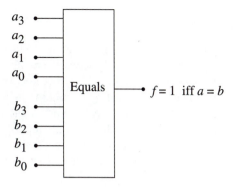

Figure 8.2 Equality detector logic block

If $(a = b)$ then
$$f = 1$$
else
$$f = 0$$

where the notation $a = b$ means that $a_n = b_n$ for $n = 0, 1, 2, 3$. Figure 8.2 shows the unit at this level, where only the inputs and outputs are important. We have assigned the name "Equals" to the block to identify its operation. Our present task will be to design the interior of the unit using basic logic gates.

As a starting point, recall that the exclusive-NOR (XNOR) function is also called the equivalence function since

$$\overline{a_n \oplus b_n} = 1 \quad \text{if} \quad a_n = b_n$$
$$= 0 \quad \text{if} \quad a_n \neq b_n \tag{8.3}$$

The XNOR characteristics are summarized in Figure 8.3 for reference. Since we want to create a circuit that detects if all bits are equal, we may construct the function f by applying the XNOR operation to each pair of bits and then ANDing the result. Explicitly, the function can be written as

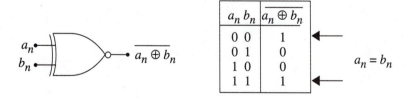

Figure 8.3 The equivalence function

$$f = (\overline{a_3 \oplus b_3}) \cdot (\overline{a_2 \oplus b_2}) \cdot (\overline{a_1 \oplus b_1}) \cdot (\overline{a_0 \oplus b_0}) \qquad \textbf{(8.4)}$$

and evaluates to 1 only if each individual XNOR operation results in a 1. The logic diagram for the Equals network may now be constructed as shown in Figure 8.4 by using a one-to-one correspondence between each term and a logic gate. The operation is easily verified by tracing the logic flow. However, the most important concept is that we have created a new logic component named Equals that has a well-defined relationship between the inputs and the outputs. Once the behavior of the component is defined, we do not need to reference the internal circuits any more. We may use the logic unit as a basic building block in more complex designs. This example shows how we may increase the hierarchical viewpoint from simple logic gates to a more complex logic network.

8.3 BCD Validity Detector

Recall from Chapter 3 that we sometimes represent the base-10 digits 0 through 9 using the binary-coded decimal (BCD) format. In this scheme, decimal numbers are encoded using the 4-bit binary word *ABCD* as shown by the table in Figure 8.5. Note, however, that *ABCD* has $2^4 = 16$ combinations, so that six of the groupings are not used; these are denoted by the "x" entries in the table.

Let us design a logic unit that detects the undefined combinations. The input to the unit will be *ABCD*, and the output will be defined as *F* such that

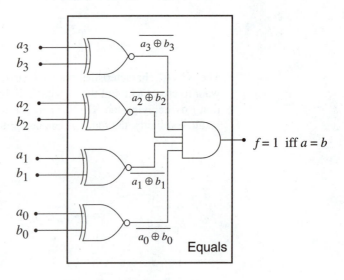

Figure 8.4 Internal circuit for equality detector

ABCD	Decimal
0000	0
0001	1
0010	2
0011	3
0100	4
0101	5
0110	6
0111	7

ABCD	Decimal
1000	8
1001	9
1010	x
1011	x
1100	x
1101	x
1110	x
1111	x

Figure 8.5 BCD-to-decimal conversion table

If ($ABCD$ is valid) then
  $F = 0$
else if ($ABCD$ is not valid) then
  $F = 1$

In this situation, F is used as a **flag** bit. If $F = 0$, then the input is accepted as a defined quantity. However, a condition of $F = 1$ "warns" the system that the input is not a valid BCD combination. In a typical application, we use the value of a flag bit to signal the network that something significant has occurred. In this case, the flag tells us that the input word is undefined. Let us name the unit the *BCD Validation Unit* and use the general symbol shown in Figure 8.6; at this level, the unit is defined entirely by the behavior described above.

The internal logic of the BCD validation unit can be determined by examining the undefined input combinations. These are

$$ABCD = 1010, 1011, 1100, 1101, 1110, \text{ and } 1111$$

so that we want $F = 1$ when any of these occur. A simple approach to finding the logic equation is to use the 4-variable Karnaugh map shown in Figure 8.7. Logic 1 entries are placed at every invalid input. Reducing gives

$$\begin{aligned} F &= A \cdot B + A \cdot C \\ &= A \cdot (B + C) \end{aligned} \tag{8.5}$$

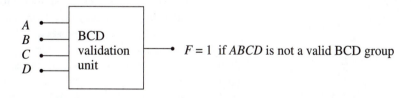

Figure 8.6 Function of the BCD validation unit

CD

AB \ 00 01 11 10

	00	01	11	10
00				
01				
11	1	1	1	1
10			1	1

Figure 8.7
Karnaugh map for undefined
input combinations

as the required logic. This is easily verified by noting the bit patterns in the unused combinations. The interior circuit for this function is shown in Figure 8.8; note that the bit D is not used in the logic. Once again, it is important to note that we may use this component as a primitive building block in a larger system without having any specific knowledge of the internal details. The operation of the component is defined entirely by the generalized function.

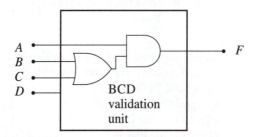

Figure 8.8 Completed design of unit

8.4 Line Decoders

A **line decoder** is a circuit that allows us to "activate" an output line by specifying a control word. An example of a general line decoder is shown in Figure 8.9(a). The 2-bit select word is denoted by S_1S_0, and the output lines are D_0, D_1, D_2, and D_3. Since there are 2 bits in the select word and 4 output lines, this is called a 2-to-4 or 2/4 decoder. In the general case, an X/Y decoder has X select bits and $Y = 2^X$ outputs.

The operation of the 2/4 decoder is summarized in the function table of Figure 8.9(b). Each combination of the select bits S_1S_0 is equivalent to a decimal number n and activates the output line D_n by setting it to a logic 1 level while the remaining lines are held at logic 0 values; this defines an **active-high** decoder. For example, the select word $S_1S_0 = 10$ corresponds to a decimal 2, giving $D_2 = 1$ while $D_0 = 0 = D_1 = D_3$. Naming this network a decoder is appropriate, as it can be viewed as taking the *encoded* binary word S_1S_0 and *decoding* it to its decimal value by activating the proper output.

Now let us turn to the problem of building a 2/4 decoder using basic logic gates. Consider each column of the function table in Figure 8.9(b). For a given output D_n,

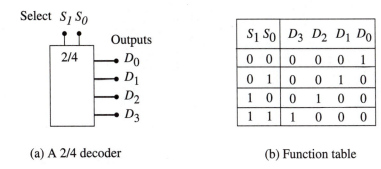

(a) A 2/4 decoder

(b) Function table

Figure 8.9 Operation of a 2/4 active-high line decoder

there is only a single 1 entry in the corresponding column. We may then write an expression for each output line, as in

$$D_0 = \overline{S_1} \cdot \overline{S_0}$$
$$D_1 = \overline{S_1} \cdot S_0$$
$$D_2 = S_1 \cdot \overline{S_0}$$
$$D_3 = S_1 \cdot S_0$$

(8.6)

This forms the basis for creating a 2/4 decoder using the AND-based logic network shown in Figure 8.10. Each output D_i is obtained by ANDing the specified control bits together, resulting in a straightforward implementation of the decoder network.

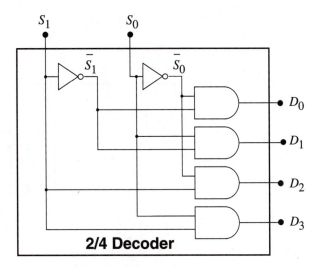

Figure 8.10 Internal circuitry for the 2/4 decoder

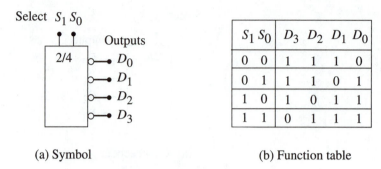

(a) Symbol

(b) Function table

S_1 S_0	D_3	D_2	D_1	D_0
0 0	1	1	1	0
0 1	1	1	0	1
1 0	1	0	1	1
1 1	0	1	1	1

Figure 8.11 Active-low 2/4 row decoder

An **active-low decoder** is exactly opposite to an active-high unit: the selected line is driven to a logic 0 level while the remaining lines are at logic 1 values. The logic symbol for an active-low decoder is shown in Figure 8.11. Proper operation is obtained by simply placing inversion bubbles at the outputs of the active-high network shown in Figure 8.10. The internal logic diagram for this unit is provided in Figure 8.12. Note that adding the bubbles at the output simply means that the output gates must be changed from ANDs into NANDs. Alternatively, we may use a more analytic description of the output as given by the expressions

$$D_0 = \overline{\overline{S_1} \cdot \overline{S_0}}$$

$$D_1 = \overline{\overline{S_1} \cdot S_0}$$

$$D_2 = \overline{S_1 \cdot \overline{S_0}}$$ (8.7)

$$D_3 = \overline{S_1 \cdot S_0}$$

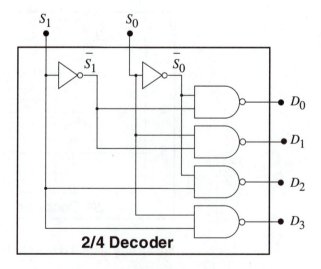

Figure 8.12 Internal circuitry for a 2/4 active-low decoder

These can be reduced to

$$D_0 = S_1 + S_0$$
$$D_1 = S_1 + \overline{S_0}$$
$$D_2 = \overline{S_1} + S_0$$
$$D_3 = \overline{S_1} + \overline{S_0}$$

(8.8)

by using the DeMorgan rules, and they show how we can replace the NAND gates by OR gates if desired. The external behavior remains the same regardless of the internal circuit details.

8.5 Multiplexors

A multiplexor (or MUX for short) is a logic component that has several inputs but only a single output. A MUX provides the capability to direct one of the inputs to the output. This is accomplished by using a control bit word that allows us to select which input will be connected to the output. Figure 8.13 illustrates the operation of a 4-to-1 multiplexor, which will also be referred to as a 4:1 MUX. The input paths are labeled P_0, P_1, P_2, P_3, and the output is denoted by F. The value of the select word $S_1 S_0$ allows us to choose one of the paths and direct it to the output, as summarized in the function table. For example, $S_1 S_0 = 00$ corresponds to decimal 0, so that P_0 is directed to the output; similarly, $S_1 S_0 = 11$ is a decimal 3, indicating that $F = P_3$ for this case. The internal operation may be visualized as using a 4-position switch that is controlled by the select word. Logically, the output may be expressed in SOP form as

$$F = \overline{S_1}\,\overline{S_0}P_0 + \overline{S_1}S_0P_1 + S_1\overline{S_0}P_2 + S_1S_0P_3$$

(8.9)

by applying the techniques from Chapter 4. We may construct the internal logic for the 4-to-1 MUX by translating this SOP expression directly into the logic circuit

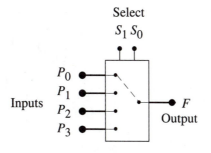

S_1	S_0	F
0	0	P_0
0	1	P_1
1	0	P_2
1	1	P_3

(a) Operation of a 4-to-1 MUX   (b) Function table

Figure 8.13 A 4-to-1 multiplexor unit

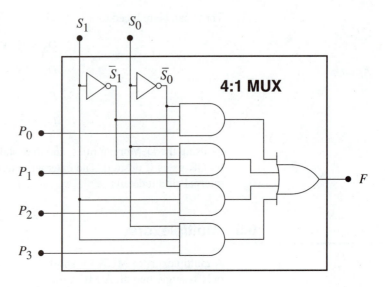

Figure 8.14 Internal circuitry for a 4:1 MUX

shown in Figure 8.14. We see that the AND-OR structure of the SOP expression provides a straightforward form to build the circuit.

The IEEE standard symbol for a 4:1 MUX is shown in Figure 8.15. The select bits S_1 and S_0 are shown with a "G-dependence" $G\frac{0}{3}$, which denotes the AND operation between the input paths P_d and the select bits S_1 and S_0. This means the following: Translate the input word S_1S_0 to its decimal value d, then AND the select word with the input path P_d to determine the output. In other words, this just generates the logic equation for the 4:1 MUX.

Larger multiplexors can be constructed in the same manner. Given n-inputs, we need to specify a select word that has y bits such that $n \neq 2^y$. For example, an 8:1 MUX requires a 3-bit select word of the form $S_2S_1S_0$ to direct one of the inputs P_0, $P_1, \ldots, P_7$ to the output. The equation for the 8:1 would have the form

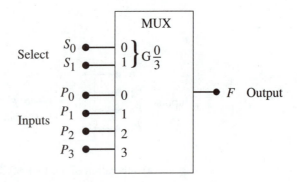

Figure 8.15 IEEE symbol for a 4:1 multiplexor

$$f = \bar{S}_2\bar{S}_1\bar{S}_0P_0 + \bar{S}_2\bar{S}_1S_0P_1 + \bar{S}_2S_1\bar{S}_0P_2 + \bar{S}_2S_1S_0P_3$$
$$+ S_2\bar{S}_1\bar{S}_0P_4 + S_2\bar{S}_1S_0P_5 + S_2S_1\bar{S}_0P_6 + S_2S_1S_0P_7 \tag{8.10}$$

such that the decimal equivalent of the control word specifies the output line. The same formalism can be used to create larger multiplexors, such as a 16:1 or 32:1. These are very useful in systems design. However, from the hardware viewpoint it is usually easier to build MUXs with a large number of inputs by using 2:1 or 4:1 networks as primitive components.

Example 8-1

An 8:1 MUX that is defined by the equation above would have the IEEE symbol shown below. The only difference between this and the 4:1 symbol is the number of inputs and the fact that the select bits $S_2S_1S_0$ provide selection from P_0 to P_7.

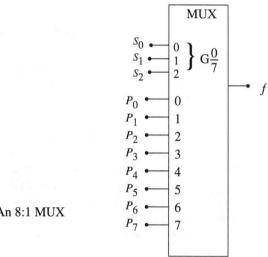

An 8:1 MUX

Example 8-2

Let us illustrate how to construct an 8:1 MUX using two 4:1 devices and one 2:1. The select word $S_2S_1S_0$ can be split into two groups: S_1S_0 to control the 4:1 MUXs and S_2 to select a desired output from the 2:1 MUX. The diagram that follows illustrates a straightforward way to accomplish this.

In this approach, devices MUX_1 and MUX_2 are used to create two output signals X_1 and X_2 defined by

$$X_1 = \bar{S}_1 \cdot \bar{S}_0 \cdot P_0 + \bar{S}_1 \cdot S_0 \cdot P_1 + S_1 \cdot \bar{S}_0 \cdot P_2 + S_1 \cdot S_0 \cdot P_3$$
$$X_2 = \bar{S}_1 \cdot \bar{S}_0 \cdot P_4 + \bar{S}_1 \cdot S_0 \cdot P_5 + S_1 \cdot \bar{S}_0 \cdot P_6 + S_1 \cdot S_0 \cdot P_7$$

Then MUX_3 provides the desired output by means of

$$f = \bar{S}_2 \cdot X_1 + S_2 \cdot X_2$$

Substituting for X_1 and X_2 shows that this produces the desired result.

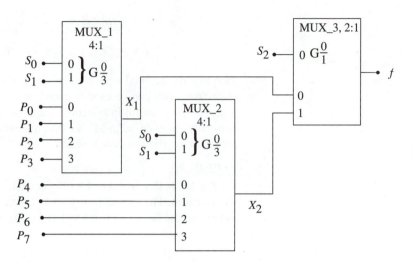

A design for an 8:1 MUX network

8.5.1 Multiplexors as Logic Elements

Consider the output equation of a 4:1 MUX rewritten in the form

$$f(x, y) = \bar{x}\bar{y} \cdot P_0 + \bar{x}y \cdot P_1 + x\bar{y} \cdot P_2 + xy \cdot P_3 \tag{8.11}$$

In arriving at this expression, we have replaced the select bits $S_1 S_0$ by the variables x and y. This expression contains all possible combinations of xy, so we can use it to create a 2-variable SOP equation by choosing the values of P_0 through P_3 to be either 1 (which makes the term contribute to f) or 0 (which eliminates the minterm).

An example is shown in Figure 8.16(a). The output of this circuit is given by

$$\begin{aligned} f(x, y) &= \bar{x}\bar{y} \cdot 0 + \bar{x}y \cdot 1 + x\bar{y} \cdot 1 + xy \cdot 1 \\ &= \bar{x}y + x\bar{y} + xy \end{aligned} \tag{8.12}$$

as indicated by the values of the inputs 0 through 3. Although this may look familiar, we have provided the Karnaugh map reduction in Figure 8.16(b) to verify that this is the same as the simple OR function

$$f = x + y \tag{8.13}$$

This idea is easily extended to other functions. Figure 8.17(a) shows the case where the output is the exclusive-OR operation that is generated by

$$\begin{aligned} f(a, b) &= \bar{a}b \cdot 1 + a\bar{b} \cdot 1 \\ &= a \oplus b \end{aligned} \tag{8.14}$$

Reversing the "1" and "0" values as in Figure 8.17(b) gives

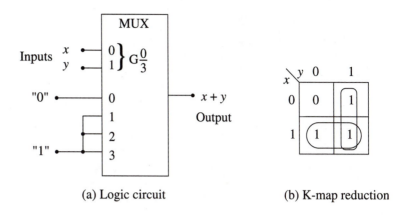

(a) Logic circuit

(b) K-map reduction

Figure 8.16 A 4:1 MUX wired to produce the OR function

$$f(a, b) = ab \cdot 1 + \bar{a}\bar{b} \cdot 1$$
$$= \overline{a \oplus b}$$

(8.15)

which is the equivalence (XNOR) operation.

The concept may be extended to larger multiplexors by following the same principles. Figure 8.18 illustrates an 8:1 MUX that uses input variables A, B, C to generate the output f. Summing the terms that have "1" entries in the paths gives the function as

$$f = \bar{A}\bar{B}C + \bar{A}BC + A\bar{B}\bar{C} + ABC$$

(8.16)

corresponding to selecting paths 1, 3, 4, and 7. Once the SOP form has been obtained, it may be reduced as needed using standard techniques.

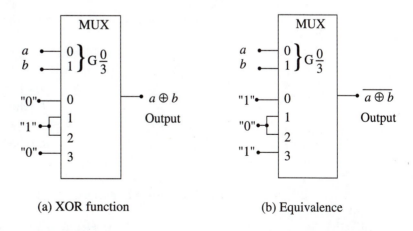

(a) XOR function

(b) Equivalence

Figure 8.17 Exclusive-OR and equivalence functions using multiplexors

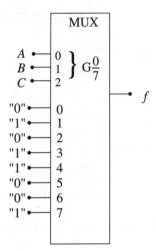

Figure 8.18
An 8:1 multiplexor used to
produce a SOP output

8.5.2 VHDL Description

Consider the 4:1 MUX shown in Figure 8.13. To write a VHDL description of this
module, we first need an entity statement such as

```
entity mux4 is
    port(
        d0, d1, d2,d3:  in bit;
        s: in bit_vector(1 downto 0);
        f: out bit);
    end mux4;
```

This defines the four data lines $d0$, $d1$, $d2$, $d3$, the 2-bit select word s, and the output
line f.

As with most logic units, the operation of the 4:1 MUX can be described in several ways. For example, we may use concurrent statements to construct

```
architecture basic of mux4 is
begin
        f < = d0 when (s = "00") else
        f < = d1 when (s = "01") else
        f < = d2 when (s = "10") else
        f < = d3 when (s = "11") ;
    end basic;
```

An equivalent listing is obtained using a sequential process in the form

```
architecture behavioral of mux4 is
begin
mux4_to_1: process ( d0, d1, d2, d3)
```

```
        begin
            if s = "00" then
            f < = d0;
            elsif s = "01" then
            f < = d1;
            elsif s = "10" then
            f < = d2;
            elsif s = "11" then
            f < = d3;
            endif;
        end process mux4_to_1;
    end behavioral;
```

In general, one may choose the VHDL construct that is most familiar, or a form that interfaces well to the rest of the network listing.

8.6 Demultiplexors

A demultiplexor (DEMUX) is the opposite of a multiplexor. It takes a single input and directs it to one of several outputs, as illustrated schematically for a 1:4 network in Figure 8.19. Choosing a value for the 2-bit select word $S_1 S_0$ uniquely determines where the input variable X is directed. Since there are four outputs, the logical description of this unit requires that we write a separate expression for each output. These are given by

$$P_0 = \overline{S_1} \cdot \overline{S_0} \cdot X$$
$$P_1 = \overline{S_1} \cdot S_0 \cdot X$$
$$P_2 = S_1 \cdot \overline{S_0} \cdot X$$
$$P_3 = S_1 \cdot S_0 \cdot X$$

$$(8.17)$$

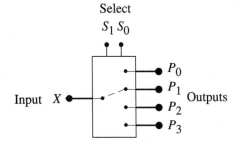

S_1 S_0	X to
0 0	P_0
0 1	P_1
1 0	P_2
1 1	P_3

(a) Operation of a 1-to-4 DEMUX   (b) Function table

Figure 8.19 Operation of a demultiplexor unit

where each line acts independently. As usual, we may construct the logic diagram using a gate-by-gate equivalent. This results in the network shown in Figure 8.20. It is important to note the similarities between the 1:4 DeMUX logic and the 4:1 MUX unit. This occurs because the two are opposites of each other.

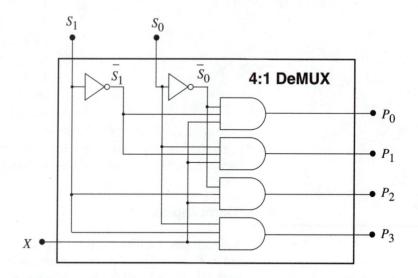

Figure 8.20 Internal logic for a 1:4 demultiplexor network

8.6.1 VHDL Description

The VHDL description of a demultiplexor is straightforward. One approach is to simply list the possibilities, as in

```
entity demux1_4 is
    port(  x : in bit;
            s : in bit;
           p0 : out bit;
           p1 : out bit;
           p2 : out bit;
           p3 : out bit);
end demux1_4;
architecture Operation of demux1_3 is
begin
        if s = "00" then
            p0 < = x;
            p1 < = '0';
            p2 < = '0';
            p3 < = '0';
        elsif s = "01" then
            p0 < = '0';
```

$$p1 <= x;$$
$$p2 <= \text{'0'};$$
$$p3 <= \text{'0'};$$
$$\text{elsif } s = \text{"10" then}$$
$$p0 <= \text{'0'};$$
$$p1 <= \text{'0'};$$
$$p2 <= x;$$
$$p3 <= \text{'0'};$$
$$\text{elsif } s = \text{"11" then}$$
$$p0 <= \text{'0'};$$
$$p1 <= \text{'0'};$$
$$p2 <= \text{'0'};$$
$$p3 <= x;$$
$$\text{end } operation;$$

An alternative listing can be created by using a vector approach or the VHDL coding may be based on the individual logic gates. Only the overall function is important in describing the logic behavior.

8.6.2 Multiplexed Transmission System

Multiplexed transmission systems allow several sources to share one line while still maintaining separate data flows. This is of great practical importance, as it helps reduce **idle time** during which the transmission line is not being used. Moreover, sharing a line reduces the hardware cost of the system.

An example of a 16-line multiplexed transmission system is shown in Figure 8.21. Sixteen separate inputs $P_0, \ldots, P_{15}$ are provided at the input of a 16:1 multiplexor network that acts as a **transmitter** network, i.e., the input of a transmission system. The 4-bit word $s_3 s_2 s_1 s_0$ is used to select which input is directed to the transmission line. The same select word is used at the **receiver** end of the transmission line to direct the demultiplexor input to the proper output channel $D_0, \ldots, D_{15}$. For example, if $s_3 s_2 s_1 s_0 = 0001$ is applied to both ends, then the input P_1 is chosen as the input,

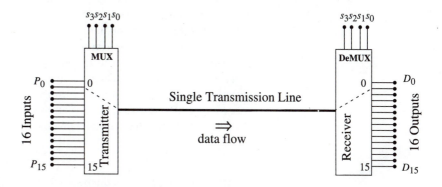

Figure 8.21 A multiplexed transmission system

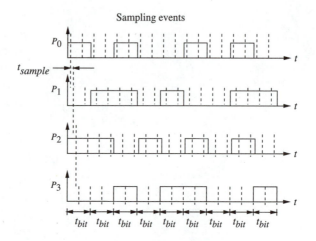

Figure 8.22 Data sampling in a 4-channel system

sent along the transmission line to the receiver, and then directed to output D_1. Incrementing the select word to a value $s_3s_2s_1s_0 = 0010$ causes the system to accept P_2 and direct it to D_2. The interesting aspect of this system is that it is possible to transmit the data simultaneously in all 16 channels using a single transmission line!

Let us examine this concept in more detail by using the example data values shown in Figure 8.22 for four input lines P_0, P_1, P_2, and P_3 that are taken to be the inputs to a 4-line MUX/DeMUX system. These are serial data channels, so that the value (0 or 1) at a particular time t is given by each plot. In this diagram, we have introduced the concept of the **bit time** t_{bit}. This is the duration of the time interval that is used to define one bit. In other words, a logic state lasts a time t_{bit}—no shorter, no longer. After this time passes, the value may change to the next value. For example, the plot for P_0 implies the bit sequence 101001010, while the plot for P_1 means 011010011, and so on. To allow us to transmit the data simultaneously in several channels along a single transmission line requires that we **sample** each line in a cyclic manner. Every dashed vertical line in the drawing indicates a sampling event, with the events spaced by a **sampling time** t_{sample}. For the 4-channel system, the data paths are sampled according the order P_0, P_1, P_2, P_3, P_0, P_1, P_2, and so on. In terms of a MUX, this means that the select word is incremented every t_{sample} interval.

Now note if the sampling time t_{sample} is much smaller than the bit time t_{bit}, then it is possible to sample and transmit the data in every channel without losing any information. This can be seen in the drawing. The samples are taken at a much higher frequency than the data are changing. This technique is called **time-division multiplexing** and is based on the **sampling theorem** in the field of digital signal processing (DSP). It is easy to see that this approach can be applied to control the data flow in large numbers of lines so long as

$$t_{sampling} \ll t_{bit} \tag{8.18}$$

is satisfied. In fact, it forms the basis of many common transmission systems such as telephone networks.

8.7 Binary Adders

Arithmetic functions such as addition and subtraction can be performed using binary numbers. These types of operations are central to building a computer.

The addition of binary numbers is based on the following identities, where "+" is used to denote the add operation:

$$
\begin{aligned}
0 + 0 &= 0 \\
0 + 1 &= 1 \\
1 + 0 &= 1 \\
1 + 1 &= 0 \quad \text{with a carry of 1}
\end{aligned}
\tag{8.19}
$$

Aside from the existence of a carry bit, we see that adding two numbers together has the same result as the exclusive-OR (XOR) logic operation. The **carry bit** that is generated when adding 1 plus 1 can be better understood if we examine the addition of two 2-bits words and their corresponding decimal values. Suppose we want to add the base-10 numbers $1_{10} + 1_{10} = 2_{10}$; the equivalent binary code designation would be

$$
\begin{array}{r}
\text{carry} \; 1 \\
01 \\
+ \; 01 \\
\hline
\text{sum} \;\; 10
\end{array}
\tag{8.20}
$$

Adding the right column gives $1 + 1 = 0$ with a carry of 1. The carry bit is then used in the left column to obtain $1 + 0 + 0 = 1$, yielding the binary sum of $10_2 = 2_{10}$ as required.

8.7.1 The Full-Adder

Let us consider the operation of a **full-adder** network that provides the operations needed to add the bits in an arbitrary column. The full-adder has three input bits: a_n and b_n, which are the data, and c_n, which is the input carry bit from the column immediately to the right. The network produces two output bits: the sum s_n and carry-out c_{n+1}. The function table shown in Figure 8.23 is derived using the basic rules of addition listed above. It is worthwhile to work through each line of the table to insure that you understand the operation of the network.

There are two expressions that may be derived from the function table, one for the sum s_n and one for the carry-out c_n. Let us start with obtaining the expression for the sum bit. There are four "1" entries in the s_n column (the second, third, fifth, and eighth rows). In standard SOP form, this gives directly that

$$
s_n = \overline{a_n} b_n \overline{c_n} + a_n \overline{b_n}\,\overline{c_n} + \overline{a_n}\,\overline{b_n} c_n + a_n b_n c_n
\tag{8.21}
$$

This can be reduced, using Boolean algebra, as follows:

$$
\begin{aligned}
s_n &= (\overline{a_n} b_n + a_n \overline{b_n})\overline{c_n} + (\overline{a_n}\,\overline{b_n} + a_n b_n\;)c_n \\
&= (a_n \oplus b_n)\overline{c_n} + (\overline{a_n \oplus b_n})c_n \\
&= a_n \oplus b_n \oplus c_n
\end{aligned}
\tag{8.22}
$$

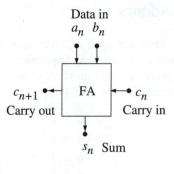

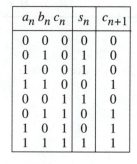

a_n b_n c_n	s_n	c_{n+1}
0 0 0	0	0
0 1 0	1	0
1 0 0	1	0
1 1 0	0	1
0 0 1	1	0
0 1 1	0	1
1 0 1	0	1
1 1 1	1	1

(a) Block diagram   (b) Function table

Figure 8.23 Full-adder characteristics

where we have made use of the XOR and XNOR equations

$$x \oplus y = \bar{x}y + x\bar{y}$$
$$\overline{x \oplus y} = xy + \bar{x}\bar{y}$$

(8.23)

The double XOR function ($a_n \oplus b_n \oplus c_n$) is also known as the **odd function**, since it evaluates to a logic 1 value if there are an odd number of 1s at the input and has a logic 0 result if there are an even number of 1s at the input.

The carry bit can be derived using the same technique. From the function table, we see that there are again four rows that yield a "1" output. Thus, we may construct the carry-out as

$$c_{n+1} = a_n b_n \bar{c}_n + \bar{a}_n b_n c_n + a_n \bar{b}_n c_n + a_n b_n c_n$$
$$= a_n b_n (\bar{c}_n + c_n) + c_n (\bar{a}_n b_n + a_n \bar{b}_n)$$
$$= a_n b_n + c_n (a_n \oplus b_n)$$

(8.24)

which gives a simple form for c_{n+1}. We note that these results could have also been obtained using the technique of Karnaugh maps.

8.7.2 Half-Adders

A special case of the full-adder is when $c_n = 0$. This results in the **half-adder** unit shown in Figure 8.24. The half-adder has only two inputs, a_n and b_n, and gives the sum s_n and carry-out c_{n+1} bits as computed from the simplified expressions

$$s_n = a_n \oplus b_n$$
$$c_{n+1} = a_n \cdot b_n$$

(8.25)

These can be verified directly using the truth table.

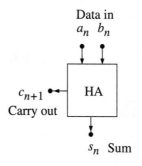

Data in
a_n b_n

c_{n+1} HA
Carry out

s_n Sum

a_n b_n	s_n	c_{n+1}
0 0	0	0
0 1	1	0
1 0	1	0
1 1	0	1

(a) Block diagram

(b) Function table

Figure 8.24 A half-adder circuit

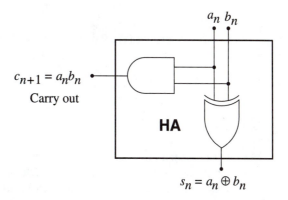

a_n b_n

$c_{n+1} = a_n b_n$
Carry out

HA

$s_n = a_n \oplus b_n$

Figure 8.25 Internal logic for a half-adder

8.7.3 Adder Circuits

Now that we have the logic expressions for the adder functions, we may construct the circuit by providing gates. Figure 8.25 shows the logic diagram for a half-adder using the XOR as a primitive logic element to generate the sum bit. This circuit is easily expanded to the full-adder network shown in Figure 8.26 by adding the carry-in bit and providing the extra logic gates.

Efficient digital logic design often requires that we go beyond the simple and obvious to improve the performance of the network. While the adder networks above are valid designs, it is possible to derive alternative expressions that are entirely equivalent to the straightforward equations but structure the logic in a different manner. Learning how to manipulate logic forms is a very powerful tool for the designer.

Consider the FA carry-out bit

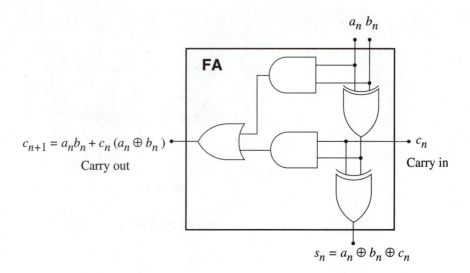

$$c_{n+1} = a_n b_n + c_n (a_n \oplus b_n)$$
Carry out

c_n
Carry in

$$s_n = a_n \oplus b_n \oplus c_n$$

Figure 8.26 Internal logic for a full-adder

$$
\begin{aligned}
c_{n+1} &= a_n b_n + c_n (a_n \oplus b_n) \\
&= a_n b_n + c_n (a_n + b_n) \\
&= a_n b_n + a_n c_n + b_n c_n
\end{aligned}
\tag{8.26}
$$

The second line is obtained by noting that the presence of the term $a_n b_n$ allows us to replace $(a_n \oplus b_n)$ with the simple OR function $(a_n + b_n)$. Now let us examine the sum bit s_n when c_{n+1} is used as an **input** bit, as in the function table provided in Figure 8.27. The SOP form is given by listing the $s_n = 1$ entries, which gives

$$
s_n = (\bar{a}_n b_n \bar{c}_n + a_n \bar{b}_n \bar{c}_n + \bar{a}_n \bar{b}_n c_n) \, \overline{c_{n+1}} + a_n b_n c_n \, c_{n+1}
\tag{8.27}
$$

a_n b_n c_n	c_{n+1}	s_n
0 0 0	0	0
0 1 0	0	1 ←
1 0 0	0	1 ←
1 1 0	1	0
0 0 1	0	1 ←
0 1 1	1	0
1 0 1	1	0
1 1 1	1	1 ←

Figure 8.27
Function table for reworking
the sum expression

This can be reduced by noting that the function table shows that when $c_{n+1} = 0$, the $s_n = 1$ if $(a_n + b_n + c_n) = 1$. If $c_{n+1} = 1$, then the only case for $s_n = 1$ occurs if $(a_n \cdot b_n \cdot c_n) = 1$. Thus, we may write

$$
s_n = a_n b_n c_n + \overline{c_{n+1}}(a_n + b_n + c_n)
\tag{8.28}
$$

as an alternative expression for the sum bit. The advantage to this algorithm is the structure of the logic equations: both c_{n+1} and s_n can be expressed in AND-OR-Invert form, as illustrated by the logic diagram in Figure 8.28.

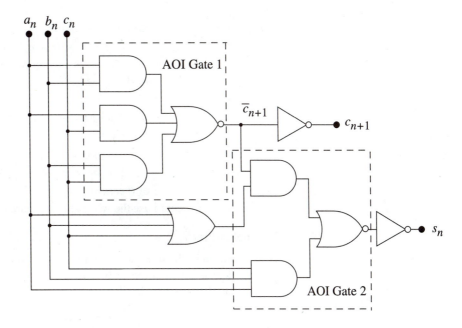

Figure 8.28 AOI full-adder logic network

8.7.4 VHDL Descriptions

It is a straightforward task to construct VHDL descriptions for the adder networks presented above.

Consider first the half-adder shown previously in Figure 8.25. The logic diagram may be used to write concurrent statements for the component description. This results in the form

```
entity half_adder is
        port(a_n, b_n:  in bit;
                s_n, c_o: out bit);
end half_adder;
```

```
architecture concurrent of half_adder is
begin
        s_n < = a_n xor b_n ;
        c_o< = a_n and b_n ;
end concurrent ;
```

Other techniques, such as dataflow, can be used, but concurrent statements yield a nice compact form.

A full adder can also be described using concurrent statements by applying the logic shown in Figure 8.26. In this case, we write

entity *full_adder* is
  port(*a_n, b_n, c_n* : in *bit*;
     s_n, c_o: out *bit*);
end *full_adder*;

architecture *concurrent* of *full_adder* is
begin
  s_n < = *a_n* xor *b_n* xor *c_n*;
  c_o < = (*a_n* and *b_n*) xor (*c_n* and (*a_n* xor *b_n*));
end *concurrent* ;

Alternatively, one could use the AOI logic in Figure 8.28 as a starting point.

8.7.5 Parallel Adders

Half-adders and full-adders are used to add individual bits together. An extension of this problem is the addition of two *n*-bit binary words. For example, suppose we want to add the two 4-bit words

$$
\begin{aligned}
A &= a_3a_2a_1a_0 \\
B &= b_3b_2b_1b_0
\end{aligned}
\tag{8.29}
$$

A **parallel adder** unit produces the sum by allowing us to input both words at the same time. Let us break down the problem explicitly by writing the addition procedure out as

$$
\begin{array}{r}
c_4 \quad c_3c_2c_1 \qquad \text{Carry bits} \\
a_3a_2a_1a_0 \\
+ \quad b_3b_2b_1b_0 \\
\hline
c_o \quad s_3s_2s_1s_0
\end{array}
\tag{8.30}
$$

This shows that we can use a single-bit adder for each column and connect the carry-out bit to the adjacent column to the left. We also note that the sum may be larger than 4 bits, so an overflow bit c_o is provided such that $c_o = 0$ means that 4 bits are sufficient to express the sum, while $c_o = 1$ implies that the sum requires more than 4 bits.

Example 8-3

To add 0101 to 0111 requires the use of the carry bits as shown:

$$
\begin{array}{r}
1\ 1\ 1 \qquad \text{Carry bits}\\
0\ 1\ 0\ 1\\
+\ \ \ \ 0\ 1\ 1\ 1\\
\hline
0\ \ \ 1\ 1\ 0\ 0
\end{array}
\tag{8.31}
$$

In this case, the sum 1100 is only 4 bits wide.

Example 8-4

Next, let's examine the sum of 1101 and 1110:

$$
\begin{array}{r}
1\ 0\ 0 \qquad \text{Carry bits}\\
1\ 1\ 0\ 1\\
+\ \ \ \ 1\ 1\ 1\ 0\\
\hline
1\ \ \ 1\ 0\ 1\ 1
\end{array}
\tag{8.32}
$$

Now the result requires 5 bits. This is because the answer (decimal 27) is larger than the maximum value (15) that can be represented by 4 bits.

Figure 8.29 shows a 4-bit parallel adder circuit that uses a **ripple-carry** scheme to transfer the carry bits. The ripple carry is characterized by having the carry-out from the nth bit used directly as the carry-in for the $(n + 1)$st bit as in the addition algorithm. Due to the need to have the carry "ripple" from the right to the left, the output bits become valid according to the time line shown in Figure 8.30(a), which assumes that the inputs are made at the time $t = 0$ and that in each full-adder circuit, the carry-out bit takes longer to compute than the sum bit. In this case, s_0 is the first bit to become valid. The next bit s_1 requires that the carry-in c_1 is valid, so that it is delayed. Similarly, s_2 is not valid until c_2 attains its final value, and so on. This clearly shows the "ripple" of the carry bits down the chain. The output of the 4-bit adder is not valid until the time t_{4b}. In Figure 8.30(b), we assume that the full-adders are constructed so that the full-adders produce the carry-out bit before the sum bit.

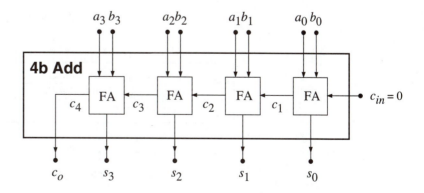

Figure 8.29 4-bit ripple-carry adder

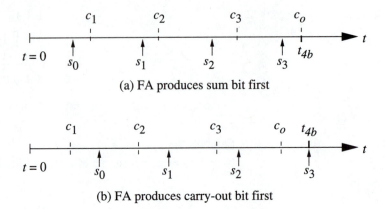

(a) FA produces sum bit first

(b) FA produces carry-out bit first

Figure 8.30 Time line for valid outputs

Note that we have designed the unit such that the 0-bit uses a full-adder (FA) circuit with $c_{in} = 0$ to add $(a_0 + b_0)$. Although a half-adder could have been used instead, the FA makes each column identical; in addition, it will be shown later that choosing a full-adder allows us to expand the functionality of the unit.

Next consider an 8-bit adder. At this point in our discussion, there are two obvious circuits that can be used. If we choose the full-adder (FA) as the primitive, then we can parallel 8 FAs to arrive at the scheme shown in Figure 8.31. This is just an extension of our earlier approach. An alternative technique involves a change in the hierarchy. Instead of viewing a single-bit full-adder as the primitive unit, suppose we instead use two 4-bit units as shown in Figure 8.32. This is entirely equivalent but uses more complex elements to build on. Of course, the two are identical in operation.

The IEEE symbol for a 4-bit adder is shown in Figure 8.33. In this convention, the inputs are denoted by P and Q, and the output sum bits are labeled as Σ; the presence of the Greek letter sigma Σ near the top of the box indicates that it is an adder unit. The carry-in and carry-out bits are shown as CI and CO, respectively. Although this example is for a 4-bit adder, it can easily be extended to an arbitrary

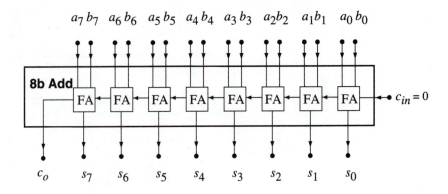

Figure 8.31 An 8-bit ripple-carry adder

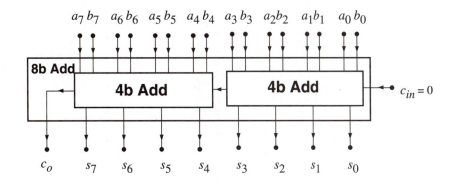

$$a_7\,b_7 \quad a_6\,b_6 \quad a_5\,b_5 \quad a_4\,b_4 \quad a_3\,b_3 \quad a_2b_2 \quad a_1b_1 \quad a_0\,b_0$$

Figure 8.32 An 8-bit adder built from two 4-bit adders

word length. It is important to note that none of the 4-bit adder symbols provide any explicit information about the internal logic or circuitry. Instead, only the overall function is important at this level of hierarchy. This allows us to concentrate on using the adder unit at the system level where we need to add n-bit words, without having to worry about the internal workings.

8.7.6 CMOS Adder Circuits

A direct approach to building a CMOS full-adder circuit is to use the basic AOI algorithms

$$s_n = a_n b_n c_n + \overline{c_{n+1}}(a_n + b_n + c_n)$$

$$c_{n+1} = a_n b_n + c_n(a_n + b_n) \tag{8.33}$$

This provides direct implementation because CMOS series-parallel gates provide a one-to-one correspondence to basic AOI-type logic gates as shown in Figure 8.34. To create the full-adder circuit, we simply use the techniques of logic formation presented in Chapter 6 to create two separate circuits. These are shown in Figure 8.35. The carry-out circuit in Figure 8.35(a) uses the inputs to produce $\overline{C_{n+1}}$ as an output. This is then used as one of the inputs to the sum gate circuit shown in Figure

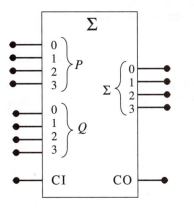

Figure 8.33
IEEE symbol for a 4-bit adder

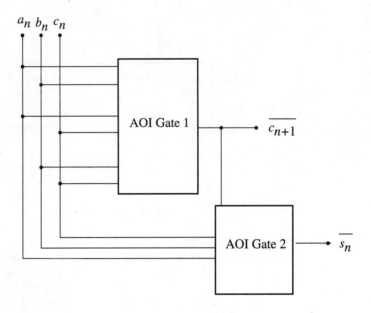

Figure 8.34 AOI logic diagram used to design a CMOS circuit

8.35(b); this produces $\overline{s_n}$. Inverters at the outputs of both gates need to be added to obtain s_n and c_{n+1}.

8.8 Subtraction

Subtraction is more complicated than addition due to the need to introduce the concept of a "borrow" into the binary system. When we subtract a smaller number from a larger number, the basic rules are

$$
\begin{aligned}
0 - 0 &= 0 \\
1 - 0 &= 1 \\
1 - 1 &= 0
\end{aligned}
\tag{8.34}
$$

However, we also need a rule for the remaining case $(0 - 1)$. This is usually written as

$$
0 - 1 = 1 \quad \text{(using a borrow of 1)} \tag{8.35}
$$

To explain this, suppose we want to calculate the base-10 difference of $2 - 1 = 1$. In base 2, this calculation would be written as

$$
\begin{array}{r}
10 \\
-01 \\
\hline
01
\end{array}
\tag{8.36}
$$

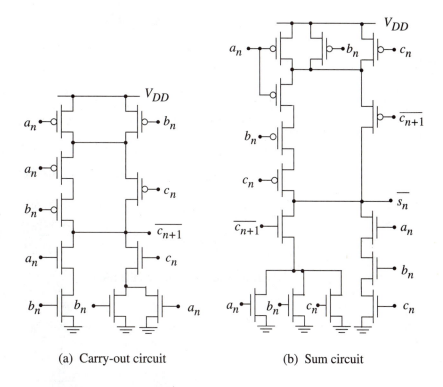

(a) Carry-out circuit   (b) Sum circuit

Figure 8.35 CMOS AOI full-adder circuit

so that the right column is indeed equivalent to writing $0 - 1 = 1$ if we borrow a 1 from the left column. This is also consistent with binary addition, since

$$
\begin{array}{r}
01 \\
+ \ 01 \\
\hline
10
\end{array}
\tag{8.37}
$$

using a carry from the right column to the left. Although it is possible to develop binary subtraction entirely from these rules, it can lead to errors in hand calculations (those that you do on paper) and is also cumbersome to implement with logic elements. It is therefore preferable to find an alternative approach to subtraction.

Let us start with the basic idea of subtraction. If we take two numbers X and Y, then computing the difference D from

$$X - Y = D \tag{8.38}$$

is equivalent to saying that

$$X + (-Y) = D \tag{8.39}$$

Consider the number $(-Y)$. This is defined such that

$$Y + (-Y) = 0 \tag{8.40}$$

Let us write $Y = y_3y_2y_1y_0$ and $(-Y) = w_3w_2w_1w_0$ and find the values of w_n such that

$$\begin{array}{c} y_3 \ y_2 \ y_1 \ y_0 \\ + \ \underline{w_3w_2w_1w_0} \\ 0 \ \ 0 \ \ 0 \ \ 0 \end{array} \qquad (8.41)$$

If we choose $w_n = \bar{y}_n$, that is,

$$w_3 = \bar{y}_3 \qquad w_2 = \bar{y}_2 \qquad w_1 = \bar{y}_1 \qquad w_0 = \bar{y}_0 \qquad (8.42)$$

then the sum in every column is automatically $(1 + 0) = 1$, so that

$$\begin{array}{c} y_3y_2y_1y_0 \\ + \ \underline{\bar{y}_3\bar{y}_2\bar{y}_1\bar{y}_0} \\ 1 \ \ 1 \ \ 1 \ \ 1 \end{array} \qquad (8.43)$$

If we add 1 to this, then the result is

$$\begin{array}{c} y_3y_2y_1y_0 \\ + \ \underline{\bar{y}_3\bar{y}_2\bar{y}_1\bar{y}_0} \\ 1 \ \ 1 \ \ 1 \ \ 1 \\ + \ \ \ \ \ \ \underline{1} \\ 1 \ \ 0 \ \ 0 \ \ 0 \ \ 0 \end{array} \qquad (8.44)$$

which has the right four bits summing to 0000 with an overflow of 1 in the fifth column. Formally, complementing each bit by $\bar{Y} = \bar{y}_3\bar{y}_2\bar{y}_1\bar{y}_0$ is referred to as taking the **1s complement** of Y; this is also referred to (in slang terms) as "flipping the bits" for an obvious reason. The **2s complement** of Y is obtained by adding 1 to the 1s complement and can be expressed as

$$\text{2s complement } (Y) = \bar{Y} + 1 \qquad (8.45)$$

Then,

$$Y + \text{2s complement } (Y) = 0 \qquad (8.46)$$

if we discard the overflow 1 in the additional column.

We are now in a position to develop an algorithm for subtraction. In order to calculate $(X - Y)$, we write

$$(X - Y) = X + \text{2s complement } (Y) \qquad (8.47)$$

If we use n-bit words, then we discard the 1 that appears in the $(n + 1)$st bit position.

Example 8-5

Suppose that $X = 1101$ and $Y = 0101$. To find the difference $(X - Y)$, we first take the 1s complement of Y by flipping the bits:

$$\text{1s complement } (Y) = 1010 \qquad (8.48)$$

Then

$$\text{2s complement } (Y) = 1010 + 1 \tag{8.49}$$
$$= 1011$$

gives the 2s complement. The difference is found by adding X to the 2s complement of Y, which gives

$$
\begin{array}{r}
1101 \\
+ \ \underline{1011} \\
1 \ \ 1000
\end{array} \tag{8.50}
$$

so discarding the 1 in the fifth column gives

$$(X - Y) = 1000 \tag{8.51}$$

As a check, we note that in decimal calculation, $X = 13$ and $Y = 5$ so $(X - Y) = 8$, which agrees with the binary calculation.

A Deeper Look at 2s Complement Subtraction

Formal justification of the mechanics of this procedure starts with a general definition of the 2s complement. Consider an n-bit binary number Y. The 2s complement of Y is defined as

$$\text{2s complement } (Y) = 2^n - Y \tag{8.52}$$

for $Y \neq 0$; in the special case where $Y = 0$, we have

$$\text{2s complement } (0) = 0 \tag{8.53}$$

by definition. Both sides of these equations are in base 2, and it is important to note that the value of the 2s complement must be calculated for a particular value of n. The definition can be rewritten by adding $0 = 1 - 1$ to the right side and regrouping to give

$$\text{2s complement } (Y) = (2^n - 1) - Y + 1 \tag{8.54}$$

The factor $(2^n - 1)$ has an interesting characteristic: it is an n-bit binary word that has all 1s. Let us check this statement for $n = 4$. In decimal, $(2^n - 1) = 16 - 1 = 15$, so in base 2,

$$(2^4 - 1) = 1111 \tag{8.55}$$

Similarly, if $n = 8$, then

$$(2^8 - 1) = 11111111 \tag{8.56}$$

and so on. When we subtract the n-bit binary word Y from $(2^n - 1)$, every column is a subtraction of either $(1 - 0) = 1$ if the Y bit is 0, or $(1 - 1) = 0$ if the Y bit is 1. This is the process of flipping the bits described above. Formally, this is called the 1s complement such that

$$1\text{s complement } (Y) = (2^n - 1) - Y \tag{8.57}$$

so that

$$2\text{s complement } (Y) = 1\text{s complement } (Y) + 1 \tag{8.58}$$

When we want to perform the subtraction $(X - Y)$, we note that

$$X + 2\text{s complement } (Y) = X + 2^n - Y$$
$$= 2^n + (X - Y) \tag{8.59}$$

Thus, since 2^n is an $(n + 1)$ bit binary word that has leading 1 as the most significant bit followed by n 0s, the difference $(X - Y)$ can be extracted by adding X plus the 2s complement of Y and discarding the overflow bit. This is, of course, the algorithm

$$(X - Y) = X + 2\text{s complement } (Y) \tag{8.60}$$

that we presented earlier. This shows that the technique can be justified by general theory.

Example 8-6

Let us illustrate the importance of the word size in this algorithm. Consider the decimal number $A = 9$. If we use a 4-bit representation, then

$$A = 1001 \tag{8.61}$$

and

$$2\text{s complement } (A) = 0110 + 1$$
$$= 0111 \tag{8.62}$$

Suppose that we let $B = 15$ and want to calculate $(B - A)$ using the 2s complement approach. Translating to binary by writing $B = 1111$ gives

$$(B - A) = 1111 + 0111$$
$$= 1\ 0110 \tag{8.63}$$

so that discarding the leading 1 yields the correct result of

$$(B - A) = 0110$$
$$= 6_{10} \tag{8.64}$$

Suppose instead that we used an 8-bit word for B. Applying the algorithm results in

$$(B - A) = 00001111 + 0111$$
$$= 00010110 \tag{8.65}$$
$$= 22_{10}$$

which is clearly incorrect! Where did we go wrong? We used the incorrect 4-bit representation for the 2s complement of A. In 8 bits, the value is

$$2\text{s complement } (A) = 00000110 + 1$$
$$= 11110111 \tag{8.66}$$

so that now

$$(B - A) = 00001111 + 11110111$$
$$= 1\ 00000110 \tag{8.67}$$
$$= 6_{10}$$

gives the correct result. This example shows that the 2s complement changes with the word size and that it is very important to maintain consistency in the calculations.

8.8.1 Subtractor Logic Circuits

To perform subtraction using the 2s complement approach, we need to construct a logic unit that accepts Y as an input and (i) calculates the 1s complement Y, and then (ii) adds 1. This operation must be performed on the entire n-bit word. Figure 8.36 shows the block diagram for generating the 2s complement of an 8-bit word $y_7y_6y_5y_4y_3y_2y_1y_0$. The 1s complement is obtained by simply inserting an inverter for each bit, while an adder is used to give the final result, which is denoted as

$$h_7h_6h_5h_4h_3h_2h_1h_0$$

This can then be used as the input word to a parallel adder.

In practical applications, both addition and subtraction are commonly required as complementary functions. A single adder/subtractor unit can be built using the ripple-carry adder discussed in the previous section. This new unit is shown in Figure 8.37. Inspecting the logic diagram shows that two basic modifications have been made to the original 8b adder. First, each b_n input line now has an XOR gate in its path before it enters the adder unit. Second, a new control bit SUB has been added to the circuit. This is connected to the input of each XOR and also to the

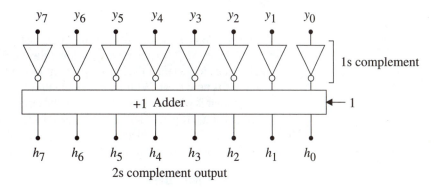

Figure 8.36 2s complement circuit

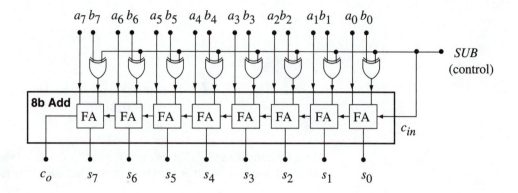

Figure 8.37 8-bit adder/subtractor unit

carry-in bit of the adder. As we will see, this circuit acts as an adder when $SUB = 0$ but changes to a 2s complement subtractor when $SUB = 1$.

To understand the modification of the b-input bits, recall the XOR logic gate and truth table shown in Figure 8.38. The inputs have been denoted as b_n and SUB.

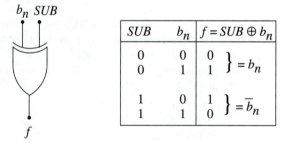

SUB	b_n	$f = SUB \oplus b_n$
0	0	0
0	1	1 $\Big\} = b_n$
1	0	1
1	1	0 $\Big\} = \overline{b}_n$

Figure 8.38 The XOR function in the adder/subtractor

When $SUB = 0$, the output is $f = b_n$, so b_n just passes through to the gate. If, on the other hand, $SUB = 1$, then $f = \overline{b}_n$. Adding the XOR gate and the SUB control bit thus allows us to generate the 1s complement of the 8-bit input B. The operation of the circuit is now clear. With $SUB = 0$, the unit functions as a normal 8-bit adder, and the output is $(A + B)$. If $SUB = 1$, then the circuit calculates the 2s complement of B using the XOR gates, and adds 1 to the results via the carry-in input. This is entirely equivalent to the subtraction algorithm, so that the output is the difference $(A - B)$. A simplified symbol for the 8-bit adder/subtractor unit is shown in Figure 8.39.

8.8.2 Negative Numbers

Up to this point, we have implicitly assumed that the difference $(A - B)$ of two numbers is positive or zero, i.e., that $A \geq B$. If $A < B$, then the result will be negative.

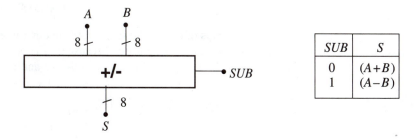

Figure 8.39 General adder/subtractor symbol

This possibility brings up the problem of how to represent a negative number in the binary numbering system.

Let us choose an 8-bit word $A = a_7a_6a_5a_4a_3a_2a_1a_0$ that has $2^8 = 256$ different combinations. If we choose to represent only positive integers, then the conversion is given by the table in Figure 8.40, representing the digits 0 through 255. If instead we want to use the 8-bit word to represent negative numbers also, then we must determine the range of values, since A has only 256 different combinations.

A simple and useful way to approach this problem is to replace the leftmost bit a_7 by a **sign bit** s and write

$$A = s \; a_6a_5a_4a_3a_2a_1a_0$$

This is interpreted as a **7-bit** word consisting of $a_6a_5a_4a_3a_2a_1a_0$ (that has $2^7 = 128$ combinations) multiplied by +1 or −1 as determined by the value of s:

$$s = 0: \text{ +1 multiplier}$$

$$s = 1: \text{ −1 multiplier}$$

This leads to the association shown in Figure 8.41, where the positive range is from 0 to +127, while the negative numbers are 0 to −127. One problem with this scheme is that we are forced to associate

Binary	Decimal
0000 0000	0
0000 0001	1
0000 0010	2
0000 0011	3
0000 0100	4
0000 0101	5
⋮	⋮
1111 1011	251
1111 1100	252
1111 1101	253
1111 1110	254
1111 1111	255

Figure 8.40
Binary-to-decimal conversion for positive integers

$$1000\ 0000 = -0$$

which is of course the same as +0. In other words, we have two different representations for 0. This arises because 256 is an even number, so that $(256/2) = 128$ implies an even split between positive and negative numbers. This is not really a

Binary	Decimal
0000 0000	0
0000 0001	1
0000 0010	2
0000 0011	3
$\vdots$	$\vdots$
0111 1110	126
0111 1111	127
1111 1111	−127
1111 1110	−126
1111 1101	−125
1111 1100	−124
$\vdots$	$\vdots$
1000 0001	−1

Figure 8.41
Sign bit convention for negative numbers

problem so long as one remembers the conversion.

An alternative approach for representing negative numbers is to use the 2s complement approach where we write the 8-bit word

$$A = s\ a_6 a_5 a_4 a_3 a_2 a_1 a_0$$

but now use the sign bit s as a negative weighted multiplier in the base-10 conversion. This means that the decimal value is

$$s \times (-2^7) + a_6 \times 2^6 + a_5 \times 2^5 + a_4 \times 2^4 + a_3 \times 2^3$$
$$+ a_2 \times 2^2 + a_1 \times 2^1 + a_0 \times 2^0$$

(8.68)

so that $s = 0$ has no effect and yields a positive number, while $s = 1$ is equivalent to adding a value of −128 to the value of the 7-bit word $a_6 a_5 a_4 a_3 a_2 a_1 a_0$, as is shown by the weighting table in Figure 8.42. This results in the translation table of Figure 8.43, which illustrates that this approach only has a single representation for zero, but the transition from 0 to −1 looks odd at first. Also note that the range of positive integers is 0 to +127, while the range for negative numbers is −1 to −128. This

Bit	s	a_6	a_5	a_4	a_3	a_2	a_1	a_0
Decimal weight	−128	64	32	16	8	4	2	1

Figure 8.42 Position weighting with sign bit convention

Binary	Decimal
0000 0000	0
0000 0001	1
0000 0010	2
0000 0011	3
⋮	⋮
0111 1110	126
0111 1111	127
1000 0000	−128
1000 0001	−127
1000 0010	−126
1000 0011	−125
⋮	⋮
1111 1111	−1

Figure 8.43
2s complement representation
for negative numbers

asymmetry is to be expected and does not cause any great problem. Moreover, since we have adopted the 2s complement approach, it is directly compatible with the adder/subtractor network.

Example 8-7

Consider the 2s complement representation

$$N = 10010111 \tag{8.69}$$

To find the base-10 value of N, we use the weight table to write

$$N = -128 + 16 + 4 + 2 + 1$$
$$= -105 \tag{8.70}$$

in this scheme.

8.9 Multiplication

Binary multiplication is defined by the following rules:

$$0 \times 0 = 0$$
$$0 \times 1 = 0$$
$$1 \times 0 = 0 \tag{8.71}$$
$$1 \times 1 = 1$$

A moment's reflection shows that this is the same as the AND operation, so that the AND gate shown in Figure 8.44 provides a bit multiplier component that we denote by the times symbol "×" to remember its usage.

A more practical problem is multiplying two n-bit words together. Let us look at the problem of creating a 2b × 2b multiplier that calculates the product of $a_1 a_0 \times$

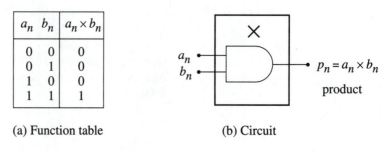

a_n	b_n	$a_n \times b_n$
0	0	0
0	1	0
1	0	0
1	1	1

(a) Function table   (b) Circuit

Figure 8.44 Binary multiplication

$b_1 b_0$. The algorithm is shown in Figure 8.45. We first multiply $a_1 a_0$ by b_0, and then multiply $a_1 a_0$ by b_1; the product word $p_3 p_2 p_1 p_0$ is 4 bits wide with

$$p_0 = a_0 \cdot b_0$$
$$p_1 = a_0 \cdot b_1 + a_1 \cdot b_0$$
$$p_2 = a_1 \cdot b_1 + c_2 \tag{8.72}$$
$$p_3 = c_3$$

and "+" meaning binary addition. In these equations, the carry bit from the right-adjacent column has been written explicitly for completeness. It is important to note that the product of two 2b words results in a 4b product. A simple logic network for the 2×2 multiplier is shown in Figure 8.46. This uses bit multipliers (components with "$\times$" in them) to perform the multiplication and half-adders (HA) to add the terms together. It is easily seen that the circuit is constructed in a manner that directly emulates the structure of the longhand algebra work shown in Figure 8.45.

We may easily extend the algorithm to calculate the product of two 4-bit words by $a_3 a_2 a_1 a_0 \times b_3 b_2 b_1 b_0$. Multiplying $a_3 a_2 a_1 a_0$ by each bit b_n for $n = 0$ to 3 and aligning each product according to its position n allows us to add the individual

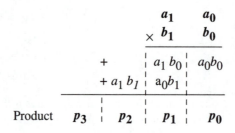

Figure 8.45 Multiplication of two 2b words

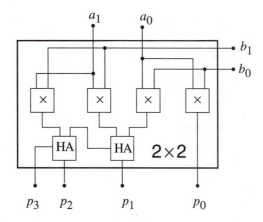

Figure 8.46 2-bit multiplier network

products to obtain the product of the two words. Mathematically, this procedure is summarized as shown in Figure 8.47. The product bits can be written as

$$
\begin{aligned}
p_0 &= a_0 \times b_0 \\
p_1 &= a_0 \times b_1 + a_1 \times b_0 \\
p_2 &= a_0 \times b_2 + a_1 \times b_1 + a_2 \times b_0 + c_2 \\
p_3 &= a_0 \times b_3 + a_1 \times b_2 + a_2 \times b_1 + a_3 \times b_0 + c_3 \\
p_4 &= a_1 \times b_3 + a_2 \times b_2 + a_3 \times b_1 + c_4 \\
p_5 &= a_2 \times b_3 + a_3 \times b_2 + c_5 \\
p_6 &= a_3 \times b_3 + c_6 \\
p_7 &= c_7
\end{aligned}
$$

(8.73)

where we have once again included carry bits from the preceding column such that c_i originates from the $(i-1)$st column and contributes to the p_i bit. A more compact form of the same set of expressions is obtained by writing

			a_3	a_2	a_1	a_0
		$\times$	b_3	b_2	b_1	b_0
			a_3b_0	a_2b_0	a_1b_0	a_0b_0
		$+ \; a_3b_1$	a_2b_1	a_1b_1	a_0b_1	
	$+ \; a_3b_2$	a_2b_2	a_1b_2	a_0b_2		
$+ \; a_3b_3$	a_2b_3	a_1b_3	a_0b_3			
Product $\; p_7$	p_6	p_5	p_4	p_3	p_2	p_1	p_0

Figure 8.47 Multiplication algorithm for two 4-bit words

$$p_i = \sum_{i = \alpha + \beta} a_\alpha \times b_\beta + c_i \qquad (8.74)$$

where the summation is over all products whose coefficients add up to i, which is the product subscript in p_i. Note that multiplying 2 n-bit words together results in a product that is $2n$ bits wide (with the extra bit being a carry from the p_6 term).

Example 8-8

Let us perform the binary multiplication of 1010×1001 using the procedure. Note that this is equivalent to the base-10 problem of $10 \times 9 = 90$.

```
        1 0 1 0
      × 1 0 0 1
        1 0 1 0
      0 0 0 0
    0 0 0 0
  1 0 1 0
  1 0 1 1 0 1 0
```

Multiplication of binary words can be performed directly using an array of basic logic elements that provides for

- bit multiplication, along with
- circuits to sum the appropriate values.

A 4-bit multiplier can be built using the 2-bit multiplier in Figure 8.46 as a basis, or it may be constructed using primitive components. The same holds true for multipliers that can handle larger-sized words. It is important to remember that a multiplier unit has inputs and outputs of specified word size and produces an output with twice the number of bits.

8.10 Transmission Gate Logic

Transmission gates (TGs) are logic-controlled switches that can be used to construct a wide variety of logic networks. The symbol for a transmission gate is shown in Figure 8.48. In the drawing, A and B are data variables, while S (and $\overline{S}$) is the switch control bit.

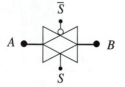

Figure 8.48
Transmission gate symbol

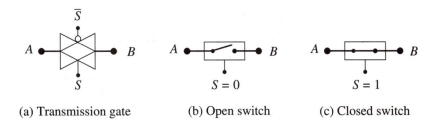

(a) Transmission gate (b) Open switch (c) Closed switch

Figure 8.49 Switch model of a transmission gate

The operation of a transmission gate is quite simple. The value of the control bit S determines whether the path between the left and right sides (A and B) is open or closed. When $S = 0$, then no path exists, and there is no relationship between A and B. However, if $S = 1$, then the TG provides a closed path between the left and right sides, and $B = A$. A switch model that illustrates this behavior is shown in Figure 8.49. Note that both S and $\overline{S}$ are required to control the transmission gate, but only S is needed in the simple switch model.

Let us now attempt to write a logic equation that describes the operation of a transmission gate. Since the value of the control bit S determines the switch position, we will start by writing

(a) $S = 0$ says that B is not determined by A

(b) $S = 1$ says that $B = A$

to describe the two cases. The first statement is obtained from the fact that $S = 0$ makes the TG behave as an open switch, so that the left and right sides are not connected to each other. Conversely, a value of $S = 1$ creates a closed data path between A and B, so that they must be equal as stated in case (b). To create a Boolean equation that describes the TG, we would like to write an expression for B in terms of A and S. Unfortunately, statement (a) for the case $S = 0$ prohibits us from accomplishing this task since A is not related to B. Owing to this situation, we will simply write an equation that is valid when $S = 1$ in the form

$$B = A \cdot S \qquad (S = 1) \tag{8.75}$$

and remember that the case where $S = 0$ is not defined. Although this may sound very restrictive at the moment, we shall see that TG networks are always designed with this limitation in mind.

8.10.1 Transmission Gate Multiplexors

As a first example, let us consider the design of the 2:1 MUX shown in Figure 8.50. The inputs are P_0 and P_1, and the switching is controlled by the variable S. When $S = 0$, transmission gate TG0 acts as a closed switch, while transmission gate TG1 acts as an open switch; in this case, the output is $F = P_0$. If $S = 1$ is applied, then TG0 is open and TG1 is closed, resulting in an output of $F = P_1$. Using the equation for a single transmission gate (where the input variable is ANDed with the signal applied to the "unbubbled" control) yields the functional expression

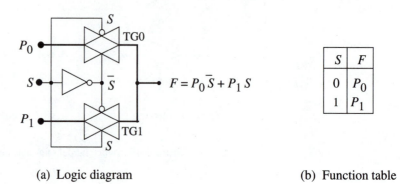

(a) Logic diagram

(b) Function table

Figure 8.50 A transmission gate 2:1 multiplexor

$$F = P_0 \cdot \bar{S} + P_1 \cdot S \qquad\qquad (8.76)$$

by noting that the top line gives a term $P_0 \cdot \bar{S}$, while the lower line yields $P_1 \cdot S$. The OR operation (+) is due to the fact that either the upper line OR the lower line is selected, depending upon the value of S.

This scheme can be applied to create multiplexors with a larger number of inputs, for example, 4:1 and 8:1 networks. Since an n-input MUX requires m select bits with $n = 2^m$, each line in the MUX will have m transmission gates between the input and the output. A 4:1 circuit is shown in Figure 8.51. Line selection is accomplished using two bits S_1 and S_0 such that the decimal value of the 2-bit word $S_1 S_0$ determines which input is routed to the output. This is accomplished by insuring that each value of $S_1 S_0$ specifies that only a single input has a complete dataflow path between the input and the output with both transmission gates closed.

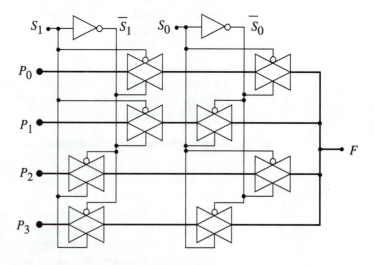

Figure 8.51 A TG-based 4:1 multiplexor

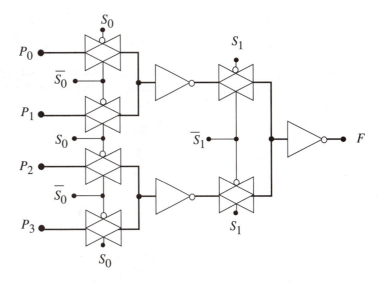

Figure 8.52 A 4:1 MUX based on 2:1 MUX units

It is possible to create larger TG-MUX networks. However, we find that the logic propagation delay through a data path increases with the number of transmission gates, and larger networks such as 8:1 and 16:1 are usually too slow for a high-speed system. It is common to use smaller cascaded MUXs with buffers (amplifiers) to overcome this problem. A 4:1 MUX created in this manner is shown in Figure 8.52. This circuit is made using three identical 2:1 MUX units. The control bit S_1 is applied to the output MUX and determines whether the upper or lower 2:1 MUX is selected. S_0 is applied to the left networks and selects which input path is directed to the output MUX. Note that each data path has two inverters, so that the correct output is achieved. This approach may be used to construct larger MUXs using 2:1 circuits as the basic units.

8.10.2 TG XOR and XNOR Gates

Transmission gates can also be used to construct various logic gates. An example is the XOR network shown in Figure 8.53(a). This is based on the 2:1 MUX circuit with the inputs replaced by variables. It is straightforward to verify that this implements the XOR function by using the rules discussed above. Consider the case where $B = 0$, so that TG0 is closed and TG1 is open. The output is $F = A$, so that this case is described by the product term $A \cdot \overline{B}$. If $B = 1$, then the situation is reversed, with TG0 open and TG1 closed; this is equivalent to the product term $\overline{A} \cdot B$. Since either the upper path OR the lower path is in operation, we may write

$$F = A \cdot \overline{B} + \overline{A} \cdot B$$
$$= A \oplus B$$

(8.77)

It is a simple matter to create the XNOR network by recalling that

$$F = A \cdot B + \bar{A} \cdot \bar{B}$$

$$= \overline{A \oplus B}$$

(8.78)

so we need only interchange B and $\bar{B}$ to change the XOR network into an XNOR circuit. The resulting logic network is shown in Figure 8.53(b).

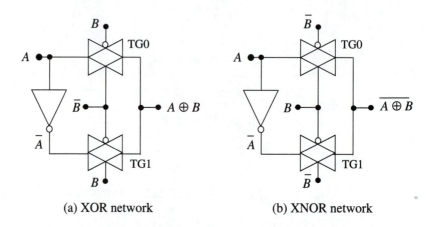

(a) XOR network   (b) XNOR network

Figure 8.53 Exclusive-OR and equivalence functions

Transmission gates can be used to create other logic functions in a similar manner. As seen from the examples above, TG-based multiplexer networks are well suited for SOP logic functions. In addition, we shall see that TGs can be used to create very useful **latch** circuits that can capture and hold logic bits.

8.10.3 CMOS Transmission Gates

Transmission gates are basic building blocks in a CMOS technology, and they are common in many logic networks. A CMOS TG is constructed by paralleling an n-channel MOSFET with a p-channel MOSFET and applying complementary control signals to the gate; this is shown in Figure 8.54. Note that the "bubbled input" of the

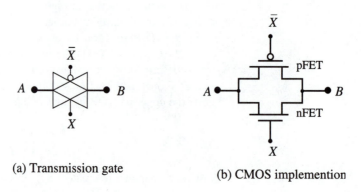

(a) Transmission gate   (b) CMOS implemention

Figure 8.54 CMOS transmission gate

TG corresponds to the gate of the pFET. The operation of the TG is based on the behavior of the MOSFETs. A value of $S = 0$ places both transistors in cutoff, and they act like open switches between the left and right sides. On the other hand, if $S = 1$, then both transistors are active, so that $B = A$. The layout of a CMOS TG requires two parallel-connected transistors, as shown in Figure 8.55. The alert reader will notice that this looks very similar to one of the inverter layouts presented in the last chapter. The transmission gate can be viewed as a complementary pair that has been wired to place the two FETs in parallel, so that the similarity between it and an inverter is no accident.

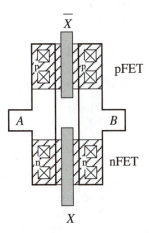

Figure 8.55
Layout of a CMOS TG

In order to create a useful transmission gate for use in a VLSI cell library, an inverter is often provided as shown in Figure 8.56. The inverter is placed at the input to provide drive current through the TG transistors, which have no direct connections to the power supply or ground. This allows us to create a silicon layout as

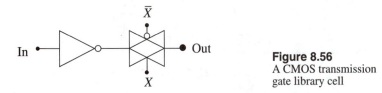

Figure 8.56
A CMOS transmission gate library cell

shown in Figure 8.57(a), which is represented by the simplified cell in Figure 8.57(b). Once this is stored in the library, it can be used as a basic building block in high-density designs. If the cell design only uses a single inverter as shown, then it will be necessary to use this characteristic in the logic equations whenever it is instanced into a particular circuit.

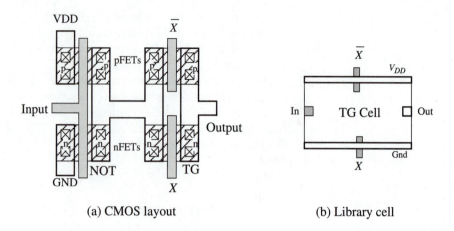

(a) CMOS layout   (b) Library cell

Figure 8.57 Layout of library cell

8.11 Summary

In this chapter we have examined the concept of building different "logic components." These digital units are more complicated than simple gates and are chosen to perform some type of useful function.

Logic components are created to help us design more complex digital networks. The most important point about our discussion is that we have moved to a relatively high level in the design hierarchy where the internal makeup of a complex logic component becomes a secondary issue. Although it may be interesting to see what is inside of the device, it is not necessary to learn all of the internal details; only the overall operation is important. This allows us to progress to very complex designs.

8.12 Problems

[8.1] Suppose that we want to compare two 4-bit words

$$a = a_3 a_2 a_1 a_0$$
$$b = b_3 b_2 b_1 b_0$$

for the case where $a_n \neq b_n$ for $n = 0, 1, 2, 3$, i.e., none of the bit positions are equal. Design the logic network that produces an output $g = 1$ when this condition is satisfied.

[8.2] Consider two 4-bit words

$$a = a_3 a_2 a_1 a_0$$
$$b = b_3 b_2 b_1 b_0$$

Design a logic component that produces an output of $h = 1$ when the two conditions $a_2 = b_2$ and $a_0 \neq b_0$ are simultaneously true.

[8.3] Consider two 4-bit words

$$a = a_3 a_2 a_1 a_0$$
$$b = b_3 b_2 b_1 b_0$$

Design a logic component that produces an output of $G = 1$ when the two conditions $\overline{a}_1 = b_2$ and $a_2 = b_3$ and $a_3 = b_0 \oplus b_2$ are simultaneously true.

[8.4] Design an active-low 2/4 decoder using OR gates for the output lines.

[8.5] Design an active-high 3/8 decoder using basic logic gates. Provide the logic diagram and an operation table.

[8.6] Design an inverting 2:1 multiplexor with inputs P_0 and P_1 that has an output

$$f = \overline{P_0 \cdot C + P_1 \cdot C}$$

using OR and NOT gates only.

[8.7] Suppose that you need a 3:1 MUX for your system. Discuss how this can be accomplished and what problems might arise in the application.

[8.8] Construct a 4:1 multiplexor using only 2:1 MUXs as the main building blocks.

[8.9] Construct an 8:1 multiplexor using only 2:1 MUXs as the main building blocks.

[8.10] Consider the 4:1 multiplexor circuit shown immediately below. Find the function $f(a, b)$ at the output.

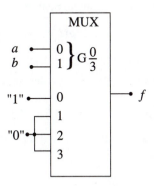

[8.11] Write the function $f(A, B, C)$ for the multiplexor network shown. Then reduce the function to its simplest form using a 3-variable Karnaugh map.

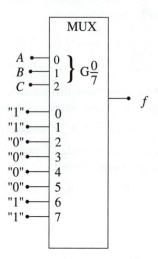

[8.12] Design a random-logic MUX network that implements the function

$$g = a \cdot b + \bar{c} \cdot d$$

[8.13] Use an 8:1 MUX to implement the function

$$F = x \cdot \bar{y} \cdot z + (\bar{x} + y) \cdot \bar{z}$$

[8.14] Use an 8:1 MUX to implement the function

$$Q = a \cdot b \cdot c + b \cdot \bar{c}$$

[8.15] Use a MUX to implement the function

$$g(u, v) = u \cdot v + \bar{u} \cdot v + u \cdot \bar{v}$$

Does the MUX technique provide an efficient design methodology for this case?

[8.16] Use a multiplexor to implement the function

$$f = a \cdot \bar{b} \cdot c + b \cdot c \cdot \bar{d} + a \cdot b \cdot c \cdot d$$

[8.17] Use a multiplexor unit to implement the function G defined by the following K-map.

x \ yz	01	11	10	00
0	1	1	1	0
1	1	0	0	1

[8.18] Use a multiplexor to implement the function $X(a, b, c, d)$ as described by the K-map below.

ab \ cd	01	11	10	00
00	1	0	1	0
01	0	0	0	0
11	1	1	0	1
10	1	0	0	1

[8.19] Design a 1:8 DeMUX network using basic logic gates.

[8.20] Add the following binary words using the rules of binary bit addition (i.e., without converting to decimal form first).

(a) $1100 + 0101$
(b) $0110 + 0110$
(c) $0011 + 0101$

[8.21] Perform the following binary additions (without using decimal equivalents).

(a) $11000110 + 01011100$
(b) $11110000 + 10101010$
(c) $10100010 + 11110001$

[8.22] Obtain the sum of the following hexadecimal numbers (denoted by 0x) by first converting to binary and then adding.

(a) 0xAE + 0x 37
(b) 0x35 + 0x F2
(c) 0xB7 + 0x 5C
(d) 0x90 + 0x 83

[8.23] Recall that the hexadecimal numbering system is based on the digits 0 through F. Perform the following hexadecimal additions without translating to binary or decimal form. State the value of the carry digit if appropriate.

(a) $7 + A = ?$
(b) $9 + B = ?$
(c) $3 + F = ?$
(d) $5 + 4 = ?$
(e) $E + C = ?$
(f) $1 + 9 = ?$

[8.24] Construct the VHDL description of the AOI full-adder network shown in Figure 8.28.

[8.25] Construct the VHDL description of a 4-bit parallel adder using the full-adders as basic modules. This will require you to use port maps as discussed in Chapter 5.

[8.26] Consider the 4-bit parallel adder shown in Figure 8.29. Each full-adder (FA) section is characterized by the following delay times: 1.5 ns for the sum bit and 1.8 ns for the carry-out bit.

(a) Calculate the total time required to evaluate the sum of two 4-bit words.

(b) Use the same technique to find the time needed to evaluate the sum for an 8-bit ripple-carry adder that is built using the same FA units.

[8.27] Consider the 4-bit parallel adder shown in Figure 8.29. Each full-adder (FA) section is characterized by the following delay times: 2.1 ns for the sum bit and 1.7 ns for the carry-out bit. Calculate the total time required to evaluate the sum of two 4-bit words.

[8.28] Find the 1s complement and the 2s complement of the following binary words.

(a) 0111
(b) 1010
(c) 1111
(d) 10110101
(e) 11001100
(f) 10100101

[8.29] Perform the following subtractions by translating into binary and applying the 2s complement approach. Use word sizes of 4b or 8b.

(a) $14 - 6$
(b) $34 - 21$
(c) $134 - 62$
(d) $196 - 118$

[8.30] Suppose that we attempt to create a 4-bit subtractor using the concept of borrowing instead of the 2s complement technique. What makes this so difficult compared to using the carry bits to design a 4-bit adder? Use drawings or equations to make your arguments.

[8.31] Consider the 4-bit word

$$sa_2a_1a_0$$

where s is the sign bit. Construct a binary-to-decimal conversion table that can be used to translate between the two.

[8.32] Perform the following binary multiplications.

(a) 1010×1011
(b) 1110×0010
(c) 1011×0111
(d) 0100×1011

[8.33] Find the product of $0x34 \times 0x17$ by converting these and performing the operation in binary form.

[8.34] Consider the multiplication of two 3-bit words $a_2 a_1 a_0 \times b_2 b_1 b_0$.

(a) Write the equations for each of the six product bits p_i.

(b) Construct the logic diagram for the 3-bit multiplier network using single-bit multiplying units (AND gates) and adders.

[8.35] Design a transmission gate (TG) network that implements the function

$$F = x \cdot \bar{y} + x \cdot z$$

[8.36] Consider the timer on a microwave oven. The main power is controlled by a safety switching signal S such that $S = 0$ turns the oven off, while $S = 1$ allows power to be applied to the microwave tube. Can you design an electrical network that can sense whether the door is OPEN or CLOSED and turn off the master power if it is OPEN?

[8.37] A full-adder circuit calculates the sum bit by means of

$$s = a \oplus b \oplus c_{in}$$

Use transmission gates to provide this function. Hint: Remember that

$$x \oplus y = x \cdot \bar{y} + \bar{x} \cdot y$$

which can be combined with the next XOR operation using algebra.

[8.38] Compare the CMOS circuit and layout of a transmission gate with that of a simple inverter (NOT gate) as discussed in Chapter 7. Discuss the similarities and differences between the two.

CHAPTER 9

Memory Elements and Arrays

Most digital systems require the ability to "remember" the values of binary variables for use in future calculations. This is accomplished by using a class of digital circuits called **memory cells** that allow the user to store the value of a binary variable and then recall the bit as needed. Although the cells can be used individually, they are most commonly found in a word-size group, called a register, or as very large arrays that can store millions of bits of data.

9.1 General Properties

A **memory element** is a circuit that can hold the value of a binary variable as required by the system. A basic memory cell is portrayed in Figure 9.1. In the most general terms, there are three operations that characterize the cell.

- **Write** In a write operation, the value of the data bit D (which is a 0 or a 1) is placed on the input and is accepted by the element.

- **Hold** The value of D that was stored in the write operation is held, even though there may be changes in the input.

- **Read** A read operation provides the value of the stored bit as the output variable Q.

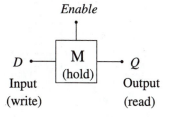

Figure 9.1
General memory cell

The *Enable* control bit may be present in certain designs such that *Enable* = 0 implies a hold, while *Enable* = 1 allows a read or a write operation.

Digital systems employ different types of memory elements whose characteristics vary with the application. One classification scheme is based on which operations are provided by the cell design. A **read/write** memory is one where the user may store values, hold them for an indefinite period of time, and read them out as needed. This is the most general type of memory possible; it is usually called **random-access memory** or **RAM**. Most read/write memories loose their contents when the power supply is disconnected. A more restrictive type is called **read-only memory (ROM)**. In a ROM, the information is permanently stored in the device before it is placed into the electronic system. A user may read the information out of a ROM but is not permitted to change the data. A variation on this is the **programmable ROM (PROM)** where the user may store the desired data, but the write procedure requires a special electronics setup and is only performed a few times.

Computers are designed around a set of memory elements that are called **registers**. A register is designed to hold an entire *n*-bit word but is otherwise identical to RAM. Regardless of the specific application, all memory elements have a common basis in digital logic.

9.2 Latches

A **latch** is a logic element that can follow data variations and transfer these changes to an output line. A simple representation is shown in Figure 9.2. In general, a latch is characterized by two main properties:

- It is **transparent** in that the output $Q(t)$ follows changes at the input at least part of the time; and,

- The storage is achieved using a **bistable circuit**, in which either $Q = 0$ or $Q = 1$ can be held in the cell.

Several types of latches can be constructed using simple logic gates.

9.2.1 The SR Latch

The **set-reset** (SR) latch is a transparent bistable element that is sensitive to changes in the inputs. An SR latch has two inputs that are labeled S and R and an output that is labeled Q; the complement of the output $\overline{Q}$ is also provided. A general block representation is shown in Figure 9.3. The name of this element is obtained from the definitions of two basic operations:

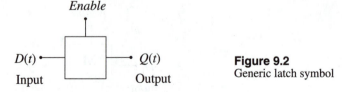

Figure 9.2
Generic latch symbol

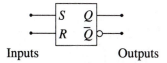

Figure 9.3
Symbol for an SR latch

Inputs   Outputs

- The **set** (S) operation, where the output is forced to a value of $Q = 1$.

- The **reset** (R) operation, where Q is forced to a value of $Q = 0$.

The terminology "set" and "reset" is quite general and is frequently used to describe conditions in a digital system. For example, if we say that we want to "set the variable A," this means that we force A to a value of 1. Similarly, resetting A means that $A \rightarrow 0$. There are two basic SR-latch circuits that will be discussed here. The first is based on NOR2 gates, while the other uses NAND2 logic. Since the NOR and the NAND are duals, the switching characteristics will be opposite of each other.

The NOR SR Latch

An SR latch can be built using two **cross-coupled** NOR gates as shown in Figure 9.4(a). The term "cross-coupled" defines the manner in which the two gates are connected; it means that the output of one NOR gate is connected to the input of the other gate, and vice versa. This is also termed a **feedback** network, as the output of each gate is "fed back" and is used as an input to the other gate. Feedback is required to create a bistable circuit that can hold the two logic values. The operation of the latch is summarized by the state table in Figure 9.4(b). Care must be exercised in interpreting the entries, as the SR latch stores bits by sensing *changes* in the S and R inputs. This means that we are really interested in $S(t)$ and $R(t)$ and their effect on the state of the latch.

Now let us analyze the SR latch function table. The first entry in the state table where $S = 0$ and $R = 0$ are maintained is called the **hold** operation; this means that there is no change in the value of Q as long as $S = 0 = R$. To understand this property, let us review the NOR truth table shown in Figure 9.5. The NOR function can be interpreted by noting that a 1 at either input gives a 0 at the output; the

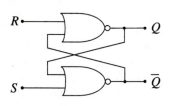

S	R	Q	$\overline{Q}$	Operation
0	0	Q	$\overline{Q}$	Hold
⊓	0	1	0	Set ($Q \rightarrow 1$)
0	⊓	0	1	Reset ($Q \rightarrow 0$)
1	1	0	0	Not used

(a) Logic diagram   (b) Function table

Figure 9.4 A NOR-based SR latch circuit

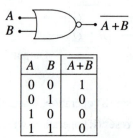

A	B	$\overline{A+B}$
0	0	1
0	1	0
1	0	0
1	1	0

Figure 9.5
NOR2 truth table

only way to obtain a 1 at the output is to have both inputs at 0 values. Now suppose that both $S = 0$ and $R = 0$ are applied to the inputs. If $Q = 0$, then both inputs to the lower NOR gate (with the S input) are 0, which means that $\overline{Q} = 1$. This is in turn applied to the upper NOR gate (with the R input), which insures that $Q = 0$ is maintained. If instead we have $Q = 1$, then the arguments are just reversed. The importance of this analysis is the fact that it is possible to hold the value of Q. The cross-coupled NOR gates form a **bistable** network that has two "stable" states. This means that the circuit can "hold" the value of $Q = 0$; the same is true for the other possibility of $Q = 1$.

To set the latch, the S input is raised from a 0 up to a 1 while $R = 0$. This is followed by S returning to 0, which places the latch back in the hold state. In the function table, this action is denoted by a pulse entry for S. With the value $S = 1$, the bottom output $\overline{Q}$ is forced to 0 as seen from the NOR truth table. The upper NOR gate then has $R = 0$ and $Q = 0$ applied to its inputs, resulting in an output of $Q = 1$. This gives the "set" action. A reset forces Q to 0 and is analyzed in the same manner.

The fourth entry in the function table is where both S and R go to logic 1 values simultaneously. We see from the truth table that this forces both outputs to 0. Since the two outputs are not complements in this case, we define this as a "not used" input combination for the SR latch. This simply means that we avoid the case where both inputs are pulsed at the same time.[1] This behavior of the SR latch must be considered when using the device in a digital system.

The NAND SR Latch

NAND gates can also be used to build an SR latch as shown in Figure 9.6. Note that the NAND gates are cross-coupled to give the necessary feedback to form a bistable network. However, the properties are opposite from the NOR-based scheme. The hold condition occurs when both $R = 1$ and $S = 1$. The latch may be set to $Q = 1$ by dropping S from a 1 to a 0 as indicated in the function table; bringing S back up to a 1 then initiates the hold. Conversely, if R is dropped from a 1 to a 0, the latch is reset to a value of $Q = 0$. Restoring R to a logic 1 level holds this value at the output. The "not used" case is when S and R are simultaneously dropped to 0.

---

[1] The condition where both inputs are 1 may cause an unstable situation where the device oscillates. This possibility depends upon the signal delays through the gates.

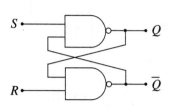

S	R	Q	$\overline{Q}$	Operation
0	0	0	0	Not used
⊔	1	1	0	Set $(Q \to 1)$
1	⊔	0	1	Reset $(Q \to 0)$
1	1	Q	$\overline{Q}$	Hold

(a) Logic diagram   (b) Function table

Figure 9.6 An SR latch built from NAND gates

We may understand why this latch acts opposite to the NOR-based circuit by studying the NAND truth table in Figure 9.7. The operation of the NAND gate may be summarized by stating that placing a logic 0 at either input results in a logic 1 output. Since the NOR gate inverts with a logic 1 input, we expect that the two latches have similar behavior, but with reversed input levels needed to induce a switching event. Owing to this observation, the symbol for this latch must be modified from that shown in Figure 9.3, where both the S and R are now designated as being active low inputs.

9.2.2 D Latch

A **D-type latch** has a single input D that acts as the input data bit. It is characterized by the ability to latch onto the value of D and track any changes. A simple way to construct a D latch is shown in Figure 9.8. This uses an SR latch with the inputs replaced by D and $\overline{D}$. The complement of D is obtained by adding the inverter. The operation of the D latch can be understood using the function table for the SR latch. When $D = 0$, the $\overline{D}$ input to the bottom NOR gate is a 1. This forces the output to $Q = 0$, which is held by the cross-coupled network. If $D = 1$, then the output of the upper NOR gate is forced to $\overline{Q} = 0$, which is the other stable state.

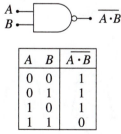

A	B	$\overline{A \cdot B}$
0	0	1
0	1	1
1	0	1
1	1	0

Figure 9.7
NAND2 truth table

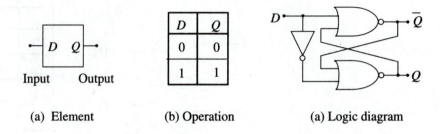

(a) Element   (b) Operation   (a) Logic diagram

Figure 9.8 Basic D-type latch

9.3 Clocks and Synchronization

In a complex digital system one must carefully control the flow of data to insure that the proper information is available to each section when it is needed. The easiest way to control the movement of data within a network is to synchronize the system operation using a well-defined reference such as a clock signal. Although clocks were introduced in Chapter 1, let us quickly review the basics.

A **clock** is a control signal that periodically makes a transition from a 0 to a 1 and then back to a 0 again, as illustrated in Figure 9.9. We will usually denote the clock by the symbol $\phi(t)$ or CLK. The most important characteristic of the clock is that it repeats every T seconds, which is known as the **period**. The **frequency** f of the clock is defined by

$$f = \frac{1}{T} \tag{9.1}$$

and gives the number of cycles (repetitions) that the clock has in 1 second. The frequency has strict units of sec^{-1}, which is called a *hertz* (abbreviated as *Hz*). In computers, the most common measurement for T is in *microseconds* (which is abbreviated as μs), where

$$1 \; \mu s = 10^{-6} \; sec$$

or in *nanoseconds* (*ns*) where

$$1 \; ns = 10^{-9} \; sec$$

With $T = 1 \; \mu s$, the frequency is

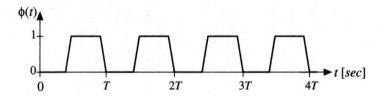

Figure 9.9 Clocking signal used for synchronization

$$f = \frac{1}{10^{-6}} = 10^6 = 1 \; MHz \qquad \textbf{(9.2)}$$

where *MHz* is read as *megahertz* and means one million cycles per second. This is the most common unit for specifying clocks in modern desktop computers.

Example 9-1

Suppose that a clock has a frequency of 500 *MHz*. The corresponding clock period *T* is computed by using the inverse relation

$$T = \frac{1}{f}$$
$$= \frac{1}{500 \times 10^6} \qquad \textbf{(9.3)}$$
$$= 2.0 \; ns$$

This is a typical order of magnitude for many modern systems.

Using a clock signal to control the operation of a memory element provides us with the ability to dictate the times when data values can be stored in the device. The resulting memory elements are characterized by the portion of the clock cycle when a data bit can be loaded. This allows for the design of very complex digital networks in which the data can be moved in a synchronous manner.

9.3.1 Clocked SR Latch

A clocked SR latch is shown in Figure 9.10(a). Comparing this with Figure 9.4, we see that the clock signal $\phi(t)$ is ANDed with both the *R* and *S* inputs. This modifies the inputs into the cross-coupled NOR circuit to have the effective values of

$$R' = R \cdot \phi(t)$$
$$S' = S \cdot \phi(t) \qquad \textbf{(9.4)}$$

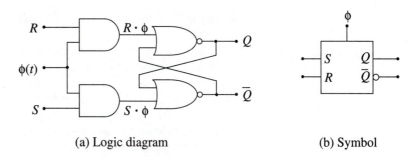

(a) Logic diagram   (b) Symbol

Figure 9.10 A clocked SR latch

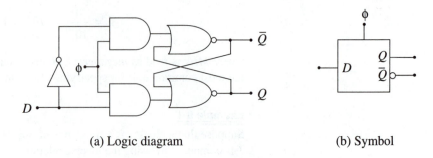

(a) Logic diagram   (b) Symbol

Figure 9.11 A clocked D-type latch

Since (0, 0) is the hold condition in the NOR latch, the latch is automatically in the hold state when $\phi = 0$; set and reset operations can only take place when $\phi = 1$. This defines a **level-triggered** or **level-sensitive** latch[2] that allows inputs anytime when the clock is at a value $\phi = 1$. The symbol for the clocked SR latch is shown in Figure 9.10(b). Note the inclusion of the clock ϕ as a control signal.

9.3.2 The D Latch

A clocked D latch may be created in the same manner, as illustrated in Figure 9.11(a). Once again, we have ANDed the clock signal ϕ with the regular inputs to obtain an effective input of

$$D' = D \cdot \phi \qquad \qquad \textbf{(9.5)}$$

This creates a level-sensitive latch with the symbol shown in Figure 9.11(b). As with the clocked SR latch above, the input is only active when $\phi = 1$. The clocked D latch is often called a transparent latch due to its behavior during this time. A clock condition of $\phi = 0$ acts to hold the values of Q and $\overline{Q}$ at the output.

Clocked latches are useful in synchronizing the data flow through a complex system. They also give more meaning to the name "latch" as they can be visualized as circuits that "latch on to" data when activated by a clocking signal.

9.4 Master-Slave and Edge-Triggered Flip-Flops

A flip-flop is a non-transparent latch that is controlled by the clock. This means that the present output value of the memory element Q is not related to the present input value A. Several different variations exist, but all have the characteristic that the output $Q(t)$ was stored at an earlier time as established by the clocking circuitry.

The basic features of a **master-slave** flip-flop network are shown in Figure 9.12. This storage element is so named because of the designation of the two internal blocks that represent two separate latching circuits. The master latch is used to

---

[2] This is also called a flip-flop in treatments that define a flip-flop as any memory clock-controlled memory element.

accept the data input bit A into the flip-flop. This value is held in the master and then transferred to the slave at a later time. Both inputs are synchronized by the clock signal $\phi(t)$.

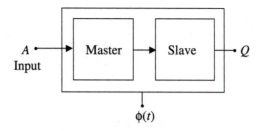

Figure 9.12 The master-slave structure

Figure 9.13 summarizes the operation of the master-slave configuration. In Figure 9.13(a), the input to the master circuit is activated and the value of A is stored there. During this time, the slave is inactive (i.e., in a hold state) and cannot accept inputs. The next operational phase is shown in Figure 9.13(b). During this time, the master is placed in a hold state while the slave is active. The value of A is then accepted by the slave and is available at the output of the flip-flop giving $Q = A$. Since the master is disabled, it continues to hold the value of A at its output even though the input has changed to B. During this time, the flip-flop is non-transparent. The important characteristic of the flip-flop is that the value of the output $Q(t)$ is not related to the present input (B) but represents the input at an **earlier** time when the master was active.

9.4.1 A Master-Slave D-Type Flip-Flop

A master-slave D-type flip-flop (DFF) is shown in Figure 9.14. This memory element is constructed by cascading two clocked D-type latches. The first latch is designated as the master circuit and is responsible for securing the input data bit D. The second D latch acts as the slave; it is used to hold the value of the data bit that it receives from the master. The operation of the DFF is easily understood once it is observed that the master and the slave circuits are controlled by opposite phases of the clock $\phi(t)$. Since the master latch has $\bar{\phi}$ applied to it, it accepts inputs when $\phi = 0$. The slave, on the other hand, uses ϕ for timing, so that it allows changes in the inputs when $\phi = 1$.

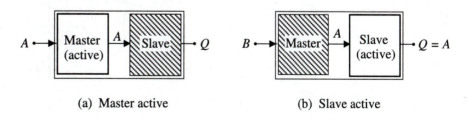

(a) Master active   (b) Slave active

Figure 9.13 Master-slave operation

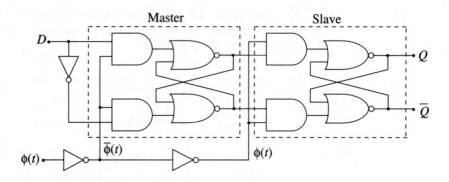

Figure 9.14 D-type flip-flop circuit

A detailed example of the DFF operation is shown in Figure 9.15. When the clock is at a value $\phi = 0$ as in Figure 9.15(a), the master latch is activated. The input data bit has a value of $D = 1$ that is accepted and stored in the first latch. During this time, the slave latch is in a hold state, so that its contents are not affected while the

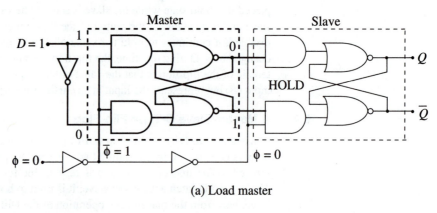

(a) Load master

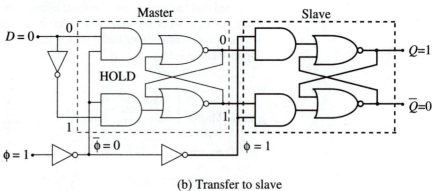

(b) Transfer to slave

Figure 9.15 DFF operation details

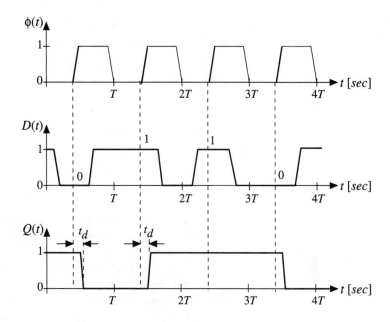

Figure 9.16 Master-slave DFF timing diagrams

master latch is loading. When the clock changes to $\phi = 1$, the master latch is placed in hold, and its output is fed to the slave circuit, which has an input that is now activated by the clock; this is shown in Figure 9.15(b). The data bit A is transferred to the slave circuit, and the output has values of $Q = 1$ and $\overline{Q} = 0$.

Let us now examine the operation of the master-slave DFF circuit using the timing diagrams in Figure 9.16. The clock signal $\phi(t)$ in Figure 9.16 provides the synchronization information such that a value of $\phi = 0$ indicates that the input D is accepted by the master, while $\phi = 1$ means that the DFF input is inactive and the data is transferred to the slave, making it available at the output $Q(t)$.

The master-slave DFF described above is designed to accept an input D during the time when the clock is at $\phi = 0$. The value that is transferred to the slave circuit (and hence, the output Q) is the value that is in the master latch when the clock makes a transition from $\phi = 0$ to $\phi = 1$. For this reason, the master-slave configuration is classified as being an **edge-sensitive** device. This means that the value of Q depends upon the contents of the master latch when the clock makes a transition from one value to the other, corresponding to an "edge" of the clock waveform.

A related memory element is the **edge-triggered flip-flop**. Functionally, this appears very similar to the edge-sensitive design. However, the edge-triggered flip-flop allows a data input D to be loaded *only* when the clock is making a transition. This is different from the master-slave circuit, in which the master latch is transparent during one half of the clock cycle. In our treatment, we will not be concerned with the internal differences between the master-slave and the edge-triggered circuits. Instead, we will take the viewpoint that any flip-flop can be used as a primitive building block like a MUX and only examine the overall characteristics of the device. The symbol for an edge-triggered DFF is shown in Figure 9.17(a). A special

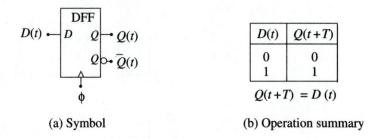

(a) Symbol   (b) Operation summary

Figure 9.17 Edge-triggered DFF operation summary

notation has been added at the clock input to indicate a **positive** edge-triggered behavior in which the input is accepted only during a clock transition from $\phi = 0$ to $\phi = 1$.

The operation of the edge-triggered DFF is described by the summary table in Figure 9.17(b); this shows that the output Q is simply the value of the input D delayed by one clock cycle. The notation in the table uses the time variable t to denote the "present" value, such as $D(t)$, where t is the time of the rising clock edge. The output, on the other hand, is written as $Q(t + T)$, where T is the clock period. This implies the value of the output variable Q during the next clock cycle, i.e., at the time $(t + T)$. The edge-triggered DFF is characterized by the equation

$$Q(t + T) = D(t) \tag{9.6}$$

which says the output is the same as the input during the previous clock cycle. This delay is an important aspect of edge-triggered D-type flip-flops, as it allows us to store the present value for use during the next clock cycle.

The delay characteristic of the edge-triggered DFF is sufficient for many applications in digital design. However, some systems require that we have the ability to hold a data bit over several clock cycles. To create a circuit that provides this, we must route the output Q back to the input D, which then "reloads" the flip-flop with the same value. A simple way to accomplish this is using the 2-to-1 MUX shown in Figure 9.18. In this scheme, the load signal LD controls the input such that $LD = 0$ holds the present value by routing the output back to the input. If $LD = 1$, then a new value d is routed to the DFF storage element.

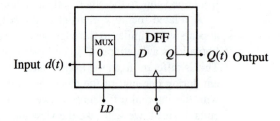

Figure 9.18 DFF with a data LOAD control

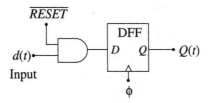

Figure 9.19 DFF with reset capabilities

Another feature that is sometimes useful is the ability to **reset** the flip-flop, i.e., force Q to 0. This can be provided by adding the AND gate as shown in Figure 9.19. In this case, the control signal is the $\overline{RESET}$ such that

$$RESET = 0: \text{Gives } D = d \text{ at the input,}$$

while

$$RESET = 1: \text{Forces } D = 0 \text{ at the input.}$$

Where the value of RESET is applied during the clock transition, this feature can also be combined with the Load operation if desired by modifying the logic.

VHDL Descriptions of D-Type Storage Elements

The D-type latch and D flip-flop are useful elements for general logic design. The VHDL descriptions introduce new ideas that are worth detailing here.

Consider first the level-triggered D latch shown in Figure 9.20. The clock input is denoted by clk but is otherwise identical to ϕ in all regards. To describe the oper-

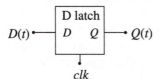

Figure 9.20
D-type latch module

ation of the latch, we must first note that the value of clk controls the behavior of the element. When $clk = 0$, the latch does not allow a change in the state Q. The value of the stored bit can only change when $clk = 1$. The description below introduces the idea of a VHDL process, which is a construct that allows timing to be included.

```
-- VHDL Description of a D Latch
entity D_latch is
    port(D : in bit;
         Q : out bit);
end D_latch;33
-- The architecture contains the process statements
architecture Level_trigger of D_latch is
```

```
begin
    process (clk, D )
    begin
        if (clk = '1') then
            Q <= D;
        end if;
    end process;
end Level_trigger;
```

The **process** keyword indicates the beginning of a sequential set of statements. This means that the commands are executed in the same order in which they are listed. The process is described by the sequence

```
process (clk, D )
begin
    if (clk = '1') then
        Q <= D;
    end if;
end process;
```

which executes exactly as it reads. In other words, a clock value of $clk = $ '1' causes D to be transferred to the output as Q.

An edge-triggered device is slightly more complicated, since we must write a process that accounts for a changing value of clk. Consider the module DFF shown in Figure 9.21 for a positive edge-triggered element. One way to describe a changing signal in VHDL is to use the 'event directive. The statement

clk 'event

is interpreted as a Boolean object with a value of TRUE or FALSE. If clk 'event is TRUE, this means that the signal clk has changed value; if clk has not changed, then this returns a FALSE outcome. To describe a positive clocking edge, we write

if (clk 'event and $clk = $ '1')

which will evaluate TRUE since clk is changing AND the value of clk is now '1'. With this in mind, the DFF can be described by the following listing.

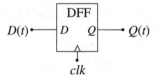

Figure 9.21
Edge-triggered DFF module

```
-- VHDL Description of an edge-triggered DFF
entity DFF is
     port(D : in bit;
          Q : out bit);
end DFF;


-- The architecture contains the process statements
architecture Positive_trigger of DFF is
begin
     process (clk, D )
     begin
          if (clk 'event and clk = '1') then
               Q <= D;
          end if;
     end process;
end Positive_trigger;
```

If we wish to describe a negative edge-triggered flip-flop, then we simply have to change the process command to reflect that the final clock value is $clk = 0$.

```
-- VHDL Description of a DFF
entity DFF is
     port(D : in bit;
          Q : out bit);
end DFF;
-- The architecture contains the process statements
architecture Negative_trigger of DFF is
begin
     process (clk, D )
     begin
          if (clk 'event and clk = '0') then
               Q <= D;
          end if;
     end process;
end Negative_trigger;
```

VHDL descriptions of all other clock-sensitive elements in this chapter can be created using the process construct.

9.4.2 Other Types of Flip-Flops

Several other flip-flops can be built using basic logic gates. Each is defined by a set of unique characteristics that may be useful to the designer, depending upon the system specifications. We will only be concerned with edge-triggered FFs where the input is accepted only during a clock edge.

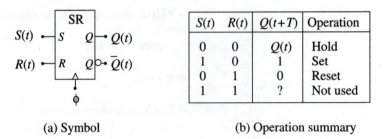

$S(t)$	$R(t)$	$Q(t+T)$	Operation
0	0	$Q(t)$	Hold
1	0	1	Set
0	1	0	Reset
1	1	?	Not used

(a) Symbol   (b) Operation summary

Figure 9.22 SR flip-flop characteristics

SR Flip-Flop

The SR flip-flop has the same basic structure as the DFF with the input circuit replaced by an SR latch. This allows two separate inputs S and R, with a hold defined by $S = 0 = R$. It is important to remember that the condition where both S and R are pulsed to 1 is not used in the operation of an SR latch, so that this case is also precluded in the SR flip-flop. The symbol for an edge-triggered SR FF is shown in Figure 9.22(a), and the corresponding operation table in Figure 9.22(b). This can be summarized by the statements

$$S(t) = 1 \rightarrow Q(t+T) = 1$$
$$R(t) = 1 \rightarrow Q(t+T) = 0$$

(9.7)

Note that the case where the inputs $S(t)$ and $R(t)$ are both equal to 1 is undefined. This is due to the use of the basic SR latch in the circuit.

JK Flip-Flop

The JK flip-flop modifies the SR FF circuit so that the case where the inputs are simultaneously pulsed to 1 induces a **toggle** operation where

$$Q \rightarrow \overline{Q}$$

(9.8)

This means if $(Q, \overline{Q}) = (1, 0)$, then the toggle changes them to $(Q, \overline{Q}) = (0, 1)$. The inputs are in the circuit, but otherwise they have the same external operation as the SR flip-flop. The characteristics are summarized by Figure 9.23. The JK flip-flop

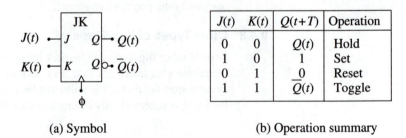

$J(t)$	$K(t)$	$Q(t+T)$	Operation
0	0	$Q(t)$	Hold
1	0	1	Set
0	1	0	Reset
1	1	$\overline{Q}(t)$	Toggle

(a) Symbol   (b) Operation summary

Figure 9.23 JK flip-flop characteristics

used to be dominant in hardware designs that were based on SSI integrated circuits, and it is still a versatile device in certain systems.

T Flip-Flop

The toggle flip-flop is a circuit that has one input T as indicated by the logic symbol in Figure 9.24. The operation of the TFF is exactly as implied by its name: the output toggles whenever T changes from 0 to 1. This is a relatively special device that does not have the versatility of the flip-flops discussed above but may still be useful.

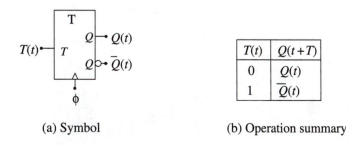

(a) Symbol (b) Operation summary

Figure 9.24 T flip-flop characteristics

9.5 Registers

A **register** is a logic element that is used to store an n-bit binary word. Almost every large digital system employs registers to hold important data, so they are worth studying in some detail. Several variations can be created, with each having its own range of applicability in digital systems design.

9.5.1 Basic Storage Register

We may build a register by simply using n single-bit storage cells in a parallel manner that allows us to read or write all cells simultaneously. Registers are usually constructed in convenient word sizes as required by the system specifications.

Let us examine an 8b register that can store a single byte of data. To create the register, we use the individual memory cell shown in Figure 9.25(a) to construct the circuit shown in Figure 9.25(b). Note that we have chosen an edge-triggered memory element as the primitive cell. For simplicity, we only indicate the input, output, and clock ports. Other control lines, such as a load (LD) signal, may be required for the element but will not be shown explicitly. Although each bit is accessible by itself, the register is used to hold an 8-bit word as a defined group. Read or write operations are thus performed simultaneously on every cell. Since edge-triggered storage cells are used throughout, writing takes place only during a clock transition. A simplified notation is shown in Figure 9.26(a) where we have simply grouped the cells together while still showing them as individual units. In Figure 9.26(b), we represent the register as a single entity with 8-bit input and output lines, and a single clock control. In many applications, it is understood that the registers are loaded according to the clocking signal ϕ (and any other conditions), and even it is eliminated in the drawings.

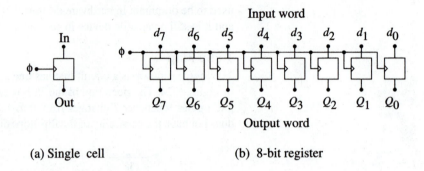

(a) Single cell   (b) 8-bit register

Figure 9.25 Creation of a register from individual storage elements

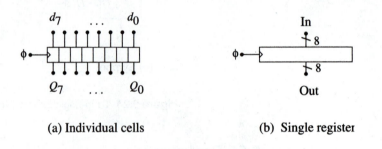

(a) Individual cells   (b) Single register

Figure 9.26 Symbols used to represent an 8b register

General-purpose registers of this type are used extensively in computer design, so that we will defer a more detailed discussion until Chapter 11 where they are introduced in the context of a computer system. The important concept to remember at this point is that the register holds an n-bit word as a group and that the register itself can be viewed as a single digital element that is made up of smaller cells.

9.5.2 Shift Registers

A shift register is designed to move bits to neighboring cells as the clock pulses are applied. To illustrate the general properties of a shift register, let us replicate the memory cell in Figure 9.27(a) to create the serial chain shown in Figure 9.27(b), where the output of each cell acts as the input for its neighbor to the right. Overall, this circuit has a single input bit D and a single output bit Q and constitutes an 8-bit shift register. Since all of the cells are controlled by the same clocking signal ϕ, every cell loads at the same time.

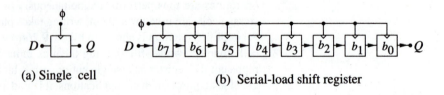

(a) Single cell   (b) Serial-load shift register

Figure 9.27 Basic shift register structure

$$D \rightarrow \boxed{b_7 \, b_6 \, b_5 \, b_4 \, b_3 \, b_2 \, b_1 \, b_0} \rightarrow Q$$

Figure 9.28 Simplified shift register symbol

The operation of the shift register is easier to understand using the simplified symbol shown in Figure 9.28. This type of drawing allows us to visually examine the contents of the cells and trace the movement of data through the structure. The inputs are still assumed to be clocked as in the original circuit, even though we do not show ϕ explicitly.

Suppose that we want to load the register with bit sequence

$$D_0 = 1, \text{ then}$$
$$D_1 = 1, \text{ then}$$
$$D_2 = 0, \text{ then}$$
$$D_3 = 0,$$

and so on, as implied by the "0,1,0,1,0,0,1,1" at the input of the shift register shown in Figure 9.29(a). When we use this type of notation for a serial input, it means that the bit on the right side is the next bit that will enter the unit or system. Since there are 8 cells in the shift register, it will take 8 clock cycles to load the entire register. In terms of the clocking diagram of Figure 9.29(b), this is the same as stating that all of the first input bit $D_0 = 1$ will be available at the output just after time $7T$.

The progression of the bits through the shift register is shown explicitly by the sequence provided in Figure 9.30. The time designation on the left side implies the contents of the register just before the next loading event takes place (since the cells are triggered by a positive clock edge). The movement of the data is straightforward to follow. During the first positive clock transition, cell d_7 is loaded with $D_0 = 1$ as shown at time T. The next positive clock edge causes $D_0 = 1$ to move to the next cell d_6, while admitting $D_1 = 1$ into cell d_7. Each subsequent positive clock transition admits the next bit and causes every stored data bit to move one cell to the right. After time $7T$, the last bit of the word is loaded into cell d_7, and the first bit $D_0 = 1$ is

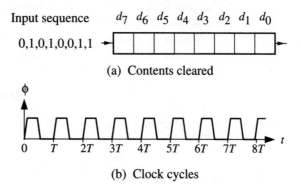

(a) Contents cleared

(b) Clock cycles

Figure 9.29 Serial data input to a shift register

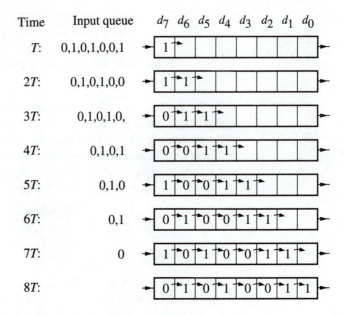

Figure 9.30 Data movement inside the shift register

available at the output of the shift register. Since D_0 was the first bit loaded and the first bit out, this is called a bit-level **FIFO** (First-In, First-Out) circuit.[3] FIFOs are very useful for certain types of storage applications.

A parallel-loading shift register allows the user to store the entire word on one clock transition, as implied by the drawing in Figure 9.31. This particular register has two shift operations that are controlled by the signals *SHR* (Shift Right) and *SHL* (Shift Left). The effects of both are illustrated in Figure 9.32. In Figure 9.32(a), the initial contents of the register are shown as the word 10101100 (as read from left to right). The Shift Right operation is initiated by setting *SHR* = 1; each bit is moved one location to the right (Figure 9.32b). Note that the leftmost bit (in italics) is automatically forced to a 0 value. Figure 9.32(c) shows the Shift Left operation that is obtained by setting the control bit *SHL* = 1. This is exactly opposite to the Shift Right in that each bit is moved left by one location. In this case, it is the rightmost bit that is forced to 0.

When a shift register is used in a complex data processing system, it is common to construct the circuits to allow shifting of more than one location. In this case, the commands are written in the form

$$\text{SHR } n$$

for shifting the contents right by *n* cells, while

$$\text{SHL } m$$

would imply an operation that shifts the contents left by *m* locations.

---

[3] FIFO is pronounced by syllables as *fi-fo*.

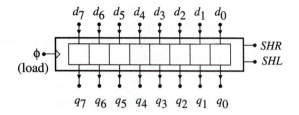

Figure 9.31 Shift register with parallel load capabilities

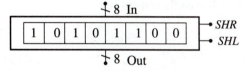

(a) Initial condition

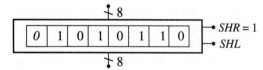

(b) After Shift Right operation

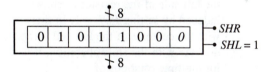

(c) After Shift Left operation

Figure 9.32 Summary of shift operations

One direct application of the shift operations is to perform multiplication and division of a binary number by a power of 2. To understand this comment, let us start with the binary word

$$N = 00011100$$

which has a base-10 value of 28. Let us perform a SHL 1 operation. This gives

$$N_{L1} = 00111000$$

which is decimal 56, i.e., $N_{L1} = 2N$. Starting with the original word and invoking a SHL 2 operation yields

$$N_{L2} = 01110000$$

which is equal to 112_{10}, so that $N_{L2} = 2^2N$. Thus, in general, the SHL n operation is equivalent to multiplying the contents of the register by 2^n.

The Shift Right operation SHR m achieves the inverse results. For example, if we start again with

$$N = 00011100$$

then SHR 1 and SHR2 operations give

$$N_{R1} = 00001110$$

$$N_{R2} = 00000111$$

It is easily verified that

$$N_{R1} = N/2$$

$$N_{R2} = N/2^2$$

corresponding to decimal values of $N_{R1} = 14$ and $N_{R2} = 7$. This demonstrates that SHR m can be used to *divide* the contents of the register by the power 2^m. [4]

In addition to the shift operations, a register may also have the capability of the rotate operations ROR (rotate right) and ROL (rotate left). These are the same as the corresponding shift movements, except that rotation does not lose a bit if it is shifted out but transfers it to the other side. The effects of both operations are summarized in Figure 9.33. In Figure 9.33(a), the contents are shown as 10101100. The rotate-right operation is initiated by setting the control bit *ROR* = 1. This moves each bit one location to the right. The rightmost bit is transferred to the left side of the register as shown. Figure 9.33(b) shows the rotate-left motion induced by setting *ROL* = 1. In this case, each bit moves one location to the left, with the contents of the left cell (originally a 1) transferred to the rightmost location. As with the shift operation, it is possible to construct a register that allows for multiple rotations of

$$\text{ROR } n \text{ and ROL } m$$

by changing the wiring of the memory cells that make up the register.

Shift and rotate operations are important in many types of digital signal processing and are included as base operations in all computers. They will be discussed in more detail in Chapter 11 in the context of computer architectures.

9.6 Random-Access Memory (RAM)

Random-access memory is the term applied to large arrays of general read/write memory cells. RAM is commonly found as the main system memory in a desktop computer unit, and it provides the ability to store all of the important data needed to run the computer, including the program and the operating system.

---

[4] This only works exactly for even-decimal integers.

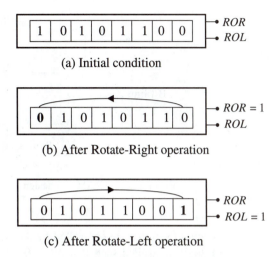

(a) Initial condition

(b) After Rotate-Right operation

(c) After Rotate-Left operation

Figure 9.33 Rotate operations

9.6.1 Static RAM Cell

Static memory is capable of holding the stored data bits so long as the power is connected to the electronic circuits.

SRAM[5] is based on the characteristics of a closed loop consisting of two cascaded inverters as shown in Figure 9.34(a). This is a bistable circuit in that either an $A = 0$ or an $A = 1$ can be held in the loop. Although the circuit may look strange at first, it is operationally equivalent to the SR latch. This can be seen by redrawing the circuit as shown in Figure 9.34(b), which clarifies the cross-coupling of the inverters. Comparing this with the NOR2 latch in Figure 9.4, we see the same structure except that single-input NOT gates are used for the logic.

A static RAM cell that stores one bit of data is constructed by embedding the cross-coupled inverters into a larger network that allows us to connect with the inverter loop. Figure 9.35 shows an SRAM cell where we have added two access

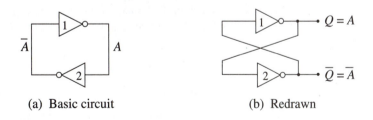

(a) Basic circuit

(b) Redrawn

Figure 9.34 Cross-coupled inverters form a bistable circuit

---

[5] SRAM is read by pronouncing the "S" and adding the word "ram" to it.

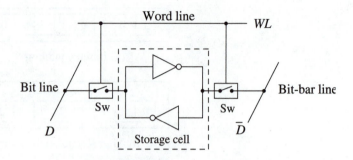

Figure 9.35 SRAM cell design

switches (Sw) that are controlled by the signal WL on the **word line**. The switches are logic-controlled such that $WL = 0$ gives open switches, while $WL = 1$ closes them. When $WL = 0$, the storage cell is isolated from external influences and holds the bit. If $WL = 1$, then both switches are closed, which connects the bit line (D) and bit-bar line ($\overline{D}$) to opposite sides of the cell. This allows writing to, or reading from, the cell. To summarize,

- $WL = 0$ is a hold state, while
- $WL = 1$ allows read and write operations.

The actual operation performed when $WL = 1$ is determined by the auxiliary circuitry.

9.6.2 SRAM Array

SRAM is usually manufactured to allow the easy storage of n-bit binary words instead of individual bits. This is a practical consideration and arises from the fact that most applications are built with this in mind. An $m \times n$ SRAM array consists of m rows, with each row containing n cells.

Let us examine how we might create a very simple memory that stores 8 words, with each word 8 bits in length, i.e., an 8×8 array. Consider first the storage of an 8-bit word. This requires 8 cells, one for each bit, as shown in Figure 9.36. We have used simplified notation for the word line connection, but the meaning remains the same: $WL = 0$ isolates the cells, while $WL = 1$ provides access and allows write or

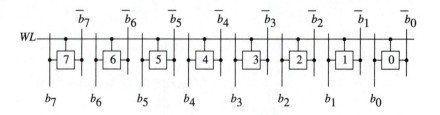

Figure 9.36 An 8-bit memory array

read operations to take place. In order to hold 8 of these words, we build seven more identical circuits, and stack them as shown in Figure 9.37. In order to select the desired row (word), we include a 3/8 active-high **row decoder** circuit. The input to the decoder is a 3-bit binary word

$$A_2A_1A_0$$

that is called the **address**, since it specifies the location of a particular word in memory. Interface circuits have been provided at the bottom of the array. These

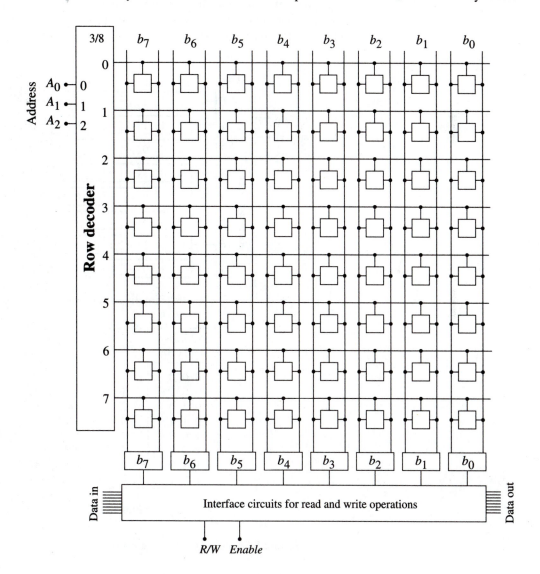

Figure 9.37 An 8 × 8 RAM array

provide the circuit needed to write to, or read from, each row in the array. *Enable* is an active-high control signal that activates the array, while *R/W* determines whether a read is taking place (with *R/W* = 1), or a write is in progress (with *R/W* = 0). The read/write control signal *R/W* is only valid when *Enable* = 1.

Let us clarify the operation of the array by studying a read operation using the simplified drawing shown in Figure 9.38(a) for the read operation. Much of the wiring has been grouped together so that we can concentrate on the data transfer without being sidetracked by too many details. The drawing shows the contents of each row; Figure 9.38(b) summarizes how the input address activates a particular work line, which in turn allows that row of data to be read out. For example, if the address is 100, then line 4 is selected, resulting in an output of 0101 0101.

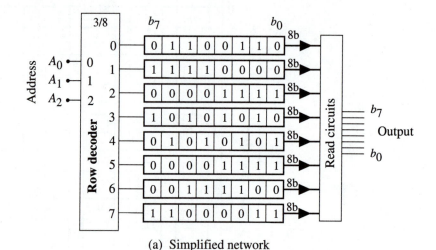

(a) Simplified network

Address $A_2 A_1 A_0$	Row	Output $b_7 \ldots b_0$
0 0 0	0	0110 0110
0 0 1	1	1111 0000
0 1 0	2	0000 1111
0 1 1	3	1010 1010
1 0 0	4	0101 0101
1 0 1	5	0000 1111
1 1 0	6	0011 1100
1 1 1	7	1100 0011

(b) Function table

Figure 9.38 Read operation in a RAM array

Simplified operational diagrams and symbols provide a concise manner in which the important characteristics of a complex digital unit can be shown explicitly without cluttering up a logic diagram. They are equivalent to viewing the unit at

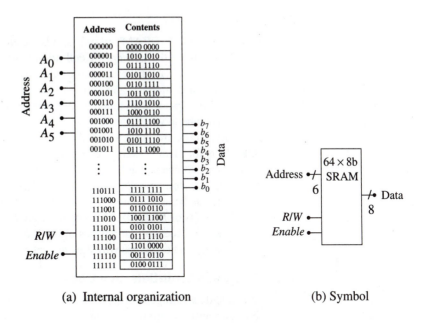

(a) Internal organization

(b) Symbol

Figure 9.39 Block diagram for a 64×8 SRAM deice

a higher level in the hierarchy and are very useful (if not mandatory) for describing a complex system such as a computer. To create a symbol, we must define all of the characteristics that are important to the "outside world." Although the internal circuits may be interesting in themselves, most are of little concern to the user.

Consider an SRAM array. At the user level, it is a device that allows us to perform the write, hold, and read operations on binary words. The capacity of the array tells us how many words can be stored, and the addressing scheme provides the manner in which we can keep track of where every word is located in the array. These properties are summarized by the block diagram in Figure 9.39(a) for a device that holds 64 8b words. The data in and out of the memory device is transferred as the 8-bit word

$$b_7 b_6 b_5 b_4 b_3 b_2 b_1 b_0$$

The address gives the location of each word and is specified by the 6-bit word

$$A_5 A_4 A_3 A_2 A_1 A_0$$

which selects a storage location. If the device is activated using *Enable* =1, then the read and write operations are selected using the *R/W* control. At this level, the internal structure of the RAM is drawn to portray the device as a set of storage locations with each location having a unique address. This allows us to illustrate the contents. A more compact symbol is shown in Figure 9.39(b). The operation of the device is identical in all regards, but none of the internal details are shown. This makes it much easier to use at the system level of the design hierarchy.

9.6.3 Dynamic RAM

A dynamic RAM (DRAM) array is similar to an SRAM array in that it allows us to store data using the concept of cell addressing. The difference between the two types of memory is in the internal design of the cells themselves. Dynamic RAM cells are much simpler and require less area on the silicon chip. This allows DRAMs to be made with much higher storage densities than SRAMs, and the cost per bit is much lower. Because of the lower price, DRAMs are used whenever large memory arrays are needed and the cost is important. The largest application of DRAM is for main system memory in desktop computers. One drawback of using "dynamic" storage is that the simpler cell design in a DRAM results in a circuit that is noticeably slower; both read and write times are increased, which then restricts how they can be used to store data in a large system.

The details of MOS DRAMs and their operation are presented in Section 9.9.2, where we will see that the physics of the bit-storing technique both gives DRAMs high density and explains why they are called "dynamic."

9.6.4 Parity and Error-Detection Codes

Reliable data storage is critical in all digital systems. Since even a single incorrect bit can cause an entire program to crash or render a database file useless, it is important to provide techniques that can detect the presence of an error and, in some cases, correct those errors. The detailed study of error-detection codes (**EDC**) and error-correction codes (**ECC**) is far beyond the scope of this book, but the fundamentals are based on some relatively simple concepts that can be introduced here.

Let us first examine the problem where we want to insure that a data byte read out of a memory location is the same as that which was written. This would require that we record some information about the data byte before it is stored and then use this information to check the byte when it is read out. A simple way to accomplish this is by introducing the concept of a **parity bit** P.

Suppose that the original data byte is written as

$$B = b_7 b_6 b_5 b_4 b_3 b_2 b_1 b_0 \tag{9.9}$$

Since each b_n is either a 1 or a 0, we can count the number of 1s in the word and use this to define a parity bit P that records whether there are an even or an odd number of 1s in the combination of the data byte *including* the parity bit P. There are two options that can be used. If we choose an **even parity** scheme, then the parity bit P_{even} has the characteristic that

P_{even} is chosen to make $(B + P_{even})$ have an even number of 1s.

For example, if $B_1 = 01101101$, then $P_{even} = 1$ since B_1 contains five 1s and this choice gives an even number. Conversely, a byte with a value of $B_2 = 1111000$ requires that $P_{even} = 0$ since B_2 already has an even number (four) of 1s. **Odd parity** is exactly opposite. In this case, we introduce the odd parity bit P_{odd} such that

$(B + P_{odd})$ has an odd number of 1s.

When applied to $B_1 = 01101101$, this would require that $P_{odd} = 0$; similarly, if $B_2 = 1111000$, then we would have $P_{odd} = 1$.

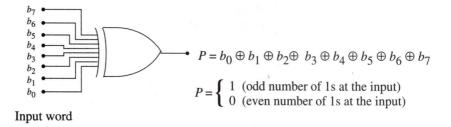

$$P = b_0 \oplus b_1 \oplus b_2 \oplus b_3 \oplus b_4 \oplus b_5 \oplus b_6 \oplus b_7$$

$$P = \begin{cases} 1 & \text{(odd number of 1s at the input)} \\ 0 & \text{(even number of 1s at the input)} \end{cases}$$

Input word

Figure 9.40 A parity generator circuit

An easy way to generate a parity bit is by using the exclusive-OR operation as shown in Figure 9.40. Recall that the XOR provides the odd function. In the logic diagram, we have defined the output as

$$P = b_7 \oplus b_6 \oplus b_5 \oplus b_4 \oplus b_3 \oplus b_2 \oplus b_1 \oplus b_0 \tag{9.10}$$

so that $P = 1$ if there are an odd number of 1s in the word, while $P = 0$ if there is an even number of 1s. With this choice,

$$P = P_{even} \tag{9.11}$$

since the total number of bits in the byte plus P has the required characteristic of even parity. If we used an XNOR gate instead, then the output would be the odd parity bit.

To apply the concept of parity to the error-detection problem, suppose that we store the original data byte plus a parity bit P as illustrated in Figure 9.41(a); from the above discussion, this choice of P represents even parity. Note that this scheme requires that the memory location be $(8 + 1) = 9$ bits wide to accommodate the

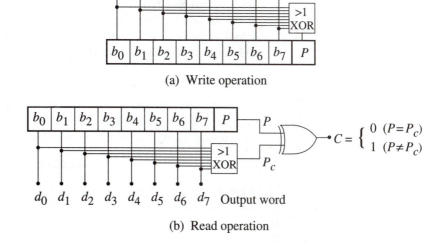

(a) Write operation

(b) Read operation

Figure 9.41 Error-detection circuitry

additional bit. The parity *bit P* provides the information on the original data. When the byte is read out of memory, as in Figure 9.41(b), we obtain the data segment

$$D = d_7 d_6 d_5 d_4 d_3 d_2 d_1 d_0 \tag{9.12}$$

which is used to calculate the new parity bit P_c as

$$P_c = d_7 \oplus d_6 \oplus d_5 \oplus d_4 \oplus d_3 \oplus d_2 \oplus d_1 \oplus d_0 \tag{9.13}$$

Next, the stored parity value P is compared to P_c using the XOR function, which results in the check bit C. If $P = P_c$, then the stored and computed parity values are equal and $C = 0$. If $P \neq P_c$, however, then two parity bits are different and $C = 1$, indicating the discrepancy between the two. Thus, a value of $C = 1$ indicates that an error has occurred, so that either the data byte or the stored parity bit is incorrect. The check bit C is usually placed in a special memory cell and called a **flag**; when $C = 1$, we say that the flag has been set and that an error has been detected. A check bit value of $C = 0$ does not guarantee that no error has occurred; it simply means that the circuit has not detected one.

9.7 Read-Only Memory (ROM)

A read-only memory is a memory array that is used for permanent data storage. Information is stored by the ROM manufacturer or the system programmer. The user has limited programming capabilities (or none at all) depending upon the specifics of the device. ROMs provide **non-volatile** storage; this means that the data remain in the memory even if the power supply is disconnected. Non-volatile storage devices are critical for many systems applications. For example, every personal computer has a ROM[6] that provides the binary codes needed to launch the system software when the power is turned on.

The symbol in Figure 9.42(a) shows the important characteristics of a small ROM. Overall, it is very similar in operation to the RAM. The read operation proceeds by first specifying the address, then enabling the chip; this results in an output of the data. The internal model shown in Figure 9.42(b) provides the necessary details of this device. Stored data words are viewed as groups of bits at specified address locations. The address word

$$A_5 A_4 A_3 A_2 A_1 A_0$$

is sent to a row decoder, which then activates the word storage locations by setting the appropriate word line high. The data word is sent to the output circuits, which are activated by the *Enable* control, resulting in the data word

$$b_7 b_6 b_5 b_4 b_3 b_2 b_1 b_0$$

at the output of the ROM. The write operation is not supported by this device, which is the distinguishing feature of a read-only memory.

---

[6] This is called the BIOS, which stands for Basic Input/Output System.

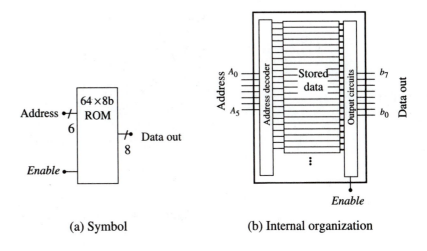

(a) Symbol   (b) Internal organization

Figure 9.42 ROM symbol and user model

9.8 CD ROM

The compact disk (CD) ROM is an interesting example of a complex digital system that incorporates electronics and optics to provide a high-density storage medium. The basic format and operation of a CD ROM are based on the CD audio standard but have more stringent requirements than those used for sound reproduction.

The CD ROM is a circular disk made out of clear polycarbonate plastic that has small indentations molded into one side. These are coated with a layer of aluminum metal, which is then covered with another layer of plastic; the label is then printed on this upper layer. As shown in Figure 9.43, the top side of the CD has the label, while the bottom clear side has the stored information. The interface between the polycarbonate plastic and the aluminum is called the **information surface**, as this is where the data is stored. The information is recorded on a single track that starts in the center of the disk and spirals outward. This is shown in Figure 9.44(a). To read the data from the disk, the disk spins over a stationary laser light source that is contained in the player mechanism. For purposes of analysis, this is most conveniently viewed as a stationary disk with a moving laser beam following a track, as portrayed in Figure 9.44(b).

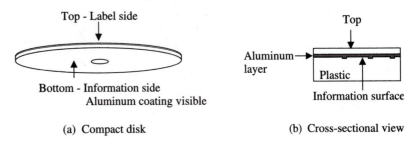

(a) Compact disk   (b) Cross-sectional view

Figure 9.43 Structure of a compact disk

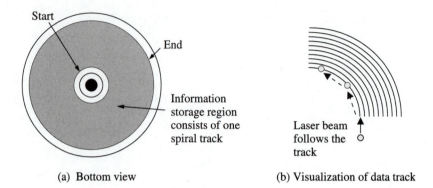

(a) Bottom view

(b) Visualization of data track

Figure 9.44 CD information tracks

Figure 9.45 shows a close-up of the information surface as seen looking through the clear plastic towards the aluminum coating. Most of the surface is flat and is referred to as **land** regions. However, a number of small indentations called **pits** are molded into the plastic and are then covered by aluminum. Data storage is achieved by providing a series of pits and lands along an implied line which is called a **track**. These tracks are placed 1.6 μm apart from one another, allowing for dense storage. A CD ROM is capable of storing about 540 MB worth of data for a 60-minute play time at normal speed. The pits and lands are "stamped" into the plastic when the disk is manufactured and cannot be changed, making this a read-only device.

Data read-out is accomplished by focusing a small beam of light from a laser onto the track as the disk spins. The beam spot follows the track at a constant velocity of $v = 1.25$ m/sec for a standard (1×) speed player. The information is read out

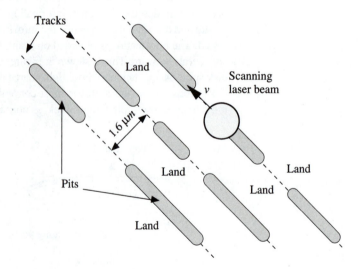

Figure 9.45 Pits and lands on the CD information surface

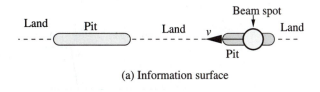

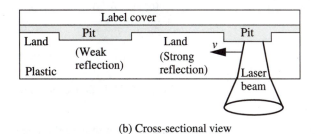

(a) Information surface

(b) Cross-sectional view

Figure 9.46 Laser reading of the information track

by tracking the light that is *reflected* off the information surface. Figure 9.46 shows a track and its cross-sectional view. When the laser beam is on a land region, the aluminum coating provides a strong reflection. If, on the other hand, the beam is scanning a pit, then the reflected beam is very weak due to optical effects known as **scattering** and **interference** of the light. Both the depth of the pit and the size of the beam spot relative to the pit width are important.

When the power P_{ref} of the reflected laser beam is monitored, we find that it varies depending upon whether the laser beam is over a pit or a land region. This is illustrated by the drawing in Figure 9.47, which shows the reflected power varying between high and low levels. Although it may seem obvious that we should use a high level for a logic 1 and a low level for a logic 0, physical considerations came into play during the design of the system, and a different **coding** scheme was chosen for the CD standard. The coding used for the CD is called **NRZ-i**. It defines logic levels in the following manner:

- **Logic 0** Constant high or low level

- **Logic 1** A transition from high-to-low or low-to high

Thus, when the laser is scanning a pit or a land region, the constant reflected power intensity is interpreted as a logic 0. A logic 1 occurs only when the beam passes from a land to a pit, or from a pit to a land. Using this coding scheme provides the logic interpretation shown in the drawing. Note that the key to this technique is defining the time interval T for one logic bit. For a 1× speed player, this gives a serial data readout rate of about 150 *KB/s* (kilobytes per second).

NRZ-i coding is used because the disk-reading process does not produce the nice well-defined reflected beams shown in the drawing. Instead, the reflected power intensity looks more like that shown in Figure 9.48 because of non-uniform laser illumination, tracking errors, vibrations, and other perturbations in the physi-

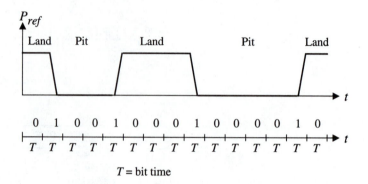

Figure 9.47 Reflected laser power as it scans the surface

cal environment. Even though the intensity varies while over a pit or land region, the CD system insures that there is a still reasonable change when the laser beam moves from a pit to a land, or vice versa. This coding scheme thus allows for a reliable output in spite of the non-ideal signal.

The use of the coding leads to other complications in the manner in which data is stored on the track. If we compare the width and length of the pits and the diameter of the laser beam (which is about 1 μm), we find that it is not possible to have two adjacent 1s on the track. In fact, this leads us to a rule that

> "1s must be separated by at least two 0s"

when recorded on the track. This implies that we cannot store "standard" coded data directly on the track of a CD! Instead, the data is subjected to a **modulation** scheme which is similar in principle to the encoding/decoding process.

Compact disks use a technique called **eight-to-fourteen modulation** (EFM) where an 8b data word is expanded to a 14b modulated equivalent such that the modulated word does not violate the spacing rule between 1s. The overall scheme is shown in Figure 9.49(a). It is the modulated bit stream that is recorded on the disk, not the original data. The actual translation is defined using a look-up table; a small section of the table has been reproduced in Figure 9.49(b). When the data is read out, it must be demodulated by simply using the table as a dictionary to produce the original 8b word.

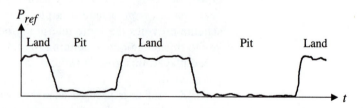

Figure 9.48 Realistic reflected signal variations

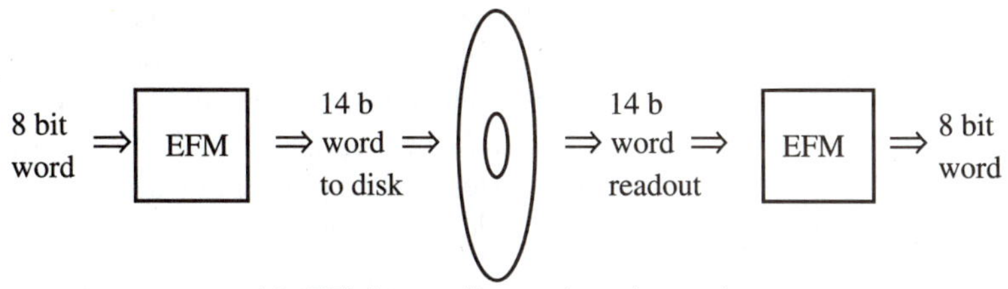

(a) EFM recording and readout scheme

8-bit word	14-bit equivalent
00000000	01001000100000
00000001	10000100000000
00000010	10010000100000
00000011	10000100010000
00000100	00000100010000
00000101	00010000100000
⋮	⋮

(b) Portion of the EFM table

Figure 9.49 CD eight-to-fourteen modulation (EFM) scheme

Another interesting aspect of the CD is its use of both error-detection codes (EDC) and error-correction codes (ECC) on the data after it is read off of the disk. The variations inherent in the CD manufacturing process lead to a relatively high probability of obtaining an error in the data stream. In general, this is specified by the **bit-error rate (BER)** such that

$$BER = \frac{N_e}{N_T} \tag{9.14}$$

where N_e = number of incorrect bits (error bits) and N_T is the total number of bits being considered. In CD audio applications (the original use of the CD standard), the *BER* is specified to have a maximum value of $BER = 10^{-9}$. Computing standards are more stringent, with $BER = 10^{-12}$ the highest acceptable value. In order to insure that the CD meets this value, a sophisticated set of EDC and ECC is used on the data before it is recorded. This results in an additional set of data that is recorded on the disk. The idea is similar to using the concept of the parity bit as in the RAM array, but it is much more powerful in that the codes allow (a) the detection of an error, and (b) the ability to correct the error by calculating what the original word was. The CD employs an algorithm embodied in a broad class of theory known as the **Reed-Solomon codes,** which work with **blocks** of data words to accomplish the task. These are so powerful that you can actually cover up a segment of track on the CD about $1mm$ long, and the processed data output will not be affected!

The compact disk system is quite complex but is an excellent example of a realistic digital system. Other approaches to optical recording and data storage have been developed. However, all tend to be bound by similar physical limitations, and each has its own unique aspects that illustrate how a digital design can solve a complex system design problem by using basic principles.

9.9 CMOS Memories

CMOS is the dominant technology for all types of memory chips. Since the MOS-FETs are small, it is possible to create memory arrays with very large capacities that are quite fast. In this section we will briefly examine some of the concepts involved in creating memory cells using CMOS techniques.

9.9.1 CMOS SRAMs

A single CMOS SRAM cell circuit is shown in Figure 9.50. It consists of two cross-coupled inverters (Mn1, Mp1 and Mn2, Mp2) for the storage cell, with access obtained using a pair of n-channel transistors MA1 and MA2. This has the same structure shown in the more generic SRAM schematic of Figure 9.35, so the operation is the same. In the literature, this is referred to as a "6T" design due to the fact that it has six transistors. The 6T design is probably the most common approach to making CMOS memories. It is also possible to create a "4T" cell that uses four MOSFETs and two resistors.

The simple scheme used to access the cell can be extended to **multiport SRAMs** where more than one pair of input/output lines can be used. A dual-port SRAM is shown in Figure 9.51. This has two independent data I/O paths that are denoted by D_1 and D_2. Cell access is controlled by two independent word lines that are labeled $WL1$ and $WL2$. Either can be used for read or write operation. One problem that arises is that we must enforce exclusivity, i.e.,

$$WL1 \cdot WL2 = 0 \tag{9.15}$$

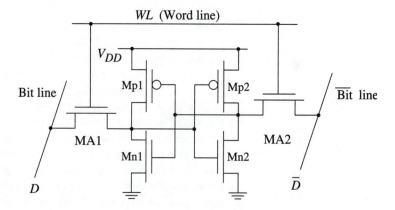

Figure 9.50 CMOS SRAM cell

to insure that one input doesn't interfere with the other. This is accomplished by using on-chip logic that arbitrates access to each cell. Multiport RAMs are useful because they allow us to hardwire the shared memory to several digital logic units at one time.

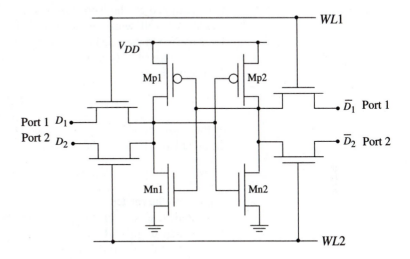

Figure 9.51 A dual-port SRAM storage cell

9.9.2 Dynamic RAM

Dynamic memories are more complicated to characterize but are used extensively for main storage because they can be built with much higher densities that SRAM cells. The descriptive adjective *dynamic* reflects the fact that the contents of a cell will change in time during a hold operation. This makes the design of dynamic memories more complicated, but modern chips usually shield the complexity from the user. The main drawback of DRAMs is that they are not as fast as SRAMs. The main advantage of DRAM is that the cost per bit is very low, making it economically attractive in system design where large arrays are required.

The circuit schematic for a DRAM cell is shown in Figure 9.52. The simplicity is apparent at first sight: the entire cell consists of a single nFET and a storage capacitor C_s. This allows for a very high integration density and makes it possible

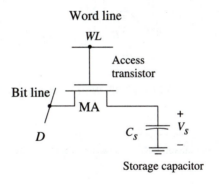

Figure 9.52
CMOS dynamic RAM (DRAM) cell design

to create a single chip that has millions of cells; 256Mb and 1Gb per chip sizes are possible with modern technology. The competitive nature of the DRAM market makes it the **technology-driver** in that many companies use their most advanced processing technology for making DRAMs.

In this basic storage circuit, the nFET acts as an access device with the conduction controlled by the word line variable WL. When $WL = 1$, the nFET acts as a closed switch allowing a write or read operation. To write to the cell, a data voltage V_D is applied to the data line, and the resulting current charges the storage capacitor C_s to a voltage V_s. This is shown in Figure 9.53. The value of the stored bit is defined by the charge

$$Q_s = C_s V_s \qquad (9.16)$$

on the capacitor. If $V_s = V_0 = 0$ V, then $Q_s = 0$ corresponding to a logic 0 value. On the other hand, if $V_s = V_1$ is a high voltage, then a logic 1 charge

$$Q_s = C_s V_1 \qquad (9.17)$$

is stored on the capacitor.

A hold state is obtained by bringing the word line to $WL = 0$, shutting off the direct conduction path between the data line and the storage capacitor. Ideally, the transistor would block all currents, and the charge Q_s would be held indefinitely. However, nFETs cannot block all of the current flow. Even with a zero gate voltage, FETs admit a small **leakage current** I_L that removes charge from the capacitor. The circuit effects of the leakage path are shown in Figure 9.54(a), while the drawing in Figure 9.54(b) shows the physical origin of the problem in the transistor where the pn junction admits a reverse current I_R that cannot be eliminated. In this case, we may write that[7]

$$I_L = I_R \qquad (9.18)$$

Since charge can leak from C_s, and the value of the bit depends upon the charge through the voltage, the data bit can only be held for a short period of time.

It is worthwhile to estimate the time interval that the value can be held. Suppose that we place a logic 1 voltage $V_s = V_1$ on the capacitor and we wish to calculate the

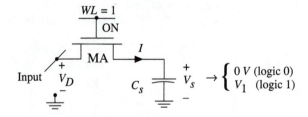

Figure 9.53 Write operation in a DRAM cell

---

[7] In realistic DRAM cells, several other leakage currents are also found to exist.

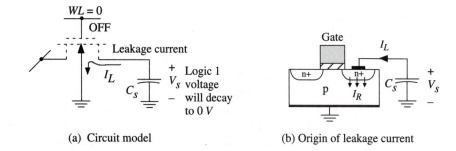

(a) Circuit model   (b) Origin of leakage current

Figure 9.54 DRAM HOLD state and leakage problem

time that the voltage can be held on the cell when the nFET is turned off. This can be accomplished by using the definition of current to write

$$I_L = -\frac{dQ_s}{dt}$$

$$= -C_s \frac{dV_s}{dt}$$

(9.19)

where we have used the fundamental relation

$$Q_s = C_s V_s$$

(9.20)

in arriving at the second line. The minus sign is used because the charge is decreasing in time. Rearranging gives

$$dt = -\left(\frac{C_s}{I_L}\right) dV_s$$

(9.21)

so that integrating both sides gives

$$\int_0^{t_H} dt = -\left(\frac{C_s}{I_L}\right) \int_{V_1}^{V_{min}} dV_s$$

(9.22)

where t_H is the **hold time**. The hold time is defined as the time interval that the capacitor can maintain the voltage above V_{min}, which is the minimum voltage that will still be interpreted as a logic 1 value. The lower limit V_1 on the right side represents the voltage of the initial stored logic 1 charge at time $t = 0$. Evaluating the integral and rearranging gives the hold time as

$$t_H = \frac{C_s}{I_L}(V_1 - V_{min})$$

(9.23)

To understand the meaning of this time interval, we note that the storage capacitor is on the order of $50\,fF = 50 \times 10^{-15}\,F$, while leakage currents are on the order of $1\,pA$ (which is $10^{-12}\,A$). Since the voltage is on the order of $1\,V$, the hold time is about

$$t_H \approx \frac{50 \times 10^{-15}}{10^{-12}} = 50 \times 10^{-3} \qquad (9.24)$$

i.e., 50 milliseconds (*ms*). This is a very short period of time. The data must be periodically updated to insure that it is valid when the system needs it. This is called a **refresh** operation and is described below.

A read operation is shown in Figure 9.55. To retrieve data from the cell, we put $WL = 1$ to turn on the access transistor, which allows current to flow to the bit line. An amplifier then provides a strengthened signal at the output. The amplifier is a critical part of the design, and much attention is devoted to it in the design process. Read circuitry must be reasonably fast in order to provide quick access to the data.

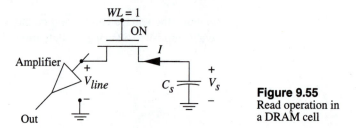

Figure 9.55
Read operation in
a DRAM cell

The refresh operation is where we periodically perform a readout of the data, amplify it, and then write it back into the cell. The refresh circuitry is included on the chip and makes it appear that the memory has long-term retention characteristics. Since the maximum hold time is on the order of 10^{-3} second, the minimum refresh frequency is on the order of

$$f_{min} = \frac{1}{t_H} \sim 1000 \; Hz \qquad (9.25)$$

Refresh rates are thus on the order of a few kilohertz. It is important to note that every cell must be periodically refreshed, so the circuitry is designed to cycle through the entire array.

Many advances in modern computer technology can be traced back to technological developments in the RAM manufacturing process. For this reason, it is a fascinating field to study even if your interests lie in other areas.

9.9.3 ROMs

CMOS technologies provide for a large variety of ROM circuits to be manufactured. Although these are classified as "Read-Only" devices, they allow user data to be entered by special circuit techniques. These usually require elevated voltage levels well above those normally used to operate the chip. Since the devices can be programmed, they are sometimes referred to as "Read-Mostly" devices.

EPROMs

EPROM is an acronym for Erasable-Programmable ROM. This type of device allows the user to both program (enter data) and erase that data.

Programming is achieved by a process in which a high voltage is used to transfer charge to a "floating" capacitor. The charge is trapped on the capacitor and cannot escape under normal circumstances. In this type of device, erasure is achieved by placing the device under an intense ultraviolet (UV) light source. Electrons absorb photons (packets of light energy), giving them energy to move back into the semiconductor. Although these were once the dominant type of user-programmable ROM, they have been replaced by devices that can be programmed and erased electrically.

E²PROMs

This type of memory is Electrically-Erasable (E^2) instead of requiring the UV technique of EPROMs. It has the advantage that the data may be erased using electrical circuitry and does not require that the chip be physically removed from the system. Because of this characteristic, it is very popular in modern digital system design.

Programming is achieved by placing a high voltage onto the pins, which causes charge to be transferred to a capacitor plate. To erase the cell, the voltage is reversed and the charge moves in the opposite direction. In a basic array, erasure is restricted to a single cell at a time. However, newer technologies allow one to erase a large number of cells simultaneously. These devices are called **flash** EPROMs, with "flash" referring to the speed at which the entire array may be erased.

9.10 **Transmission Gate Circuits**

Transmission gates can be used to create CMOS memory elements in integrated circuit designs. They are useful because a TG behaves almost like an ideal switch that can be open and closed. Using a TG at the input to a storage circuit allows one to control the latching action using the TG control mechanism.

9.10.1 **Basic Latch**

Figure 9.56 shows a basic latch circuit that uses two transmission gates TG0 and TG1 to control the data flow; the two inverters provide the bistable storage. The operation of the circuit is accomplished by the load control bit LD and its complement $\overline{LD}$. Note that the control signals to transmission gates TG0 and TG1 are applied in an opposite manner; when one TG is closed, the other is open.

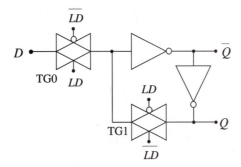

Figure 9.56 TG latch circuit

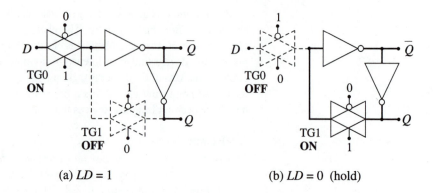

(a) $LD = 1$ (b) $LD = 0$ (hold)

Figure 9.57 TG latch operation

To understand the operation of the latch, let us first examine the case where LD = 1. This closes TG0 and opens TG1, giving the conditions portrayed in Figure 9.57(a). The input data bit D can enter the circuit and gives $Q = D$ at the output of the second inverter; this allows us to load (write) a data bit into the latch. Note that the latch is transparent so long as $LD = 1$, since the output will track changes at the input. The hold condition is achieved by having $LD = 0$ as shown in Figure 9.57(b). This turns off TG0, which in turn blocks the entry of new data. Transmission gate TG1 is a closed switch, yielding the bistable cross-coupled inverter circuit that can maintain the value of Q.

The basic TG latch is classified as being a level-sensitive device since it accepts inputs when $LD = 1$. It can be made into a clock-controlled latch by incorporating a clock signal ϕ into the control set. In Figure 9.58, the LD (and $\overline{LD}$) signal has been replaced by ϕ (and $\overline{\phi}$) everywhere; this gives a latch that accepts new data whenever the clock is in a state $\phi = 1$.

Another approach to clocking the input is shown in Figure 9.59. In this case, the input transmission gate TG0 is controlled by the composite signals $LD \cdot \phi$ and

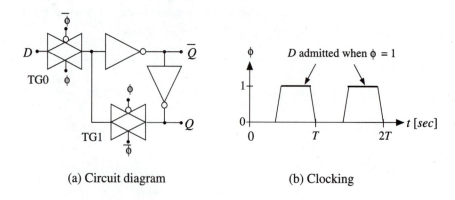

(a) Circuit diagram   (b) Clocking

Figure 9.58 Clocked TG latch circuit

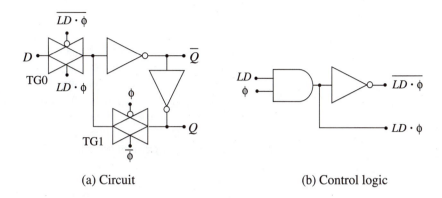

(a) Circuit   (b) Control logic

Figure 9.59 Clocked latch with a controlled load

$\overline{LD \cdot \phi}$. New data is admitted only when both $LD = 1$ **and** $\phi = 1$ are true. The circuit thus acts similarly to the clocked version in Figure 9.58, except that LD provides an additional level of control over the flow of data. An example of the timing for this circuit is illustrated by the waveforms in Figure 9.60. As always, the clocking signal $\phi(t)$ is used as the timing reference. In this case, the load signal LD is high only during the first and fourth clock cycles. The composite signal $LD \cdot \phi$ thus follows the clock during these times but is zero in the middle two clock cycles when $LD = 0$. The result of this observation is that the latch will only accept data during the first and fourth cycles.

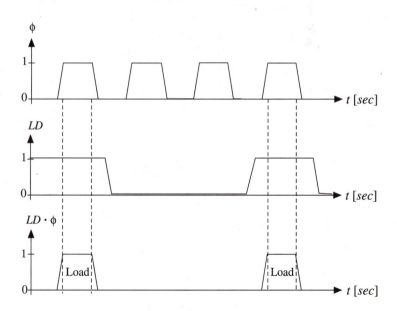

Figure 9.60 Timing diagrams for the clocked TG latch with load control

9.10.2 TG Flip-Flop

A master-slave flip-flop can be created by cascading two oppositely phased TG latches as shown in Figure 9.61. The first latch (Stage 1) acts as the master, while the second latch (Stage 2) is the slave circuit. This can be seen by noting that the input transmission gates are controlled by opposite clock phases. The master is active and accepts data when $\phi = 0$, while the input to the slave requires a clock value of $\phi = 1$ to admit new information. This is identical in operation to a positive edge-sensitive DFF. A negative edge-sensitive DFF can be obtained by simply reversing the clocking signals ϕ and $\bar{\phi}$. This circuit is illustrated in Figure 9.62.

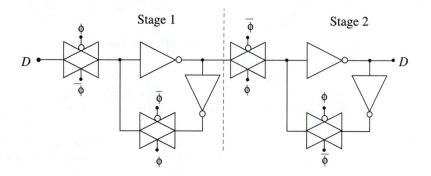

Figure 9.61 Master-slave DFF circuit

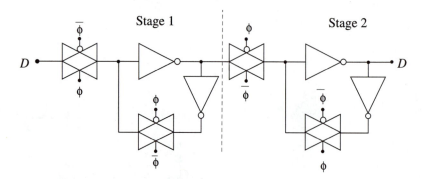

Figure 9.62 Negative edge master-slave DFF circuit

9.11 Problems

[9.1] The set (*S*) and reset (*R*) signals below are applied to an SR latch. Sketch the output $Q(t)$ for the device.

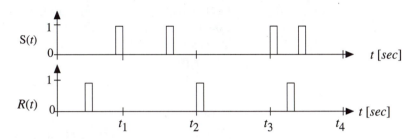

[9.2] The signals below are applied to the clocked SR latch in Figure 9.10. Sketch the output $Q(t)$.

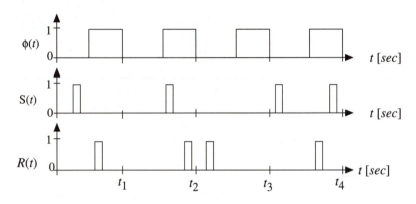

[9.3] The data signal $D(t)$ shown below is applied to the input of a positive edge-triggered DFF. Sketch the output $Q(t)$ for the device.

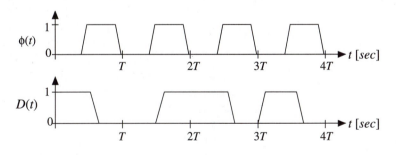

[9.4] Redo the previous problem for the case where $D(t)$ is applied to the input of a negative edge-triggered DFF.

[9.5] The reset (R) input to a NOR-based SR latch is pulsed to a logic 1 while $S = 0$. Draw the circuit and label the logic value (0 or 1) at every node when this occurs.

[9.6] Consider the NAND-based SR latch. What are the outputs when both S and R are at 0 levels? Why is this the "not used" case?

[9.7] Consider an 8-bit shift register that initially has the binary word 01011100 stored in it. What are the contents of the register after the following operations have been performed? For each case, assume the initial state given above.

(a) SHR 2
(b) SHL 1
(c) ROL 3
(d) ROR 2

[9.8] An 8-bit shift register has the binary equivalent of the decimal number 86 stored in it. What are the base-10 equivalent contents of the register after the following operations have been performed? For each case, assume the same initial state given.

(a) SHR 1
(b) SHL 1
(c) SHR 2
(d) ROR 2

[9.9] A 16-bit shift register contains the binary equivalent of 0x 4A35. Find the base-10 equivalent contents of the register after the following operations have been performed. Assume the same initial contents for each case.

(a) SHR 6
(b) SHR 3
(c) SHL 8
(d) SHL 16

[9.10] Consider the 8×8 SRAM array in Figure 9.37. The contents of each row is given by the following list:

Row 0: 1010 1111
Row 1: 1010 1111
Row 2: 1010 1111
Row 3: 1010 1111
Row 4: 1010 1111
Row 5: 1010 1111
Row 6: 1010 1111
Row 7: 1010 1111

What is the decimal value of the output when the following addresses are applied to the row decoder?

(a) $A_2A_1A_0 = 101$
(b) $A_2A_1A_0 = 011$
(c) $A_2A_1A_0 = 100$
(d) $A_2A_1A_0 = 110$

[9.11] Draw the block diagram for a 16×4 SRAM using the drawing in Figure 9.38 as a basis.

[9.12] Consider the 64×8 SRAM shown in Figure 9.39. This device is to be used as a basis for creating a larger memory array with a size of 64b $\times$ 32b. Construct the block diagram for the large memory.

[9.13] Consider the 64×8 SRAM shown in Figure 9.39. You would like to use this device to design a memory array that holds 256 32-bit words. Construct the block diagram for this memory array. You may want to consider adding a **chip select input** CS to each device that controls each chip such that $CS = 0$ disables the device, while $CS = 1$ enables it, allowing data to be written and read.

[9.14] Search the computer advertisements in a recent newspaper or a computer magazine and find the best prices you can on SRAM and DRAM chips or modules. Then compute the price per bit for each type. What is the percentage difference in price between the two types of memory?

[9.15] Study the computer ads in a magazine or a newspaper. How many specific applications of SRAM in a personal computer can you find?

[9.16] Consider the 3-bit data word $b_2 b_1 b_0$. The bits are used to generate an even parity bit p. The $3 + 1 = 4$ bits are then stored in memory. When the data is read from the memory, we obtain the results listed below. Determine if the data meets the parity check, or if there is an error.

(a) $p\ b_2 b_1 b_0 = 0\ 111$
(b) $p\ b_2 b_1 b_0 = 1\ 101$
(c) $p\ b_2 b_1 b_0 = 0\ 100$
(d) $p\ b_2 b_1 b_0 = 1\ 101$
(e) $p\ b_2 b_1 b_0 = 0\ 000$

[9.17] Suppose that we want to store 1024 16-bit data words on the surface of a compact disk. How many bits will be actually stored on the information surface of the disk?

[9.18] The powerful error-correction capabilities of a CD ROM are obtained by recording a lot of data that contains enough information to reconstruct any erroneous outputs. Since this is not user-accessible data, it is called **overhead** and it reduces the amount of useful storage.

Information on a CD is divided into **blocks** of data that simplifies the directory. Each block contains 2352 bytes, of which 2048 bytes are actual data. The longest possible "play time" of a standard CD is 74 minutes, corresponding to 333,000 blocks.

(a) Calculate the total number of user bytes for a 74-minute play time.
(b) What percentage of the total stored data is overhead?
(c) How many user bytes are available if the play time is reduced to 60 minutes? (The CD has a linear relationship between storage and play time.)

[9.19] A data transmission system is specified to have BER of 10^{-10}. Suppose that the system transmits data at a rate of 100 Mb per second. What is the average time that the system can transmit data without an error being expected? How could we combat this problem in a real-world environment?

[9.20] Consider a dynamic RAM (DRAM) cell that is in a hold state. The storage capacitance has a value of $C_s = 60\,fF$, and the leakage current is estimated to be $I_L = 0.2\,pA$. The voltage characteristics are known to be $V_1 = 3\,V$ and $V_{min} = 1.2\,V$ for a logic 1 state. Find the hold time t_H of the cell.

[9.21] A dynamic RAM (DRAM) cell has a storage capacitance with $C_s = 55\,F$, and the leakage current is estimated to be $I_L = 0.14\,pA$. The maximum voltage across the capacitor is $V_1 = 3\,V$ and the minimum logic 1 voltage is $V_{min} = 1\,V$.

(a) How long can the cell hold a logic 0 state?
(b) What is the hold time for a logic 1 state?

[9.22] A **soft error** in a DRAM is the situation where an alpha particle strikes the silicon and causes charges to flow. This can change the contents of the cell. Can you think of a DRAM environment where this may be of particular importance?

CHAPTER 10

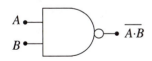

Sequential Logic Networks

One of the most powerful aspects of a digital network is the ability to perform operations in sequence. In a simple design, the order of the steps remains the same, and the operation just loops on itself. However, it is much more useful to examine systems where the sequence depends upon the results of earlier steps.

10.1 The Concept of a Sequential Network

A **sequential network** is a digital system where the output is determined by both the present input **and** the result of a previous event. This type of digital system is very common in the everyday world. In fact, a desktop computer system can be classified as a complex sequential network.

Let us examine the concept of sequencing by means of a simple example: the automatic teller machine that allows you to perform banking transactions from a remote location. To operate the machine, you must follow a set of operations in the proper sequence. The first steps are common to every transaction:

1. Insert your banking card, then

2. Enter your personal identification number (PIN).

After this, you must choose the transaction and respond to the prompts. For example, suppose that you choose

3. Withdrawal from checking account.

Then, the machine will prompt you to enter the amount, to which you might type

4. $20.00

You will then receive the money and be prompted for the next transaction.

This everyday procedure illustrates the concept of a sequential process. Once the system sequence is initiated, different transactions are possible. You make a choice by providing inputs in a specific sequence. The result of an entry is used to determine the next step. In this example, the sequencing is a result of the software that is programmed into the computer.

A sequential network is based on the same idea, except that the sequence is determined by the wiring of logic blocks and memory elements. Given a set of inputs, we may perform operations using logic gates networks. When the outputs are computed, we will save some (or all) of the "present" results for use in the "next" calculation. The next calculation depends upon the result of the present one.

Let us look at a digital unit called a **counter** to illustrate some of the characteristics of sequential logic. The simple counter shown in Figure 10.1 uses a pulse train that is fed to the input X. The unit is designed to count the pulses as they arrive at the input; the number of received pulses is indicated by the 3-bit binary word $n_2 n_1 n_0$. Since this is limited to the decimal range 0-7, we will provide an output R that is defined such that

$$R = 0 \quad \text{Count is in range 0-6}$$
$$R = 1 \quad \text{Counter to 7; output will be reset to 0 on next pulse}$$

(10.1)

This cycles the value of $n_2 n_1 n_0$ according to the sequence

$$000 \to 001 \to 010 \to 011 \to 100 \to 101 \to 110 \to 111 \to 000 \ldots \quad \textbf{(10.2)}$$

Note that the counter will be defined to reset to 0 automatically at the end of the counting sequence. As seen from the diagram, the output R is given by

$$R = n_2 \cdot n_1 \cdot n_0 \quad \textbf{(10.3)}$$

so that it detects the 111 value.

Now let us introduce some formal terminology. We define each possible value of the word $n_2 n_1 n_0$ as a distinct **state** of the **machine**. This means that there are eight possible states that the network can be in. To describe the operation of the circuit, we introduce the use of drawings that are called **state diagrams.** These show (1) each individual state of a machine, and (2) all possible sequences for the circuit changing from one state to another. The basic element of a state diagram is shown in Figure 10.2. It consists of a circle that lists the state, with arrows in and out to

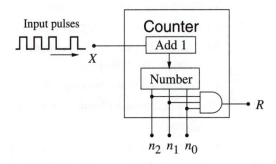

Figure 10.1 A simple counter network

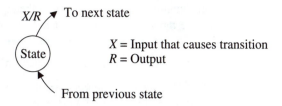

Figure 10.2 Symbolic representation of a state

show transitions to the state and away from the state, respectively. A state is considered as stable unless an action causes the circuit to change to another state. In the notation of the drawing, we use X/R to denote the value of the input X that causes the transition in the direction shown, and the value of the output R when the transition occurs. Every bubble must have lines that show what happens for every possible value of X.

Let us now construct the state diagram for the counter. Since the sequence is well defined, we note that a pulse input of $X = 1$ causes a transition to the next state; the output R is 0 unless the count reaches a value of $n_2n_1n_0 = 111$, which gives $R = 1$. This leads to the diagram shown in Figure 10.3. To interpret the diagram, simply start at any state, and follow the arrow out of the state every time a pulse is received ($X = 1$). The eighth pulse returns you to the starting point, and the cycle repeats. The diagram thus provides all of the information needed to describe the behavior of the network. It may also be used as a starting point to design the logic network itself.

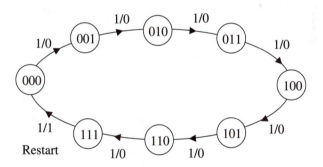

Figure 10.3 Simplified state diagram for the counter

10.1.1 Sequential Network Requirements

To create a sequential network, we need at least

- A combination logic network to perform Boolean operations, and
- A memory element that stores the result of an earlier operation.

The output of the memory is fed to the input of the combinational logic network to provide the sequential dependence from one state to the next. We will restrict our discussion to **synchronous logic networks**[1] where we require

- A clock signal $\phi(t)$ to synchronize the events.

Figure 10.4 shows the important features of the clock $\phi(t)$. The period is denoted by T with units of seconds, and the frequency of the clock is calculated from

$$f = \frac{1}{T} \ Hz \tag{10.4}$$

as discussed in Chapter 9. Since the clock is characterized by a repeating waveform, it provides a convenient *synchronization* reference. Note in particular that any point on the clock repeats every T seconds. As shown on the waveform, we have chosen the time t when the data bit is loaded for use as a reference that will be termed the **present (clock) cycle**. Anything at $(t - T)$ will be referred to as happening in the **previous cycle**, while events a $(t + T)$ are said to occur during the **next cycle**. To simplify the notation, we will write these times as $t - 1$ and $t + 1$, respectively.

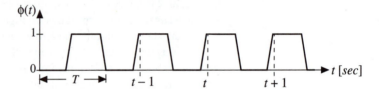

Figure 10.4 Clocking signal convention

We will achieve the system synchronization by applying the clock $\phi(t)$ to the memory and create what will be termed a **state element**; a state element is used to hold the state. For simplicity, we will generally employ positive edge-triggered state elements as shown in Figure 10.5. This means that the value of the data bit D is admitted to the memory element only during the time that the clock is making a transition from a 0 to a 1 state. This means that the state element is opaque (non-transparent) at all other times, and the present value of the input D does not change the stored state or the output.

The DFF has a very nice characteristic when used as a state element. This is summarized in Figure 10.6. We have labeled the input as D_Q corresponding to the output Q; this is convenient notation that will be used throughout our discussion. Now then, from the timing diagram we see that the present value of D_Q at time t is destined to become the next value of Q at time $(t + 1)$. Thus,

$$Q(t + 1) = D_Q(t) \tag{10.5}$$

---

[1] A non-synchronous, or **event-driven**, network is much more complicated to analyze and will not be discussed here.

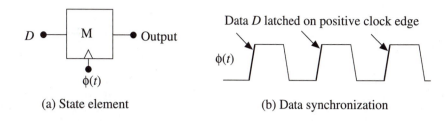

(a) State element   (b) Data synchronization

Figure 10.5 A positive edge-triggered DFF

describes the memory characteristic of the DFF. An equivalent statement is

$$Q(t) = D_Q(t-1) \tag{10.6}$$

since this says that the present value of the output $Q(t)$ was the previous value of the input data bit. This simple property provides us with the power that we need to analyze circuits that use D flip-flops as state elements.

10.1.2 A General Sequential Network

There are several types of sequential networks that can be built. A basic configuration that illustrates many of the important features is portrayed in the block diagram of Figure 10.7. There are two main elements of the network: two combinational logic blocks numbered as #1 and #2, and the state element (a clocked memory) labeled as "M." All sequential circuits have memory elements, but the placement and connection of the logic blocks will probably be different than the generalized drawing shows.

The externally applied inputs A, B, C are the inputs to the network. These are fed to both combinational (random or structured) logic networks. Logic block #1 generates the internal signals x and y, which are then stored in the memory unit M during the next rising clock edge. The outputs of the memory unit are denoted as a and b and represent the previous values of x and y, i.e.,

$$a(t) = x(t-1)$$
$$b(t) = y(t-1) \tag{10.7}$$

	DFF		Present Input = Next Output

D_Q Input — D Q — Q Output $\phi(t)$

$$D_Q(t) = Q(t+1)$$

(a) State element   (b) State behavior

Figure 10.6 DFF characteristics

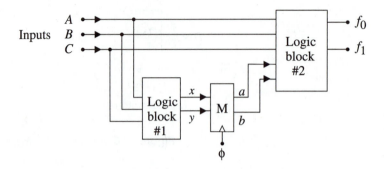

Figure 10.7 An example of a sequential network

These are used along with A, B, and C as inputs to the logic block #2, which yields the network output functions f_0 and f_1. The key to understanding the behavior of a sequential network is to recognize that the output states f_0 and f_1 are in general functions of both the external inputs and the values provided by the memory:

$$f_0 = f_0(A, B, C; a, b)$$
$$f_1 = f_1(A, B, C; a, b)$$

(10.8)

This illustrates how the previous inputs (signals x and y) will influence the next state (through a and b). In other words, the events that take place during a given clock cycle will help determine what action is performed during the next clock cycle. This is why we use the adjective *sequential* to describe this type of digital system.

As a final comment, we note that the memory block in this example must be capable of storing two bits x and y (or equivalently a and b) using two state elements (FFs). The number of possible states for this machine is thus calculated $2^2 = 4$, and these states are designated by the listing

$$(ab) = (00), (01), (10), (11)$$

(10.9)

A state diagram for this network would consist of four state bubbles, one for each possible state value.

Figure 10.8 shows another general form for a sequential network. In this configuration, a single block of combinational logic is used to provide both the outputs g_0 and g_1 and the state inputs x and y. The memory block M has state outputs a and b that act with the primary inputs A, B, and C in determining the outputs and the next state. It is possible to draw a few other general circuits, but all have similar characteristics, namely, combinational logic blocks and one or more state elements.

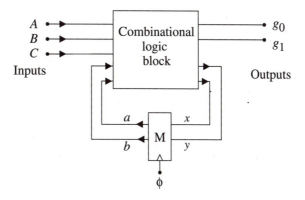

Figure 10.8 Another example of a state machine

10.2 Analysis of Sequential Networks

We will examine sequential logic circuits by first analyzing a given sequential circuit to build an understanding of the operation and important concepts. Once this is completed, we may then turn to the problem of designing a sequential logic circuit from a set of specifications.

10.2.1 Single-State Variable Circuits

Let us examine a simple network to illustrate the important characteristics. At the system level, this may be portrayed as the block shown in Figure 10.9. The input is denoted by c and the output is T. The state variable of the circuit has been denoted by X, but it is not directly observable at this level. Since only one state variable is defined, only two values are possible: $X = 0$ and $X = 1$.

The details of the state machine must be extracted from the logic diagram, which is shown in Figure 10.10(a) for this example. In terms of the general structure portrayed in Figure 10.8 above, the XOR gate is the logic block. The first step towards analyzing a sequential circuit is to write all of the relations that describe the logic flow. The value of the output T at the present time t is given by

$$T(t) = c(t) \oplus X(t) \tag{10.10}$$

since both are connected to the input of the XOR gate. The input to the DFF is denoted by $D_X(t)$. This defines the value of $X(t + 1)$ of the next state as

$$X(t + 1) = D_X(t) = T(t) \tag{10.11}$$

since the output $T(t)$ also acts as the input to the DFF.

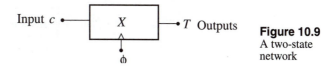

Input c —— X —— T Outputs

ϕ

Figure 10.9
A two-state
network

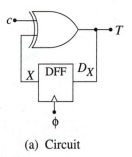

Present Values			Next Value
$c(t)$	$X(t)$	$T(t)$	$X(t+1)$
0	0	0	0
0	1	1	1
1	0	1	1
1	1	0	0

(a) Circuit (b) State table

Figure 10.10 A two-state sequential circuit

To describe the operation of the network, we will first use these equations to derive the **state table**. A state table is just a listing of the present values (at time t) of the input, the output, and the state, and the next value of the state at time $(t + 1)$. The state table for this network is provided in Figure 10.10(b). The present values of $c(t)$ and $X(t)$ allow us to calculate the present value of $T(t)$ using the XOR relation. The "Next Value" $X(t + 1)$ is simply equal to the present value of $T(t)$ due to the DFF equation. The state table tells us how the present state $X(t)$ is related to the next state $X(t + 1)$.

The state diagram is shown in Figure 10.11. It is constructed using the information in the state table concerning the present and next state values. Since there are only two possible states, $X = 0$ and $X = 1$, only two possibilities are needed. Let us assume a present state of $X = 0$. The state table shows that the network remains in the same state if $c = 0$ but changes to $X = 1$ if $c = 1$. Similarly, if the network is initially in the state $X = 1$, then it remains there if $c = 0$ but makes a transition back to $X = 0$ if $c = 1$ is applied. These four possibilities are indicated (along with the output value of T) in the drawings with the notation c/T. State diagrams are useful for visualizing the overall function of a network. For example, this shows that the variable c acts as a control bit such that $c = 0$ causes the system to stay in the same state, while a value of $c = 1$ induces a change to the other state, i.e., a toggle.

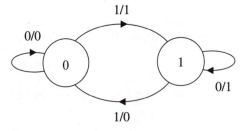

Figure 10.11 State diagram for the circuit in Figure 10.10

For our next example, consider the circuit shown in Figure 10.12. This network has two inputs x and y, and one state element with the state denoted by A. Tracing the network output f gives the relation

$$f = \overline{x \cdot A + y} \tag{10.12}$$

as seen by analyzing the logic gates. The present input to the DFF is $D_A(t)$ such that

$$D_A = \bar{f} = x \cdot A + y \tag{10.13}$$

The next state $A\,(t + 1)$ is given by the DFF relation

$$A(t + 1) = D_A(t) \tag{10.14}$$

These equations provide the formal description of the network.

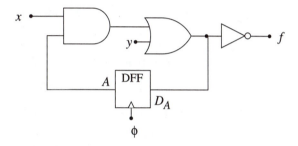

Figure 10.12 A two-input, single-state sequential circuit

To use this information, we use the present inputs (at time t) to define the present values of $D_A(t)$ and the output $f(t)$. It is then possible to use the equations to create the state table shown in Figure 10.13. As in the previous example, this listing provides the link between the present values of the inputs and the state variable $A(t)$, and output $f(t)$, and also the next value $A\,(t + 1)$ of the state variable. The entries

Present State	Present Inputs		Present Output	Next State
$A(t)$	$x(t)$	$y(t)$	$f(t)$	$A(t+1)$
0	0	0	1	0
0	0	1	0	1
0	1	0	1	0
0	1	1	0	1
1	0	0	1	0
1	0	1	0	1
1	1	0	0	1
1	1	1	0	1

Figure 10.13 State table for the circuit in Figure 10.12

have been calculated using the equations above, and they are easily verified for each line using the explicit forms

$$f(t) = \overline{x(t) \cdot A(t) + y(t)}$$
$$D_A(t) = x(t) \cdot A(t) + y(t) = A(t+1) \tag{10.15}$$

Including the time variable helps us to distinguish between the present and next state values.

The state table can now be used to construct the state diagram. Let us start with the state $A = 0$ and determine the effect of each input combination. Since the network only has one state variable, there are only two possibilities: either we remain at $A = 0$, or the inputs induce a transition to $A = 1$. These can be determined by comparing the listing for $A(t)$ with that for $A(t+1)$. The input combinations

$$(xy) = (00),\ (10) \tag{10.16}$$

correspond to the case where both $A(t) = 0$ and $A(t+1) = 0$, while

$$(xy) = (01),\ (11) \tag{10.17}$$

induce a transition to $A(t+1) = 1$. Similarly, if we start with $A(t) = 1$, then

$$(xy) = (01),\ (11),\ (10) \tag{10.18}$$

results in a value of $A(t+1) = 1$, while

$$(xy) = (00) \tag{10.19}$$

gives a final state of $A(t+1) = 1$. Combining this information yields the state diagram shown in Figure 10.14. Note that the value of the output f has been included in the diagram, so that it indeed contains exactly the same information as the state table. It does have the advantage that it is somewhat easier to follow.

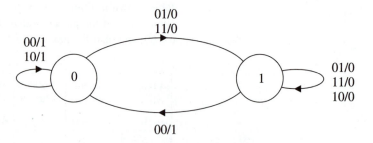

Figure 10.14 State diagram for the circuit in Figure 10.12

For our third example, let us analyze the logic network shown in Figure 10.15. The inputs are designated as u and v, and the output is g. The state variable at the output of the DFF has been defined as R. The logic equation for the output is given by

$$g = (u + v) \oplus R \tag{10.20}$$

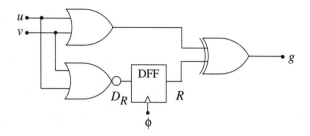

Figure 10.15 A three-gate sequential circuit

where all of the quantities are present time (i.e., at t) values. The next state condition is obtained from the DFF relation

$$D_R(t) = \overline{u(t) + v(t)}$$

$$= R(t+1)$$

(10.21)

Using the equations allows us to construct the state table in Figure 10.16. The entries can be easily verified by listing all possible values of u, v, and R, and computing the value of g. Similarly, the next state condition is obtained directly from calculating $\overline{u+v}$ using the present input values. The table provides the basis for the state diagram shown in Figure 10.17. It is worth repeating the fact that the state table and the state diagram contain the same information and are equivalent representations.

	Present Values			Next State
$R(t)$	$u(t)$	$v(t)$	$g(t)$	$R(t+1)$
0	0	0	0	1
0	0	1	1	0
0	1	0	1	0
0	1	1	1	0
1	0	0	1	1
1	0	1	0	0
1	1	0	0	0
1	1	1	0	0

Figure 10.16 State table for the circuit in Figure 10.15

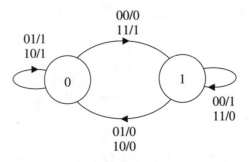

Figure 10.17 State diagram for the circuit in Figure 10.15

10.2.2 Multi-State Variable Networks

Increasing the number of state elements gives a larger number of possible states. For example, if we use two flip-flops, then the circuit can have four states, a circuit with three FFs has eight states, and so on. This allows us to create logic networks that can deal with very complex situations. Of course, the circuits themselves can become quite complicated and must often be analyzed by computer toolsets. However, the analysis approach remains the same regardless of the number of state variables.

Consider the network shown in Figure 10.18. This circuit has one input S, two outputs a_x and a_y, and two state variables x and y. The states of the circuit are thus given by

$$(xy) = (00), (01), (10), (11) \tag{10.22}$$

The state variable x is used as an input into both XOR gates, while the variable y only feeds back to the lower gate. The first step in analyzing this circuit is to write the equations of the outputs. These are

$$
\begin{aligned}
a_x(t) &= S(t) \oplus x(t) \\
a_y(t) &= S(t) \oplus x(t) \oplus y(t)
\end{aligned}
\tag{10.23}
$$

where we show the dependence on the present time t explicitly. The relationship between the present state and the next state is summarized by the DFF expressions

$$
\begin{aligned}
x(t+1) &= D_x(t) = a_x(t) \\
y(t+1) &= D_y(t) = a_y(t)
\end{aligned}
\tag{10.24}
$$

Combining these equations allows us to construct the state table in Figure 10.19. As always, we list all possible combinations of S, x, and y to find the present outputs, and use the flip-flop equations to find the next state. Both calculations are easily accomplished, especially when it is recognized that a_y is just the odd operation that we encountered when studying adder circuits in Chapter 8. The final step in the analysis is to create the state diagram from the state table. This requires that we provide a state bubble for each of the four possibilities and then use the information in

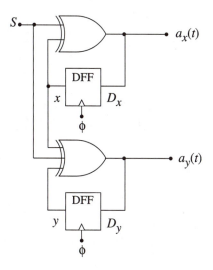

Figure 10.18 Sequential circuit with two state variables

the state table for each entry to find the details of the transition. Each line in the table shows the present state and next state, which allows us to determine the changes. The resulting state diagram is shown in Figure 10.20 with the transitions denoted by $S/a_x a_y$ (which denotes input/output). This particular example brings out an important point. Note that there is no direct transition path between the (00) and (01) states. To get from (00) to (01), the system must first go through the (11) state. In the reverse path, the sequence is (01) → (10) → (00). This is an excellent example of what is meant by a *sequence* in digital logic. Also note that this circuit has a single input of S and that the behavior of each state is defined for both values of the input, $S = 0$ and $S = 1$. This characteristic is very important and it defines the action for every possibility.

The utility of state diagrams is clear at this point: they provide a visual summary of the system operation. As was mentioned several times above, the information

Present Values				Next State	
$S(t)$	$x(t)$	$y(t)$	$a_x(t)$ $a_y(t)$	$x(t+1)$	$y(t+1)$
0	0	0	0 0	0	0
0	0	1	0 1	0	1
0	1	0	1 1	1	1
0	1	1	1 0	1	0
1	0	0	1 1	1	1
1	0	1	1 0	1	0
1	1	0	0 0	0	0
1	1	1	0 1	0	1

Figure 10.19 State table for the two-state network

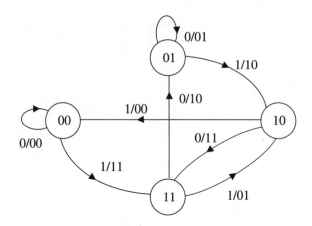

Figure 10.20 State diagram for the two-state network

contained in state tables and state diagrams is identical. Since it is often more convenient to formulate a design solution by starting with a state diagram, it is worthwhile learning how to use the diagram to construct a state table.

Consider the state diagram shown in Figure 10.21. This describes a system that has two state variables (since there are four bubbles) that we will arbitrarily denote as A and B. The transitions are determined by two input variables that we will call X and Y. There are no outputs specified; this implies that the state variables A and B are themselves the outputs of the network. To construct the state table, we list each state and determine the effects of all possible input combinations. Since there are two inputs X and Y, every state bubble has four possible inputs. This information is used to construct the first two columns of the state table in Figure 10.22. The third column is determined by the next state value for each line. As an example, consider the first group of entries in the table where the present state is given as (00). Each possible input combination $(XY) = (00)$, (01), (10), (11) is listed, and the Next State

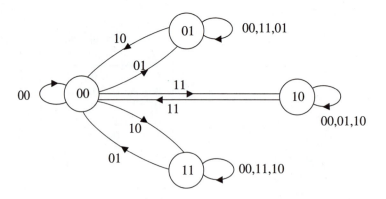

Figure 10.21 Network defined by a state diagram

Inputs	Present State t	Next State $(t + 1)$
X Y	A B	A B
0 0	0 0	0 0
0 1	0 0	0 1
1 0	0 0	1 1
1 1	0 0	1 0
0 0	1 1	1 1
0 1	1 1	0 0
1 0	1 1	1 1
1 1	1 1	1 1
0 0	0 1	0 1
0 1	0 1	0 1
1 0	0 1	0 0
1 1	0 1	0 1
0 0	1 0	1 0
0 1	1 0	1 0
1 0	1 0	1 0
1 1	1 0	0 0

Figure 10.22 State table corresponding to state diagram

entries of (AB) = (00), (01), (10), (10) are respectively determined by tracing the four paths out of the (00) bubble. This procedure can be used to translate any state diagram into a state table.

10.2.3 General Characteristics

All of the examples above have similar characteristics. The important lessons to be learned from the analyses are that

- The state of the circuit is determined by the inputs and the present state.

- The system may remain in a given state or change to another, depending upon the conditions.

- State tables and state diagrams are simply different representations of the same information.

These characteristics apply to all sequential networks, regardless of their complexity.

10.3 Sequential Network Design

Up to this point we have analyzed circuits for their behavior. Systems design, on the other hand, deals with the opposite problem, where we start with a listing of the desired transition characteristics and use this information to create the network.

A key component in the design of a sequential circuit is the state element that provides memory. In our study, we have concentrated on using the edge-triggered D flip-flop with an input $D(t)$ and an output $Q(t)$. When analyzing a circuit that contains a DFF, we calculate the value of the output for a given input. The design problem requires that we ask the opposite question:

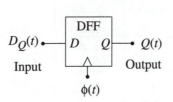

To get from $Q(t)$	To the next state $Q(t+1)$	Requires an input of $D_Q(t)$
0	0	0
0	1	1
1	0	0
1	1	1

Figure 10.23 DFF excitation table

Given the output, what was the input that caused the transition?

This information is provided in the DFF equation

$$Q(t+1) = D(t) \tag{10.25}$$

and is summarized using the **excitation table** shown in Figure 10.23. The table summarizes the behavior of the FF by listing every possible transition and the input that induced the change. Armed with this viewpoint, we may now progress to the design problem.

Suppose that we want to design the logic network that has the behavior shown in the state diagram of Figure 10.24. We use this drawing to deduce the necessary characteristics of the circuit. These are

- There are two states "0" and "1" shown, so that the network has a single state variable that we will call $A(t)$.

- Since the transitions are of the form "0/1," etc., there are a single input variable $x(t)$ and a single output function $f(t)$, where the labels are arbitrary.

This defines the overall structure of the system.

Once we have defined all of the Boolean quantities, we may create the state table from the diagram. The result is shown in Figure 10.25(a). All present state values have been listed, along with the next state $A(t+1)$. We have assumed the use of a D flip-flop and added one more column in the table that tells us the value of the

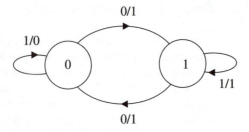

Figure 10.24 State diagram defining the design problem

| Present Values | | Next | FF In |
A(t)	x(t)	f(t)	A(t+1)	$D_A(t)$
0	0	1	1	1
0	1	0	0	0
1	0	1	0	0
1	1	1	1	1

(a) State table (b) K-map for $f(t)$ (c) K-map for $D_A(t)$

Figure 10.25 Logic design information deduced from the state diagram

present input $D_A(t)$ to the FF needed to induce the transition from the present state $A(t)$ to the next state $A(t + 1)$.

The last step in the design process is to construct the circuit that behaves according to this table. We may view the network as being made up of two functional logic blocks that respectively have outputs of $f(t)$ and $D_A(t)$. The first is obvious, since $f(t)$ is the output of the network. The second is more subtle. We need to find a logic expression for $D_A(t)$ so that we know how to wire the circuit to insure that the system will make the correct transitions. This is contained in the excitation table that was used to construct the $D_A(t)$ column.

Now that the philosophy has been explained, let us complete the details. The logical expressions for $f(t)$ and $D_A(t)$ can be found using the standard techniques presented in Chapter 3. One approach is to create the Karnaugh maps shown in Figures 10.25(b) and 10.25(c) for each function, which respectively yield

$$f = A + \bar{x}$$
$$D_A = \overline{A \oplus x}$$

(10.26)

as the necessary operations. Since the network input is x and the output is f, we may wire the interior of the circuit as shown in Figure 10.26; the state variable A is an internal variable that has no outside connection. The circuit has been constructed by just wiring the two gates as indicated by the equations and then connecting them together via the common state variable A. It is useful to note that we may check our design by invoking the analysis procedure to check that the logic network we designed does in fact give the correct state diagram.

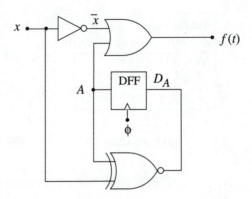

Figure 10.26 Completed network

10.4 Binary Counters

A binary counter has an output that increments its decimal-equivalent value by 1 with every clock pulse. Counters do not need any inputs other than the clock ϕ, although it is possible to add control inputs.

Consider the circuit shown in Figure 10.27. This has two outputs that form the 2-bit word $A_1 A_0$. As will be seen in the analysis, this constitutes a 2-bit cyclic counter such that $A_1 A_0$ cycles from 00 to 11 and then restarts automatically. The first step in determining the operation is to write the logic equations. Tracing the inputs to the state elements gives

$$
\begin{aligned}
D_0 &= \bar{A}_0 \\
D_1 &= A_1 \oplus A_0
\end{aligned}
\tag{10.27}
$$

while the flip-flops are characterized by the usual expressions

$$
\begin{aligned}
A_1(t+1) &= D_1(t) \\
A_0(t+1) &= D_0(t)
\end{aligned}
\tag{10.28}
$$

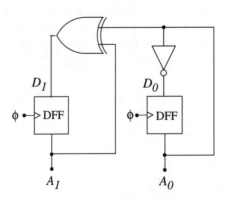

Figure 10.27 A simple 2-bit counter

Present $A_1 A_0$	Next $A_1 A_0$	Inputs $D_1 D_0$
0 0	0 1	0 1
0 1	1 0	1 0
1 0	1 1	1 1
1 1	0 0	0 0

(a) State table

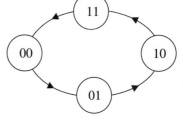

(b) State diagram

Figure 10.28 Counter characteristics

These allow us to construct the state table shown in Figure 10.28(a). If we compare the present value of $A_1 A_0$ to the next value, we see that the outputs indeed cycle in the sequence

$$00 \rightarrow 01 \rightarrow 10 \rightarrow 11 \rightarrow 00 \ldots \qquad (10.29)$$

corresponding to the decimal sequence

$$0 \rightarrow 1 \rightarrow 2 \rightarrow 3 \rightarrow 0 \ldots \qquad (10.30)$$

as advertised. The state diagram for the counter is shown in Figure 10.28(b). Since there are no inputs (other than the clock) or outputs, no designating labels are needed on the transition lines. Instead, the system is implied to follow the sequence indicated by changing one state with every clock period.

Let us extend the concept and examine the design of a 3-bit counter. Defining the output word to be $A_2 A_1 A_0$, our program will be to design a circuit that has the characteristics described by the state diagram shown in Figure 10.29. This says that the state of the circuit changes according to

$$000 \rightarrow 001 \rightarrow 010 \rightarrow 011 \rightarrow 100 \rightarrow 101 \rightarrow 110 \rightarrow 111 \rightarrow 000 \ldots \qquad (10.31)$$

corresponding to the decimal values 0 through 7 and back.

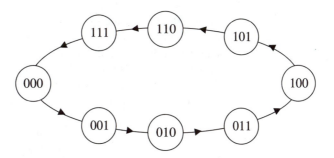

Figure 10.29 State diagram for a 3-bit counter

The design process for a counter is the same as for any sequential network. Since there are three variables, we need to use three D-type flip-flops for state elements. Denoting the inputs by D_2, D_1, and D_0, corresponding to outputs A_2, A_1, and A_0, respectively, we may write

$$A_2(t+1) = D_2(t)$$
$$A_1(t+1) = D_1(t) \qquad \textbf{(10.32)}$$
$$A_0(t+1) = D_0(t)$$

as usual. Combining the DFF excitation equations with the state diagram gives us sufficient information to construct the state table shown in Figure 10.30. This lists the present and next states and the DFF values that induce the transitions.

Present $A_2\,A_1\,A_0$	Next $A_2\,A_1\,A_0$	Inputs $D_2\,D_1\,D_0$
0 0 0	0 0 1	0 | 0 | 1
0 0 1	0 1 0	0 | 1 | 0
0 1 0	0 1 1	0 | 1 | 1
0 1 1	1 0 0	1 | 0 | 0
1 0 0	1 0 1	1 | 0 | 1
1 0 1	1 1 0	1 | 1 | 0
1 1 0	1 1 1	1 | 1 | 1
1 1 1	0 0 0	0 | 0 | 0

Figure 10.30 State table for the 3-bit counter

The next step in the design process is to find the logical expression for each of the flip-flop inputs D_2, D_1, and D_0 in terms of the present values of A_2, A_1, and A_0. These can be read directly off the table in the SOP forms

$$D_2 = \bar{A}_2 A_1 A_0 + A_2 \bar{A}_1 \bar{A}_0 + A_2 \bar{A}_1 A_0 + A_2 A_1 \bar{A}_0$$
$$D_1 = \bar{A}_2 \bar{A}_1 A_0 + \bar{A}_2 A_1 \bar{A}_0 + A_2 \bar{A}_1 A_0 + A_2 A_1 \bar{A}_0 \qquad \textbf{(10.33)}$$
$$D_0 = \bar{A}_2 \bar{A}_1 \bar{A}_0 + \bar{A}_2 A_1 \bar{A}_0 + A_2 \bar{A}_1 \bar{A}_0 + A_2 A_1 \bar{A}_0$$

by including all of the "1" entries in each column. These expressions provide the structure of the combinational logic blocks needed to construct the network. One approach to finalizing the design is to reduce these to simplest form and then use basic logic gates. This procedure can be done using Boolean algebra directly or the Karnaugh map approach. Choosing the latter gives the three maps shown in Figure 10.31 for the FF inputs. With the groupings shown, the functions reduce to

$$D_2 = A_2 \bar{A}_0 + A_0(A_1 \oplus A_2)$$
$$D_2 = A_0 \oplus A_1 \qquad \textbf{(10.34)}$$
$$D_0 = \bar{A}_0$$

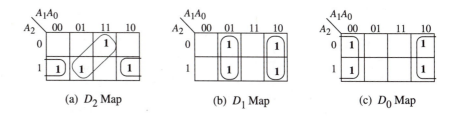

(a) D_2 Map (b) D_1 Map (c) D_0 Map

Figure 10.31 Karnaugh map reductions

by applying the usual rules. Note that the diagonal grouping in D_2 is not a formal reduction but corresponds to the XOR operation $A_1 \oplus A_2$. Finally, we may use these equations to construct the logic network shown in Figure 10.32, completing the design process.

An alternative approach to implementing the circuit is to use a programmable logic array (PLA) to create the SOP expressions directly. The general form of the PLA-based network based on AND-OR arrays is shown in Figure 10.33. The important characteristic of the sequential circuit is that the outputs $A_2A_1A_0$ are fed through the logic arrays and induce the changes in the flip-flop inputs $D_2D_1D_0$. The details of the internal wiring are shown in Figure 10.34. The design has been based on a generic array in which the AND logic creates all possible minterms m_0 through m_7, even though some combinations do not appear in the formation of the D_n flip-flop inputs. Once the minterms are calculated, the OR array allows the designer to choose the desired terms for the summation by providing a wired connection.

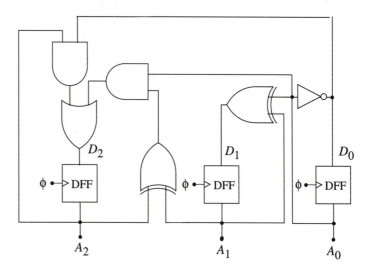

Figure 10.32 3-bit counter logic network

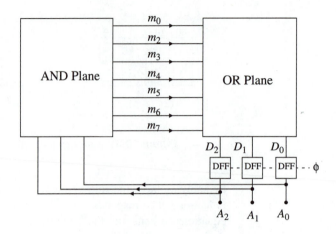

Figure 10.33 PLA-based sequential logic network

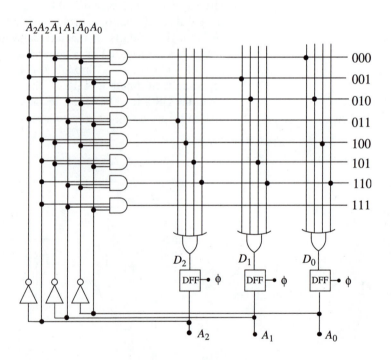

Figure 10.34 PLA wiring for the 3-bit counter network

10.5   The Importance of State Machines

Sequential networks provide much of the power that is generally associated with digital techniques. Since state machines can be designed to react to any input and provide an appropriate response, they can be placed in "stand-alone" situations where they do not need any help from humans. They truly are the basis of modern automation. The list of applications of sequential logic networks is limited only by one's imagination. For example, a traffic light controller that must sequence in the proper order, the control of a piece of equipment in a manufacturing line, and a soft drink dispensing machine can all be designed using dedicated sequential logic networks.

The computer revolution has, however, brought about a noticeable change in the scope and application of sequential circuits. The cost of general-purpose computing chips has fallen dramatically to the point where they are used almost without thought. The early generations of computer chips that initiated the widespread availability of desktop computers can now be purchased for a few dollars each. Although they may seem weak and archaic compared to the latest high-speed processors, they are extremely powerful devices that can be easily programmed to perform tasks as needed. Moreover, specialized chips called **microcontrollers** are found in many "embedded applications" such as laser printers, automobiles, and microwave ovens. A microcontroller is simply a computer chip that has been retrofitted to be hardwired into a control-type circuit instead of a general-purpose computer. Chips such as these often provide a more cost-effective design by allowing the sequential nature of the network to be programmed with software instead of requiring the building of dedicated circuits.

Even when a hardwire solution is preferred over a computer-type approach, engineers do not generally resort to gate-level designs except in simple cases. Instead, state-of-the-art general logic devices such as field-programmable gate arrays (FPGAs) allow one to design and implement complex sequential circuits using sophisticated CAD tools. A good toolset has the ability to link state diagrams with state tables and then determine the necessary connections in the FPGA.

The main point of this discussion is that the modern world of sequential network design tends to revolve around software solutions and advanced-capability logic devices. Both are involved fields and require many hours of study to master. More detailed discussions are well beyond the scope of this book, but many excellent books have been written on them.

Rather than pursue a deeper study of general sequential network design, we will instead choose to concentrate on a single type of state machine: the computer. This important structure can be considered a complex sequential circuit in every regard. It accepts inputs, reacts to them, and produces outputs that may affect the next set of operations. However, since a computer is quite large and complicated, we will tend to steer our discussions to the high end of the design hierarchy and concentrate on system-level considerations.

10.6 Problems

[10.1] Consider the sequential circuit shown in Figure P10.1. In this circuit, p is the input, f is the output, and A is the state variable.

(a) Determine the equations for $f(t)$ and $D_A(t)$.
(b) Construct the state table for the circuit.
(c) Use the state table to draw the state diagram.

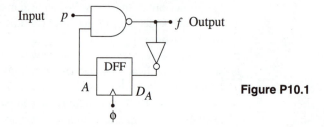

Figure P10.1

[10.2] Consider the sequential circuit shown in Figure 10.10(a) in the main body of the text. Suppose that the XOR2 gate is replaced by an OR2 gate, but everything else remains the same.

(a) Construct the state table for the modified circuit.
(b) Create the state diagram from your new state table.

[10.3] Construct the state diagram from the information provided in the table of Figure P10.2.

Present Values		Next
$A(t)$	$x(t)$	$A(t+1)$
0	0	0
0	1	1
1	0	1
1	1	0

Figure P10.2

[10.4] Construct the state diagram for the system described in the state table of Figure P10.3. Note that x is the input and A and B are the state variables.

Present Values			Next State	
$x(t)$	$A(t)$	$B(t)$	$A(t+1)$	$B(t+1)$
0	0	1	0	0
0	0	0	1	1
0	1	1	1	0
0	1	0	0	1
1	0	1	1	1
1	0	0	1	0
1	1	0	0	1
1	1	1	0	0

Figure P10.3

[10.5] Construct the state diagram for the system described by the information in the table provided in Figure P10.4.

Present $a_2\,a_1\,a_0$			Next $a_2\,a_1\,a_0$		
0	0	0	1	0	0
0	0	1	1	0	1
0	1	0	1	1	0
0	1	1	1	1	1
1	0	0	0	0	0
1	0	1	0	0	1
1	1	0	0	1	0
1	1	1	0	1	1

Figure P10.4

[10.6] Construct the state table for the network described by the state diagram in Figure P10.5.

(a) Denote the state variable as Q and the input word as $d_1 d_0$.
(b) Design the circuit using the information from your state table.

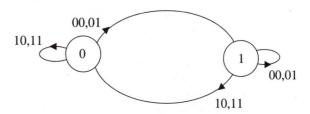

Figure P10.5

[10.7] The state diagram in Figure P10.6 describes a system with state variable $S_1 S_0$ and input x. Construct the state table for the network.

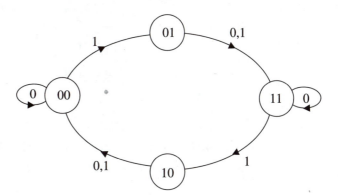

Figure P10.6

[10.8] Use the information in the state diagram shown in Figure P10.7 to construct the state table. Denote the state as A_1A_0 and the input variable as X.

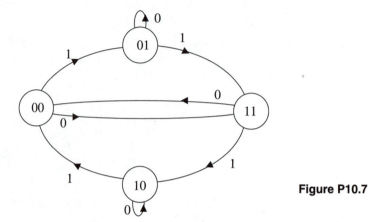

Figure P10.7

[10.9] Consider a 4-bit binary counter that increments on every clock pulse.

(a) Construct the state diagram for a counter that has an output made up of the state variable word $A_3A_2A_1A_0$.

(b) Construct the state table by assuming that the circuit consists of four D-type flip-flops with the inputs D_3, D_2, D_1, D_0 corresponding to the outputs A_3, A_2, A_1, A_0, respectively.

(c) Determine the equations for the FF inputs as functions of the state variables A_3, A_2, A_1, A_0, respectively.

(d) Design the PLA-based circuit for this counter.

[10.10] Consider the sequential circuit shown in Figure P10.8.

(a) Construct the state table.

(b) Use the table to create the state diagram.

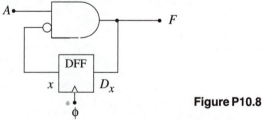

Figure P10.8

[10.11] Examine the sequential circuit in Figure P10.9.

(a) Construct the state table.

(b) Construct the state diagram.

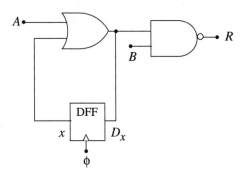

Figure P10.9

[10.12]   Examine the sequential circuit in Figure P10.10.

(a) Construct the state table.
(b) Construct the state diagram.

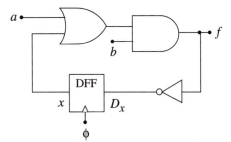

Figure P10.10

[10.13]   Consider the four-state sequential circuit shown in Figure P10.11. Study the circuit and then construct the state table that describes its operation.

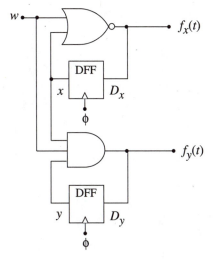

Figure P10.11

[10.14] The state diagram in Figure P10.12 is an example of a **sequence detector**. In general, a sequence detector is used to look at an input stream of bits and detect a particular sequence. Study the diagram, and remember that the notation X/Y is interpreted as (Input Value)/(Output Value). Let d be the input variable and F the output, and perform the following tasks.

(a) Construct the state table for the network.

(b) What sequence of bits does this network detect? How do you know when the sequence has been detected?

(c) Construct the state diagram for a circuit that detects the input sequence $d = 0$, then $d = 0$, then $d = 0$, then $d = 1$, and then starts over again.

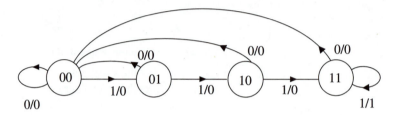

Figure P10.12

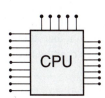

CHAPTER 11

Computer Basics

A computer is one of the most familiar examples of a digital network. The study of computer architecture deals with how the basic digital sections that make up a computer are interconnected to form the system characteristics. In this chapter, we will examine the components that are used to build a computer and how they act in unison to provide basic functions.

11.1   An Overview of Computer Operations

To understand the basic units that comprise a computer, let us think about the operations that are used when working with a typical desktop personal computer.

Turning on the computer initiates a program called the **operating system**; common examples are Unix and the Microsoft Windows® variants. An operating system provides the commands that are required to keep the system functional from the user viewpoint. It contains routines to scan the keyboard and mouse continuously to detect inputs, retrieve and store data on disk drives, send data to the video memory for display on the monitor, and launch application programs such as word processors and logic CAD toolsets.

Launching an applications program requires first loading the program from the disk drive (or other source) into the main system memory and then specifying the memory address of the first instruction. The operating system finds the first command, retrieves and executes it, then sequences to the second command, and so on. Once this is accomplished, the program takes control of the sequencing, and the application is running. The program itself consists of a set of sequential commands that tell the computer to move data around in the system (such as in a database program) or perform some operation such as addition.

This simple overview is sufficient to illustrate the major logic units that are used to build a computer. First, we must have some sort of storage device to hold the program and data. As we will see, computers usually provide this in two forms, large system memory and a small set of fast registers. Second, we must have a logic unit that can perform basic binary operations, such as those needed for arithmetic and logic functions. Finally, there must be a logic network that can accept a command and translate it into sets of binary signals that control the operation of the system.

11.1.1 Major Components of a Computer

A classical computer can be divided into five major sections, with each section performing a set of specific tasks. The block diagram in Figure 11.1 illustrates the main sections from a general viewpoint. To the outside user, the interface between our world and the binary environment is through a set of circuits that allow us to enter information and retrieve the results. These actions are governed by two major types of circuits.

1. **Input networks**. These circuits allows devices such as the keyboard, a mouse, a disk drive, a CD ROM, or a scanner to provide input data to the computer.

2. **Output networks**. Since the information is processed in binary, it must be decoded for use in the everyday world. Examples of output devices are the monitor and a disk drive in write mode.

Some devices can be used for both input and output functions. For example, a modem allows the user to both send and receive data.

Although the input/output (**I/O**) circuits are integral parts of a computer, the internal circuits tend to be the most interesting to the newcomer, for it is here that the inner workings of the machine are found. The internal circuits consist of three major sections.

3. **Memory.** Memory provides the storage for programs, data, and other necessary items such as the operating system.

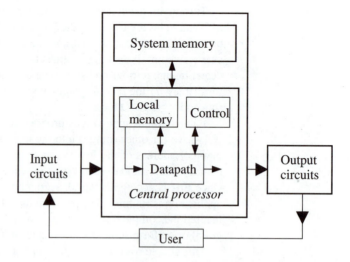

Figure 11.1 Major components of a computer

4. **The datapath**. As implied by its name, this represents the paths that the data follow during the processing events. As we shall see, the structure of the datapath determines the operations that a computer can perform. This includes items such as algebraic manipulations (addition, multiplication, etc.) and data movement to and from the memory.

5. **Control**. A datapath is designed to provide many different operations. Each operation has its own requirements as to how the data is processed. For example, addition is accomplished using an adder, so that the input words must be directed to the adder circuitry. The control unit is responsible for insuring that the data is sent to the correct set of circuits.

The datapath and control units are usually grouped together to form the **central processing unit** (CPU). In Figure 11.1, we have included a block labeled as local memory in the central processor. Local memory, called **cache** memory, allows the central processor to function without always having to retrieve data from the main memory.

11.1.2 What Can a Computer Do?

Although a computer can tackle extremely complex tasks, the internal operational modes are surprisingly limited. In general, a computer provides only two basic types of operations:

- data movement

and

- performing binary operations.

Data movement is exactly as it sounds: moving data from one point to another within the system. This includes storing and retrieving information from the memory unit, in addition to moving data around within the datapath circuits in the CPU.

The set of binary operations that are built into a computer are usually grouped into two main groups. First, logic functions such as NOT and XNOR allow us to perform Boolean operations on the binary words that pass through the datapath. When used correctly, this provides the machine with decision-making capabilities. The second group of operations are the arithmetic functions. These include addition and subtraction, as well as multiplication and division in the more advanced designs.

Every operation that a computer can perform is called an **instruction**. The group of instructions is called the **instruction set**. The number and types of instructions are determined by the structure of the datapath circuits. For example, a computer cannot perform addition unless the datapath has adders in it. Every computer is distinguished from all others by the specific details of the instruction set.

11.1.3 The von Neumann Model

Until recently, most computers were based on the model developed by John von Neumann that describes the sequential execution of each instruction line in a program. Figure 11.2 provides a simple visualization aid for understanding the main characteristics of the von Neumann machine.

There are two main sections portrayed in the block diagram. The memory unit is used for storage of both the **program** and the **data**. In general, these two types of information are kept in different portions of the memory unit, as implied by the drawing. The program consists of a sequence of instructions that are labeled

Instruction 0, Instruction 1, Instruction 2, . . .

and so on. The program is sequential in that the instruction number indicates the usual order of execution. It is the responsibility of the control unit to determine the next instruction. The term *data* is used to indicate the information that is being processed. For example, *data* might include items such as names and addresses.

The central processing unit (CPU) is the other main section of the von Neumann model. It consists of the **datapath** and the **control unit**. The datapath block represents the logic circuits that are needed to implement every instruction, such as addition or logic operations. Since the datapath network is capable of providing several different operations, we must send signals that specify which circuits are to be used. This is the responsibility of the control unit; it sends signals to the datapath unit to activate the correct circuits. A separate storage area in the control unit is used to hold the current instruction word; this is shown as a small rectangle in the drawing and is called the **instruction register** (**IR**).

The von Neumann model of a computer is based upon a repeating four-cycle procedure to execute a program. The cycles are named and described as follows.

1. **Instruction Fetch**. During the first cycle, the central processor sends a signal to the memory unit, telling it which instruction is needed. The memory responds by sending the instruction to the CPU, where it is held in the control unit.

2. **Instruction Decode**. Decoding is the process of interpreting the binary instruction and determining what needs to be done within the CPU to implement this operation. This information is sent to the datapath from the control unit.

3. **Instruction Execute**. After the datapath receives the information from the control unit, the instruction may be executed. The datapath receives the necessary input data, either from memory or from a local storage within the datapath itself, and outputs the results.

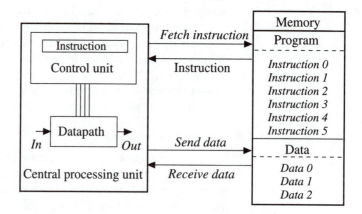

Figure 11.2 The von Neumann model of a programmable computer

4. **Storage**. The final phase of the cycle is where the results are stored back in the memory.

Every instruction in the program is treated using the same sequence of procedures.

One important observation about this description of a computer is that the system repeats the same four cycles so long as the program is running. The main difference among the instructions is handled by changing the function performed by the datapath network. As we will see later, the datapath circuitry provides different operations by "steering" the data through various combinations of logic cascades using multiplexors and other switching elements.

It is also useful to examine the speed of a computer in this model. If the computer is built in hardware, then one instruction will require a total time

$$t_{Inst} = t_{IF} + t_{ID} + t_{EX} + t_S \tag{11.1}$$

to complete. In this equation,

t_{IF} = time needed to perform the instruction fetch operation,

t_{ID} = time needed to perform the instruction decoding,

t_{EX} = execution time in the datapath, and

t_S = time required to store the result.

Obviously, a small value of t_{Inst} implies a faster computer since more instructions can be completed in one second. In the von Neumann model, the speed of a computer is increased by decreasing the time needed to perform each instruction. Since the timing is usually controlled by an externally applied clock signal $\phi(t)$, the higher the clock frequency f, the faster the computer runs.

The key to understanding the intrinsic computational power of a computer is based on combining the sequence above with the concept of the instruction set. Every instruction either moves data from one point to another or takes a binary input, performs a logic or arithmetic operation, and returns a result. By itself, a single operation does not accomplish much. However, if the instruction time t_{Inst} is small, then the computer can perform many instructions in a very short time. As an example, even a "slow" CPU has an instruction time t_{Inst} that is less than 0.1 μs[1] implying that the computer can perform about 10 **million** instructions **per** second (which is known as 10 MIPS). Thus, even though a single operation is limited, it is the ability of a computer to execute millions of instructions that provides the computing power.

11.1.4 Programming

We are all familiar with the concept of programming a computer using a high-level language such as Pascal or C. A **program** is an ordered list of commands that tell the sequence of operations to accomplish a specific task. Each line (or group of lines) in the program is constructed using special words that initiate certain actions. The specific manner in which the commands are constructed is called the **syntax**; this includes the order and use of specific command words, and miscellaneous

---

[1] Recall that a microsecond is defined by 1 $\mu s = 10^{-6}$ *sec*.

aspects such as the use of delimiters (periods, commas, semicolons, etc.) that must be adhered to. Every high-level language is defined in this manner.

Since digital systems are based on the binary number system, a computer can only deal with binary words that consist solely of 0s and 1s. This is called the **machine language** level, and it is quite cumbersome for humans to use. The binary logic circuits of a computer are designed to accept commands in a specified binary format and cause the desired actions to take place. The **word size n** of a computer refers to the number of bits that are used for the data segments. In many systems, the same word size is used for the instructions. Although n can be any power of 2, most modern systems use word sizes that are even multiples of 16 bits. For example, desktop computers currently employ 32- or 64-bit word sizes.

Let us now turn to the concept of how we would program a digital computer that is built using logic networks. Programs that are written in a high-level language must be translated to machine language using a special program called a **compiler**. This process is illustrated schematically in Figure 11.3. A compiler accepts a program that is written in a high-level language and produces a sequence of machine-language instructions that the computer can understand. Machine language is quite difficult to read and interpret, so we introduce yet another type of language, called **assembly language,** to help us understand the operations inside of the computer. Assembler commands have a one-to-one correspondence with each binary machine-language command, but they use mnemonics such as "add" and "sub" to make them easier to interpret. We will tend to state computer instructions using assembly language. Although it may seem a bit odd at first, it is easy to learn since the mnemonics tend to be well chosen for ease of memorization. More important, assembly language allows you to see the details of how a computer operates.

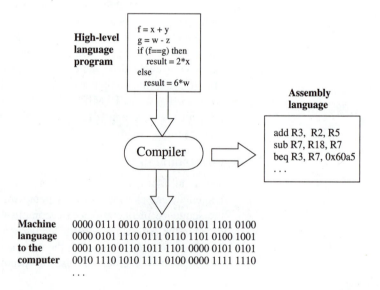

Figure 11.3 Levels of programming languages

11.1.5 Computer Registers

A register is just a set of individual memory cells that are grouped together to store an entire word. Registers are used in several different ways in a computer, but all registers have the same basic function: they can be used to store a binary word. Although we have already examined registers in Chapter 9, we will move to a more concise viewpoint here to help aid our understanding.

When discussing dataflow in a computer, we will represent registers using simplified symbols. These are shown in Figure 11.4. A single memory cell that can store one bit is shown in Figure 11.4(a); note that the details of the internal circuitry are ignored, since only the storage characteristics are important at this level of the hierarchy. To create a 32-bit register, we simply combine 32 individual cells as shown in Figure 11.4(b) and label the bits as d_{31} to d_0. Although this provides all of the information needed, it is cumbersome to draw because of the 32 input and 32 output lines. This leads us to the simpler drawing of Figure 11.4(c), which is the preferred symbol for our discussion. The input and output lines have been replaced by single **data bus** lines that represent 32 individual bit lines. The **width** of the bus is denoted by the "slash" (/) across the bus, with the number of bits that are carried by the bus (32). This simplified symbol is sufficient for our drawings as long as we remember that it represents a digital element that stores a 32b word. Sometimes, it will be useful to show explicitly the contents of the register (the word that is stored inside) as in Figure 11.4(d).

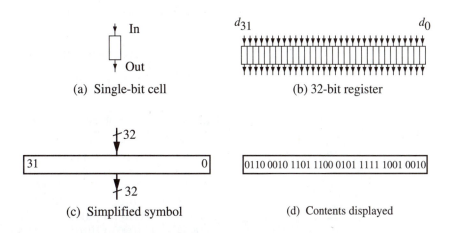

(a) Single-bit cell   (b) 32-bit register

(c) Simplified symbol   (d) Contents displayed

Figure 11.4 Schematic symbols used to represent registers

11.2 The Central Processing Unit: A First Look

The heart of a computer is the central processing unit. This group of circuits determines the operations that the computer can provide and usually dictates the performance of the entire system. As one might expect, the CPU is relatively complicated to study in detail; this is especially true if the entire CPU is examined as a single unit. However, the concept of hierarchical design can be used to break the CPU into several distinct sections, each of which can be studied independently of the others.

We will examine the basic operations of a "typical" modern computer system. Although the discussion is kept quite general, our study will include some of the details using rather specific references. These have been chosen to provide a sound background for those who choose to study computers in more detail in the future.[2]

11.2.1 The Instruction Fetch Network

To execute a program that is stored in memory, we must bring the instruction into the processor. This is achieved using the **instruction fetch** network, which is a subsystem within the control unit.

A program is a sequential listing of binary words, with every word providing the information needed by the logic networks to perform a specific operation. When the program is stored in memory, every binary word is assigned a unique address that corresponds to its location in the program. This is illustrated by the example provided in Figure 11.5. The instruction sequence is given by the number such that *Inst* 0 is first, *Inst* 1 is next, and so on. Every instruction is represented by a binary word. We have used 32-bit words in the drawing, but the actual size depends on the computer.

It is common to partition the memory in a manner that allows us to specify the address of every 8-bit byte. A 32-bit word consists of 4 bytes, so the addresses for two in-order instructions differ by 4. Let us examine how the data is stored by referring to the values in the drawing. *Inst* 0 is located at decimal address 0400 and has a binary value of

$$01101100\ 11111010\ 11110000\ 11110000$$

Address	Binary Instruction	Order
0400	01101100 11111010 11110000 11110000	*Inst* 0
0404	11000101 10110101 00001111 11110000	*Inst* 1
0408	01000110 10011111 10101010 10101010	*Inst* 2
0412	10010011 01101110 00110011 00110011	*Inst* 3
0416	10101001 01000101 11100011 11100011	*Inst* 4
0420	10001000 10001101 10011001 10011001	*Inst* 5
0424	11100010 10101001 11100010 10101010	*Inst* 6
0428	00100111 01101010 00110010 01011000	*Inst* 7
0432	10010010 11010011 10010011 01001001	*Inst* 8

Figure 11.5 A program sequence stored in memory

---

[2] The architecture chosen here, as explained later, is an example of what is called a RISC design.

The next instruction is *Inst* 1, which is located at decimal address 0404 with a binary value of

$$11000101 \ 10110101 \ 00001111 \ 11110000$$

and so on.

A basic instruction fetch logic network is described by the block diagram in Figure 11.6. The main section of the instruction fetch network contains two registers. The **instruction register** (IR) holds the binary instruction word for the current operation, such as an addition, that is specified by the program. The **program counter** (PC) is used to maintain the flow of the program by counting the instructions as they are executed. It contains the memory address for the next instruction that is to be retrieved and executed by the computer.

The operation of the instruction fetch network is dictated by the fact that the program listing is stored in the memory unit. The program counter is a register that stores the address of the current instruction being fetched from memory. This is fed to the memory, and the corresponding data word is transferred to the instruction register. For example, if we wish to obtain *Inst* 0 from memory, then we load the PC with the binary equivalent of 0400; this results in the desired transfer to the IR. To obtain the next instruction (*Inst* 1), an adder unit is provided to increment the contents of the PC according to

$$PC = PC + X$$

where $X = 4$ is the spacing needed to get to the next instruction. In the present example, this means that

$$PC = 0400 + 4$$

which causes *Inst 1* to be transferred to the IR. Once in the IR, the instruction is decoded by the other circuitry in the computer, which causes the instruction to be carried out. When the procedure is finished, the instruction is **retired**, and the next instruction is fetched from the memory and transferred to the IR. This procedure allows us to execute the program lines in order.

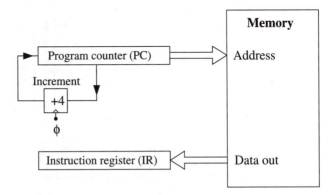

Figure 11.6 Operation of the instruction fetch (IF) network

11.2.2 Concept of the Datapath

Now that we have seen how instructions can be fetched from memory and held in the instruction register, let us enlarge our view of the central processor to that shown in Figure 11.7. This block diagram portrays the main sections of the CPU, but only a small amount of the actual wiring among the sections has been included. The lack of detail is intentional, for the interconnection schemes are best understood later when we examine the architectural features that give the programming features of the computer.

The instruction register is shown in the heart of the control unit. Once an instruction is loaded in the IR, the bits in the instruction word act as inputs into the control logic network shown in the drawing. The control logic network consists of logic gates that provide outputs known collectively as **control signals** that are fed to the datapath on **control lines**. The control signals are used to tell the logic circuits the proper settings for digital components such as multiplexor units. In some cases, the output from the control logic unit will consist of data bits from the instruction itself.

Now consider the section labeled as the **Datapath**. The datapath circuits provide the logic needed for every instruction that can be performed by the computer. The flow of the data is denoted by the large arrows contained within the datapath block, which indicate the routes that the data word may take. Multiplexors and demultiplexors are represented by simple junctions; their meaning is clarified in Figure 11.8. This choice of data flow provides for a large class of basic computer instructions, including general movement of data within the system. In general, the data-

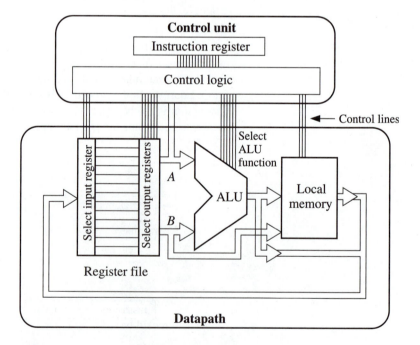

Figure 11.7 The central processor consists of the datapath and the control unit

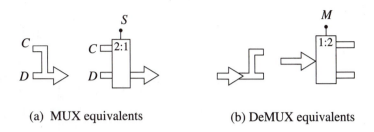

(a) MUX equivalents   (b) DeMUX equivalents

Figure 11.8 Simplified MUX and DeMUX symbols

path network can be broken down into three main groups of circuits: the **register file**, the **ALU**, and the **local memory** block.

The register file is a group of general-purpose storage registers that are used to store data words for use in the current chain of calculations. Switching networks allow us to select the input register for data storage and also to select the output registers used in a read operation. The register file portrayed in the diagram allows simultaneous outputs from two different registers that are then fed to the inputs A and B of the next logic block.

The arithmetic and logic unit (**ALU**) is a group of circuits that provide all of the arithmetic operations (such as addition and subtraction) and logic functions (such as NOT and OR). The unique aspect of the ALU is that it is a single circuit that can provide several different operations depending on the control signal applied to it. In the drawing, these signals originate from the control logic section and are labeled by "Select ALU function."

The remaining block shown in the drawing is labeled as **local memory**. This is a small section of memory that is called **cache memory**. It is included in the CPU to provide fast read and write operations that will not slow down the operation of the computer. Although it is shown as being isolated within the CPU, in reality it communicates with both the CPU components and the main system memory.

11.2.3   Datapath Operations

Now let us examine the datapath to find out how data can actually move within the unit. As drawn, the basic CPU allows three main operations to be performed. These are grouped according to the origin and final destination of the data.

Register-to-Register Operations

This type of operation takes data words from the register file and uses them as inputs into the ALU. The result is then stored back into the register file.

The equivalent datapath for a register-to-register operation is shown in Figure 11.9. In this example, the data words x and y are copied from the register file and act as the ALU inputs A and B, respectively. The ALU then produces a result $R(A, B) = z$ that is directed back to the register file and stored there. Since this type of operation allows the data to flow through the ALU, it is used to perform all of the arithmetic and logic operations in the CPU. For example, instructions such as

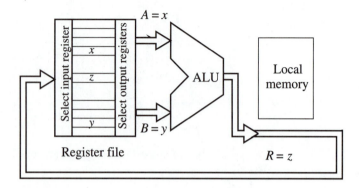

Figure 11.9 Datapath for a register-to-register operation where data originates from the registers and the result is stored back into the registers

$$R = A + B$$
$$R = A - B$$
$$R = A \cdot B$$
$$R = \overline{A}$$

(11.2)

can all be implemented using this type of datapath operation. Note, however, that the data must first be in the registers before an instruction can be executed. This makes it mandatory to provide operations that allow data transfer between the registers and the memory.

Load

When a program is launched, data is usually loaded only into the memory of the computer. To implement a register-to-register operation in our system, data must be moved from the memory to the register file before the ALU can access it. This operation is called a **load**, and it gives the **load word** instruction mnemonic **lw**. Figure 11.10 shows the portion of the datapath network that implements a load operation. Since we wish to take a word from memory, we must specify the address where the word is located. This is accomplished by the ALU input A, which is taken from the instruction register and control unit; the actual value of A is part of the program. Once the address is given to the memory unit, the data d is transferred from the specified location to one of the registers.

Store

A **store** operation gives the **store word** (**sw**) instruction and is the opposite of a load. It allows us to move a data word from a register and write it to the memory array. The data flow path in Figure 11.11 shows the important aspects of this operation. The address A is passed through the ALU and specifies the location where the word will be placed. The data x is then taken from the register file and placed in memory.

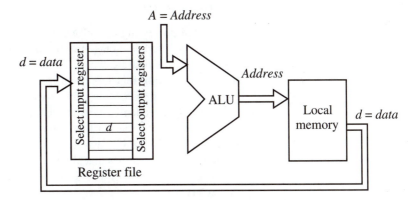

Figure 11.10 The load word instruction allows us to move a data word from the memory to a particular register

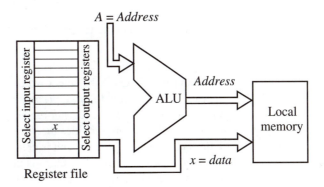

Figure 11.11 The store word instruction allows us to move a data word from a register and store it in the memory

Although our view of a computer is highly simplified at the moment, these three types of operations illustrate the most important data flow paths that are used to create the instruction set of a modern computer.

11.3 Datapath Components

Now that we have seen the characteristics of the basic data flow paths in a computer, let us study the datapath components in more detail. Each major unit is examined in this section; our goal is to gain a more thorough description of the system-level operation of each component. After we have completed the detailed study, we will discuss how the connection of the logic units creates the instruction set of the computer.

11.3.1 The Register File

The central processing unit of a computer provides a way to store data words that can be easily used as inputs into the ALU. These storage locations are made up of several registers that are wired into the datapath in a convenient manner. Collectively, the group of registers is called the **register file**. It may be thought of as the electronic analog of a filing cabinet where each register is like a drawer that can hold items independently of what is contained in all of the other drawers.

An example of a register file is shown in Figure 11.12. It consists of 32 registers labeled *R0* through *R31*; each register is taken to be 32 bits wide. The register file has a single 32b input bus that allows the user to store one word at a time into a register in the file. The input word is written to the register specified by the 5b **destination select** word

$$d_4 d_3 d_2 d_1 d_0$$

Two 32b outputs from the register file denoted by *A* and *B* are shown on the right side. The data words transferred to these lines depend upon the **source select** word

$$s_4 s_3 s_2 s_1 s_0$$

(for choosing *A*) and the **target select** word

$$t_4 t_3 t_2 t_1 t_0$$

(for choosing *B*). Once the select words are specified, the contents of the selected register are transferred to the appropriate output. This is a non-destructive operation in that the contents of the register remain intact after the read operation.

The block diagram provides the system-level description of the register file, but it does not illustrate how the unit is created from smaller building blocks. Although knowledge of the internal workings is not needed to understand how the unit functions in the computer, it is worthwhile to study how the register file is created using simpler components. Let us drop one level in the hierarchy and describe the input and output select operations using the drawing in Figure 11.13. At this level, the

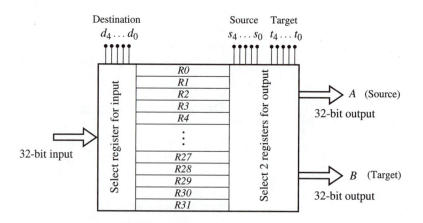

Figure 11.12 Structure of the register file

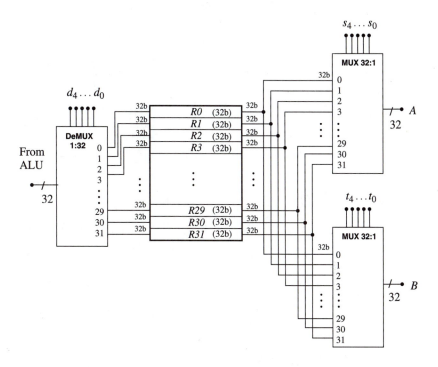

Figure 11.13 A unit-level view of the register file structure

register is still viewed as a single 32b structure with 32b input and output lines. Selection of the registers for the outputs A and B is achieved using two independent 32:1 multiplexors with 32b inputs. Both of the MUX units have the same inputs. This means, for example, that the output of register $R0$ is connected to the 0 inputs in both MUX units, the output of register $R1$ is connected the MUX 1 inputs, and so on. The output registers are chosen by the 5-bit MUX control words $s_4 s_3 s_2 s_1 s_0$ and $t_4 t_3 t_2 t_1 t_0$. For example, if

$$s_4 s_3 s_2 s_1 s_0 = 00010 \quad \text{and} \quad t_4 t_3 t_2 t_1 t_0 = 11100$$

then A is the contents of register $R2$ while B is equal to the contents of register $R28$. Once again, note that both A and B are 32-bit words in this drawing.

The input (destination) register is determined by the word $d_4 d_3 d_2 d_1 d_0$ that controls the 1:32 demultiplexor unit at the input side of the register file. As with the output side, the value of $d_4 d_3 d_2 d_1 d_0$ determines which register the data is sent to. For example, a value of

$$d_4 d_3 d_2 d_1 d_0 = 00001$$

stores the input word in register $R1$,

$$d_4 d_3 d_2 d_1 d_0 = 00010$$

sends the word to $R2$, etc., allowing us to dictate the destination as needed.

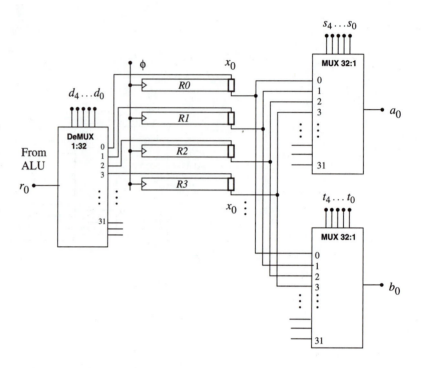

Figure 11.14 A bit-level logic diagram of the register file

Since logic circuits are created at the single-bit level, it is useful to drop one level further in the system hierarchy to arrive at the simplified circuit in Figure 11.14. This illustrates the same system as in the previous drawing but portrays only the 0th bit b_0 of each register. The behavior is the same for every remaining bit b_1 through b_{31}, so that showing a single bit is sufficient. Note that the clocking signal ϕ has been included explicitly here but was implied in the higher-level views.

This view illustrates the basic component-level view in the hierarchy. To actually build the complete register file, we would take 32 identical circuits and form the system-level structure. Conversely, if we want to implement this in hardware, we must go down in the hierarchy. The next step would be to use logic gates to construct the components, and then we would go to the transistor level. The final step would be the physical design of the silicon chip.

11.3.2 The Arithmetic and Logic Unit

The arithmetic and logic unit (ALU) is the section of the computer that provides the required arithmetic functions, such as ADD and SUB, and all logic operations, such as NOT, NAND, and XOR. Although the specific functions vary with the processor, most existing computers have a large group of common operations.

The symbol for an ALU is shown in Figure 11.15. This unit has two data inputs A and B, and an output R as the result of the operation. Although the discussion is quite general, we will assume that A, B, and R are all 32b wide to make the architecture more realistic.

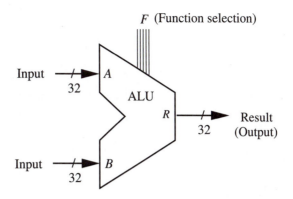

Figure 11.15 Symbol for the arithmetic and logic unit (ALU)

At the system level, the operation of the ALU is controlled by the **function select** word F; the number of bits f_i in F depends upon the number of operations that are built into the unit. The ALU is designed to provide different results depending upon the value of F. This can be stated in equation form by writing that

$$R = R(A, B, F) \tag{11.3}$$

where A and B are the inputs, while F acts as a control select word. This is interpreted as follows. F can assume different values F_0, F_1, F_2, etc.; each value of F specifies a different operation and hence a distinct result R_α. This can be stated by saying

$$
\begin{aligned}
R &= R_a(A, B) &&\text{if } F = F_0 \\
&= R_b(A, B) &&\text{if } F = F_1 \\
&= R_c(A, B) &&\text{if } F = F_2
\end{aligned}
\tag{11.4}
$$

$$\cdots$$

which illustrates the important point that the ALU can provide different functions R_a, R_b, R_c, and so on.

The basic construction of the ALU is determined by the operations that are to be built into the processor. In order to understand the construction of the unit, let us assume that F is the 3-bit word

$$F = f_2 f_1 f_0 \tag{11.5}$$

which has eight possible values. In this case, there are a maximum of eight different results R_a to R_h (where the subscripts range over the first eight letters of the alphabet). Since only one result can be selected at a time, we may write R as

$$
\begin{aligned}
R = R_a \cdot (\overline{f_2}\,\overline{f_1}\,\overline{f_0}) + R_b \cdot (\overline{f_2}\,\overline{f_1}\,f_0) + R_c \cdot (\overline{f_2}\,f_1\,\overline{f_0}) + R_d \cdot (\overline{f_2}\,f_1\,f_0) \\
+ R_e \cdot (f_2\,\overline{f_1}\,\overline{f_0}) + R_f \cdot (f_2\,\overline{f_1}\,f_0) + R_g \cdot (f_2\,f_1\,\overline{f_0}) + R_h \cdot (f_2\,f_1\,f_0)
\end{aligned}
\tag{11.6}
$$

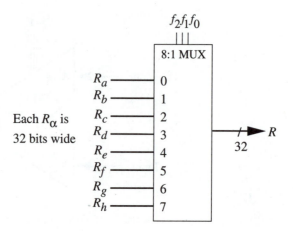

Figure 11.16 A multiplexor used to select one of several results

This chooses the output depending upon the value of $f_2 f_1 f_0$. A moment's reflection will verify that this is simply the equation for an 8:1 MUX with inputs R_a to R_h, with selection accomplished by the value of $f_2 f_1 f_0$. The multiplexor corresponding to this equation is shown in Figure 11.16. To construct the ALU, we need to provide the various results $R_a, \ldots, R_h$ that are used as inputs into the MUX.

Before proceeding further, let us examine the parallel structure of the ALU. Each of the data lines A, B, and R is 32 bits wide. Denoting the individual bits as

$$A = a_{31} a_{30} a_{29} \ldots a_1 a_0$$
$$B = b_{31} b_{30} b_{29} \ldots b_1 b_0 \qquad \textbf{(11.7)}$$
$$R = r_{31} r_{30} r_{29} \ldots r_1 r_0$$

allows us to visualize the entire ALU as shown in Figure 11.17(a). Since logic circuits can only deal with individual bits, the ALU is created by using a parallel grouping of 32 identical networks. If we examine the nth bit, where n is any bit in

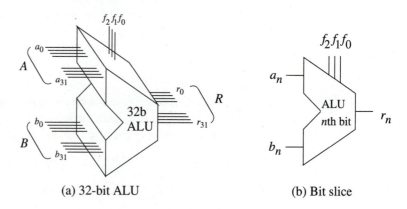

(a) 32-bit ALU (b) Bit slice

Figure 11.17 Concept of a bit slice of the ALU

the range 0 through 31, then we arrive at the **bit slice** shown in Figure 11.17(b). This is a typical circuit that is obtained by "slicing" the 32-bit structure into individual bits, then selecting one of the slices to study. The utility of this concept is easy to understand. Once we have characterized the circuits for the nth bit, we can construct the 32-bit ALU by just paralleling 32 of the circuits. It is not necessary to introduce the full complexity of the parallel network.

Let us now return to the problem of structuring the ALU. Figure 11.18 shows the bit slice circuit that is implied by characteristics we have discussed. At this level, the inputs are denoted by a_n and b_n. These are used to generate the different results r_a, r_b, ... via a function network, and the results are fed into a multiplexor unit. The function selection word $f_2 f_1 f_0$ determines which input is fed to the output r_n. This diagram makes it clear that the ALU can only perform the operations provided within the function network. It is therefore important to choose the most useful operations.

Figure 11.19(a) provides an example of an ALU bit slice. We have included the algebraic operations of addition and subtraction, and the logic functions AND, OR, XNOR (equivalence), NOT, and "pass through." The operation table in Figure 11.20(b) is obtained by first examining the result applied to each MUX input and then applying the switching characteristics of the MUX itself. It is important to note that the ALU has the circuitry to provide all eight functions, but only one is directed to the output.[3]

As an example of how this ALU operates, suppose that we want to add the two bits a_n and b_n. The input bits a_n and b_n pass through the add/subtract unit denoted by "+/–" and produce the sum s_n. Choosing the ALU control bits as

$$f_2 f_1 f_0 = 000$$

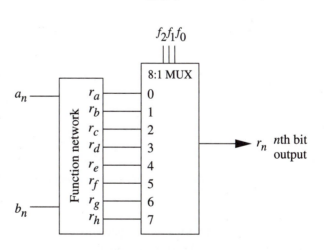

Figure 11.18 Unit-level structures of a bit slice

---

[3] Note that this example is for illustration purposes only and does not correspond to any real-world processor.

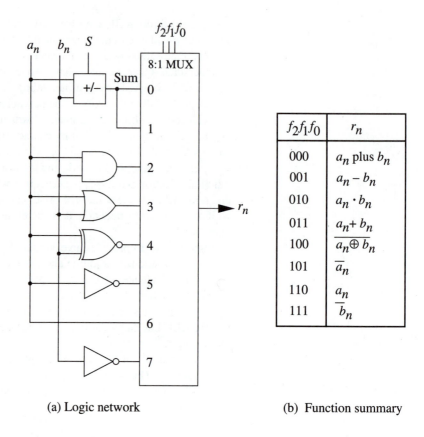

(a) Logic network   (b) Function summary

Figure 11.19 Example of an ALU bit-slice design

selects the MUX input 0 for transmission to the output y_n. This, of course, corresponds to the sum and gives an output of

$$y_n = a_n + b_n$$

If instead we had used

$$f_2 f_1 f_0 = 001$$

then MUX input 1 would have been selected, yielding an output of the difference

$$y_n = a_n - b_n$$

We have assumed that the add/sub select bit S can be provided with the proper value using additional circuitry that is not shown in the drawing.

Now then, since we know how to design one bit slice of the ALU, a circuit n bits wide can be obtained by

- replicating this bit slice $(n - 1)$ times, and then
- wiring the n circuits in parallel.

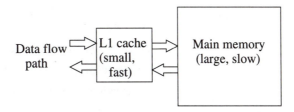

Figure 11.20 Concept of local memory

One important aspect that needs to be remembered is that the carry-out bit from the add/sub unit numbered j must be used as the carry-in bit to the add/sub unit that is numbered $(j + 1)$ to insure that the word addition and subtractions are correct.

11.3.3 The Local Memory

In our example, a local memory unit has been provided in the datapath. In this context, it would be classified as an **L1** (level 1) if it is on the same integrated circuit as the CPU circuitry. In a realistic computer, the L1 memory provides fast read/write storage but is very small when compared to the size of the main memory. This type of memory is used to provide fast load and store operations, and it is generically called a **cache**. A cache is connected between the data flow path and the main memory unit as shown in Figure 11.20. In the present discussion, the L1 cache just acts as a general memory unit, providing storage with read and write operations. The details of cache memory are examined in more depth in Chapter 12.

The operation of the cache will be assumed to be operationally identical to a general read/write memory array that can be modeled at the system level using the simplified block shown in Figure 11.21. The location in the array is specified using the address input. The value of the read/write control signal R/W specifies whether a read/write operation takes place or the memory is rendered inactive.

11.4 Instructions and the Datapath

The datapath consists of three main sections: the register file, the ALU, and the memory unit. We can summarize our study of CPU components thus far as follows:

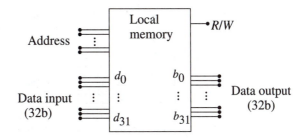

Figure 11.21 Operational model for the local memory

- The ALU functions determine the type of arithmetic and logic operations that can be performed.
- The register file provides a set of fast local storage locations.
- The cache memory allows us access to the large system memory.

Since the interconnection scheme establishes the allowed data flow paths, we can state that

- The instructions that can be implemented on a given computer are determined by the properties of each unit and how they are connected to form the system.

This is called the **architecture** of the computer since it tells us how the system is built. The last statement merely recognizes the fact that to include a specific instruction, we must (a) include any necessary logic in the ALU, and (b) provide a data flow path so that the operation can be completed. For example, if we want to include ADD in our repertoire, then we must use adders in the ALU and also insure that we can select the desired data words to steer through the ALU.

To illustrate this point in more detail, we will start with the basic units and then show how the datapath can be constructed. At this level, our goal is to design a single unit that has multiple datapaths, each supporting a specific type of data flow path. The path is selected by the instruction word and the associated control logic, and it will appear as a set of control signals at this level.

Let us first create a datapath that will allow us to perform the following:

- Retrieve a data word from a register RX.
- Retrieve a data word from a register RY.
- Perform a logic or arithmetic operation on these two words.
- Then store the result in register RZ.

This can be accomplished by the data flow path shown in Figure 11.22. The register file outputs are interfaced directly to the inputs A and B of the ALU. The result R appears at the output of the ALU and is fed back to the register file where it can be stored. As mentioned earlier in the chapter, this is called a **register-to-register instruction** since it uses data from the register file and stores the result there.

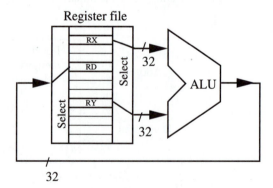

Figure 11.22 Datapath for register-to-register operations

Once the datapath has been determined, the actual operations that can result with A and B as input depend upon the characteristic functions built into the ALU. At this level, the individual instructions are most easily described using **assembly language**. As mentioned earlier, assembly language uses mnemonics to describe the operations; a mnemonic is a shortened version of a complete word that is easy to remember. The format of an assembly-language instruction defines the order and meaning of each symbol. In our treatment, a register-to-register instruction that adds two numbers together will be written as

add RZ, RX, RY   # Add contents of RX to contents of RY
                      # and place the result in RZ

where everything to the right of the pound sign "#" is considered a comment statement and has no effect on the operation. Similarly, subtraction will be written as

sub R12, R7, R8   # R7-R8 $\rightarrow$ R12

where the comment statement means "subtract the contents of register R8 from the contents of register R7 and place the result in register R12" in a descriptive form.

Other register-to-register operations that are commonly found in a computer are logic functions such as

and RC, R1, R2   # AND(R1,R2) $\rightarrow$ RC

and

or R4, R22, R1   # OR(R22,R1) $\rightarrow$ R4

Although the operations are on 32b words, the logic operations are defined to act on the aligned bits. For example, if we have

$$X = 1111\ 0000\ 1111\ 0000\ 1111\ 0000\ 1111\ 0000$$

$$Y = 1010\ 1010\ 1010\ 1010\ 1010\ 1010\ 1010\ 1010$$

then

$$\text{AND}(X,Y) = 1010\ 0000\ 1010\ 0000\ 1010\ 0000\ 1010\ 0000$$

while

$$\text{OR}(X,Y) = 1111\ 1010\ 1111\ 1010\ 1111\ 1010\ 1111\ 1010$$

These can be justified directly from the manner in which we paralleled identical bit slice circuits to create the 32-bit ALU. As can be seen from our discussion, the types of operations are restricted only by the capabilities of the ALU.

The next datapath we will examine will allow us to perform the store operation by the following steps:

- Obtain the memory address from the instruction.
- Send the data from register RX to the memory.
- Then store the word in the proper location.

The data flow path for this type of operation is drawn in Figure 11.23. The address is sent directly to ALU input A, while the contents of RX are routed to the memory unit. This is a store word instruction and is written in our assembly language as

sw RX, ADDRESS   #Store contents of RX at ADDRESS

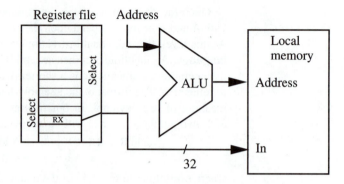

Figure 11.23 Datapath for the store word operation

For example, suppose that R4 contained the hexadecimal word 0x 2AF529B1 that we wish to store at memory location 0x 4385. The command would be

sw R4, 0x 4385

which would result in 0x 2AF529B1 being stored at the proper address.

The last datapath we will examine here is the load word instruction that allows us to copy a word from the memory and place it in a register RX. The instruction is written as

lw RX, ADDRESS   #Copy contents of memory at ADDRESS
                     # into register RX

Figure 11.24 illustrates the datapath for this type of instruction. Once again, the address is passed through the ALU to the memory unit, where it specifies the location to be read. The data word is then transmitted from the memory unit to the register file and stored in the specified register RX.

Now that we know the characteristics of the individual data flow paths, we turn to the problem of designing a single unit that can implement all three possibilities

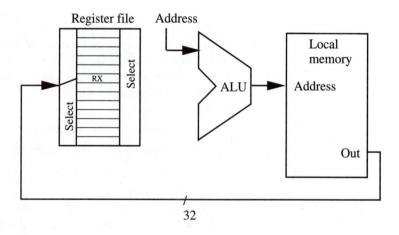

Figure 11.24 Datapath for the load word operation

by varying a set of control signals c_0 and c_1 that specify which path is to be used. Multiplexors provide this function in a direct manner, leading us to construct the datapath shown in Figure 11.25. The interconnections among the units are constants, but the actual path chosen for the data flow depends upon the setting of the control signals applied to the multiplexors. Also note that the ALU operation is selected by the control bits $f_2f_1f_0$ in the manner discussed in Section 11.3.2. A complete datapath is thus specified by the two MUX control bits c_0 and c_1 and the ALU select word $f_2f_1f_0$.

Let us clarify the general description of the datapath by examining how the datapath is changed using the control bits. The A input to the ALU is determined by the value of c_0 such that

If $c_0 = 0$, then A is an address (sw or lw operation).

If $c_0 = 1$, then A is the contents of a register.

The other control bit c_1 determines whether the output of the ALU is sent to the address input of the memory unit or back to the register file. This is seen by noting that

If $c_1 = 0$, then the ALU output is directed to the memory address port,

while

If $c_1 = 1$, then the ALU output is directed to the register file.

The first case corresponds to a sw or lw operation, while the second is for a register-to-register operation.

It is important to note that any instruction that reads from, or writes to, the register file must also provide information that allows the particular register to be chosen. Since we have assumed 32 registers R0 through R31, this means it takes 5b to select a register. A register-to-register operation such as

        xor R2, R4, R19   # XOR(R4,R19) → R2

requires that $3 \times 5 = 15$ register select bits be specified, while the memory access operation of

        lw R3, 0x5A28

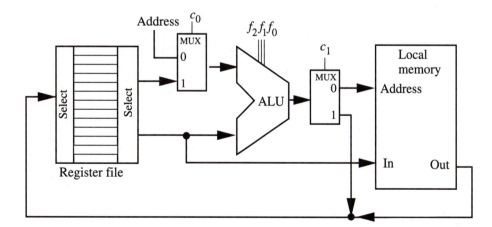

Figure 11.25 Datapath that allows register-to-register, load word, and store word operations

needs to have 5 bits to select the destination register R3. These are the bit groups that were called $s_4 \, s_3 \, s_2 \, s_1 \, s_0$ (source), $t_4 \, t_3 \, t_2 \, t_1 \, t_0$ (target), and $d_4 \, d_3 \, d_2 \, d_1 d_0$ (destination) in our discussion of the register file.

Variations in the Instruction Set

The example above illustrates the important fact that the instruction set of a computer is determined by the architecture of the system. The computer can only implement those instructions that are allowed by the datapath.

As an example of the relationship between the physical architecture and the instruction set, note that the load word (lw) and store word (sw) are restricted to addresses that are specified in the instruction itself. In other words, if we want to store the contents of R5 into memory at an address of 0x 6A45, we write

        sw R5, 0x 6A45   # Copy the contents of R5 into memory at
                              # address 0x 6A45

so that only the immediate value of the address is allowed. This is one type of **addressing mode**, but several other addressing modes enhance the programming capabilities.

Let us consider the case where register R30 contains the value of decimal 4500; we denote this by writing

$$[R30]=4500.$$

Suppose that we would like to use 4500 as a reference in our program and store words at memory addresses 4500, 4504, 4508, 4512, and so on.[4] The present architecture does not allow this type of instruction. However, the modified datapath shown in Figure 11.26 provides the necessary data flow paths. A quick comparison

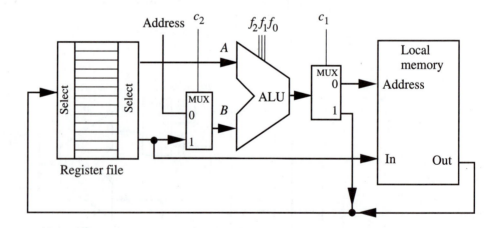

Figure 11.26 A modified datapath that allows the implementation of more instructions

---

[4] Remember that we are using the convention of addressing 8b bytes such that it takes 4B to store a 32b word. This means that the word addresses increment by 4.

of the two networks shows that we have changed the address input to where it is multiplexed with B as an ALU input. The new control signal is denoted by c_2 such that

> If $c_2 = 0$, then the ALU input B is an address, while
>
> If $c_2 = 1$, then the ALU input B is from a register.

To understand how this changes the instruction set, note that, in addition to the instructions already discussed, we can now perform the operation

> sw R2, 4(R30)   # Copy the contents of R2 into memory at
> # address ([R30]+4) = M(R30+4)

where the shorthand notation M(R30+4) means the memory location at the address computed by adding 4 to the contents of R30. This is permitted by

- Selecting register R30 as the A input to the ALU,

- Placing $c_2 = 0$ to use the value of 4 as the B input to the ALU,

- Instructing the ALU to add to produce the address ([R30]+4) = 4504, and

- Routing the contents of register R2 to be directed to the memory input (around the multiplexor).

This instruction is shown explicitly by the highlighted data paths in Figure 11.27.

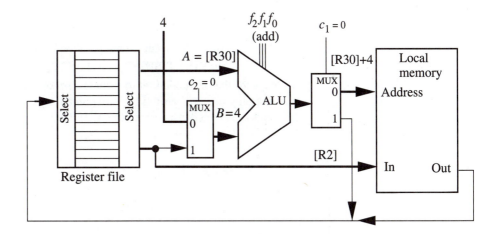

Figure 11.27 Datapath for the sw R2, 4(R30) instruction

To store the registers R3, R4, and R5 sequentially into the next higher addresses, we use the instructions

> sw R3, 8(R30)   # [R3]$\rightarrow$ M[R30+8]
> sw R4, 12(R30)   # [R4]$\rightarrow$ M[R30+12]
> sw R5, 16(R30)   # [R5]$\rightarrow$ M[R30+16]

which place data in memory at addresses 4508, 4512, and 4516, respectively. This datapath allows us to implement the analogous load word instructions that have the form

lw RX, A(RY)   # M[RY+A]→ RX

which means that the word in memory at address (RY + A) is copied to register RX.

It is clear from this example that the instruction set of a computer is determined by

- The characteristics of each unit, and

- The manner in which they are interconnected

as stated previously. The dependence of the instruction set on the logic and hardware illustrates the importance of viewing the system at various levels in the hierarchy.

As another example, suppose that we add a parallel-load shift register at the output of the ALU as shown in Figure 11.28. The shift operations are specified by the SHR (shift right) and SHL (shift left) control words, with the number of required control bits determined by the capabilities of the shift register itself. Adding this section to the datapath allows us to execute instructions such as

shl R14, R8, 3   # Shift contents of R8 left 3 bits
and place results in register R14

or, in general form,

shl RD, RT, SHIFT   # shift(RT) left by SHIFT→ RD

with the same format for a shift right operation:

shr RD, RT, SHIFT   # shift(RT) right by SHIFT→ RD

This expands the instruction set to include a set of shift (and rotate) operations that are commonly found in various types of computing algorithms. In most designs, the shift register is included as part of the ALU itself.

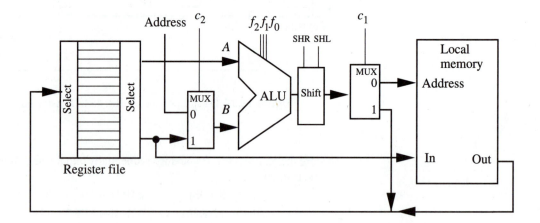

Figure 11.28 Addition of a shift unit to enhance the instruction set

11.5 The Control Unit

Now that we have seen how the datapath is created, let us examine how the binary instruction actually controls the datapath. Once an instruction arrives from memory, it is placed in the instruction register (IR) as shown in Figure 11.29(a), with a simplified block symbol shown in Figure 11.29(b). At this level, the instruction is cryptic since it appears as just thirty-two 1s and 0s in a row. Machine code has that characteristic, making it difficult to follow.

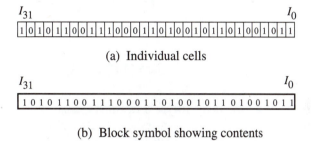

(a) Individual cells

I_{31} _____ I_0

| 1 0 1 0 1 1 0 0 1 1 1 0 0 0 1 1 0 1 0 0 1 0 1 1 0 1 0 0 1 0 1 1 |

(b) Block symbol showing contents

Figure 11.29 A machine-language instruction in the IR

The key to translating a machine-language instruction into something that makes more sense is to introduce the concept of an **instruction format**. An instruction format divides the 32 bits into smaller groups of bits called **fields**. The location of each field in the word is well defined and is assigned a particular meaning. For example, we may allocate a section of bits to be the binary word that specifies a particular register. Although a real computer has several instruction formats, we will only need two for the simple machine described here. One will be for register-to-register operations and will be called an **R-type** format. The other will be used for sw and lw operations and will be called an **I-type** instruction since it uses immediate data, i.e., data that is provided in the instruction itself and does not have to be obtained from another source.[5]

An R-type instruction format is shown in Figure 11.30(a). The 32-bit word has been divided into six different fields with the designated number of bits shown. The key to understanding the format is that each field has an independent meaning and usage. This is clarified by the equivalent viewpoint shown in Figure 11.30(b) where the fields have been designated by names.

Consider a register-to-register instruction of the form

        and RD, RS, RT   # AND(RS,RT) → RD

---

[5] It is worthwhile to mention that although R- and I-type instruction formats are specific to the MIPS processor, most RISC designs have similar (or identical) instruction types.

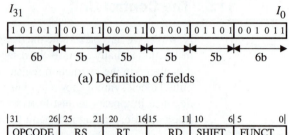

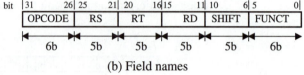

(a) Definition of fields

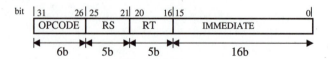

(b) Field names

Figure 11.30 An R-type instruction format

In the R-type instruction format, each register is specified by the appropriate field. Assuming that there are 32 registers R0 through R31 in the register file, we must allocate 5 bits for each. These are shown as the fields RS, RT, and RD in the drawing. The logical AND operation is specified by two of the fields in this example: the 6-bit OPCODE (short for operation code) field and the 6-bit FUNCT (short for function) field. In a large computer, these are used to generate the function select word $f_2 f_1 f_0$ for the ALU function, all multiplexor settings (like c_0, c_1, and c_2 in the previous datapath examples), and any other necessary control signals in the system. The last field shown in the R-type instruction format is the 5-bit SHIFT field that can be used to specify shift and rotate operations.

Next, consider the load word and store word operations. Since both move data words between the register file and memory, they can be described by the same instruction format. These instructions are based on **immediate** data (that supplied in the instruction itself) and are represented by the I-type format in Figure 11.31. Note that the two left bytes (bits 16–31) have the same fields as those defined for the R-type. In this case, the registers are defined by instructions such as

$$\text{sw RT, 48(RS)} \qquad \text{# [RT]}\rightarrow \text{M[RS+48]}$$

and

$$\text{lw RT, 100(RS)} \qquad \text{#M[RS+100]} \rightarrow \text{RT}$$

where the source register RS holds the base address, while the target register represents the origin or destination of the data segment being moved. The right 16 bits of the I-type instruction (bits 15–0) are labeled IMMEDIATE and are used for data or specifying an address. In the two example instructions above, the immediate values are 48 and 100, respectively.

Figure 11.31 Fields in an I-type instruction format

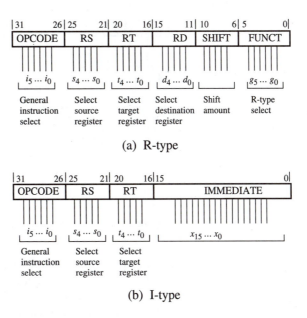

Figure 11.32 Comparison of R-type and I-type formats

Now let us turn to the overall problem of controlling the datapath. Every binary instruction must contain all of the information needed to perform the operation. This includes specification of the registers, the ALU function used in the operation, and datapath selection via the control signals. Datapath control is achieved by placing the instruction into the IR, and then using the different fields to provide the necessary information. To understand the connection, consider the R-type and I-type instruction formats as they relate to the controls signals sent to the datapath. In Figure 11.32(a), each bit in the instruction register has been associated with an output line. The 6 OPCODE bits have been labeled i_5 through i_0 and are used to specify the instruction grouping. Each of the register fields RS, RT, and RD has 5 bits and can be used directly to select registers as in Figure 11.12; similarly, the SHIFT field can be applied to the shift register. The FUNCT field has 6 bits and combines with the OPCODE field to generate the ALU control bits $f_2 f_1 f_0$. The I-type format is shown in Figure 11.32(b). By inspection it is easily seen that the left 16 bits are defined in the same manner as in the R-type format. The right bits 15 through 0, on the other hand, are used for IMMEDIATE data or addresses. The overlap of the left fields allows us to construct a single wiring diagram as shown in Figure 11.33. In order to accommodate both R-type and I-type formats, the 16 bits on the right side are wired to provide parallel control signals that can be used for either instruction type.

Specifying the instruction formats and the associated wiring allows us to obtain the complete view of the central processing unit shown in Figure 11.34. This illustrates how the IR is interfaced to the datapath to control the operations. The register selection fields for RS, RD, and RT can be directly connected to the register file. The remaining control signals that are needed for the multiplexors and choosing the ALU function are obtained using a logic network that is contained within the section labeled as control logic. This accepts inputs from the OPCODE, SHIFT, and

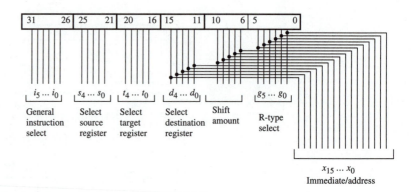

Figure 11.33 General wiring from the instruction register

FUNCT fields and generates the necessary outputs. The key to understanding this process is noting that the fields are encoded to correspond to specific instructions. Every set of input bits generates a specific set of control signals that provide the proper setting for the datapath components. Note that the control bit c_3 determines which register field is to be used to specify the destination. This is required because an R-type instruction uses RD, while an I-type does not have RD but employs the target field RT instead for its destination.

The control signals are determined by the **OPCODE map** for the processor. This is the encoding that is used for the instruction field OPCODE; an R-type instruction also uses the FUNCT field. When the computer is designed, every instruction is defined by a unique set of binary values in the appropriate field(s). This is called a "mapping" since it defines the relationship between an instruction and a unique binary word. The control logic network is designed around this mapping. The OPCODE and FUNCT act as primary inputs and determine all values of the control bits needed for the hardware to execute a particular instruction. As an example of an OPCODE mapping, three entries used in the MIPS processor design are

$$OPCODE = 000000 \Rightarrow \text{register-to-register operations}$$
$$OPCODE = 100011 \Rightarrow \text{load word (lw)}$$
$$OPCODE = 101011 \Rightarrow \text{store word (sw)}$$

Once an R-type (register-to-register) instruction is specified in OPCODE, the FUNCT field is used to determine which operation to perform. For example, the MIPS set gives

$$OPCODE = 000000 \quad \text{and} \quad FUNCT = 100000 \Rightarrow ADD$$
$$OPCODE = 000000 \quad \text{and} \quad FUNCT = 100010 \Rightarrow SUB$$

and so on. This provides the critical connection between machine-language instructions that consist of 32b binary words, and the actual operations that are being executed by the processor.

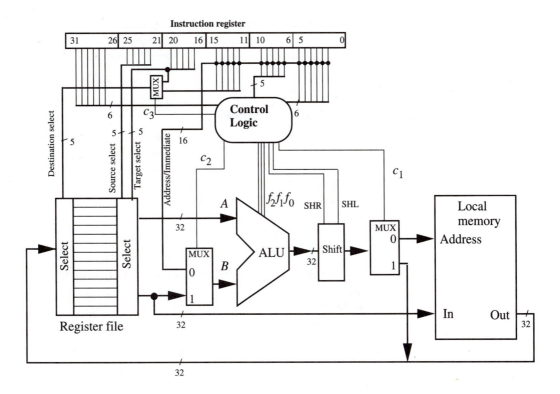

Figure 11.34 Wiring diagram showing both the datapath and control signals

11.6 CISC and RISC Architectures

Let us now examine a classification of computers that is based on the number and types of instructions that are built into the system. We have seen that every instruction type must be allowed by the architecture of the datapath. The larger the variety of the instructions, the more data flow variations must be built into the processor, and the more complicated the system becomes.

If we track the early historical evolution of computer designs, we find that every successive generation was generally more sophisticated than the preceding one. Improved technology allowed for a wider variety of instruction types and addressing modes, and many data flow possibilities were included. These evolved to what are now called **complex instruction set computers** (**CISC** machines). Compilers were written to make use of the expanded instruction set, and when assembly code was used, programming was easier due to the large number of available instructions. This philosophy dominated the computer industry for many decades. However, research studies in the 1970s laid the foundations for a modified approach to designing and building computers.

This approach centers on designing a **reduced instruction set computer** (**RISC**) that has a small number of instructions (as compared with a CISC machine). At the hardware level, the design emphasizes running the instructions that are provided as fast as possible. The target of a RISC architecture computer is to achieve fast computation using a relatively simple datapath. To appreciate the RISC philosophy, let us first examine the CISC architecture in more detail and use this as a basis for comparison.

11.6.1 CISC and Microprogramming

CISC computers obtain their large instruction set by using a technique called **microprogramming** that embeds a sequential logic network inside of the control unit. It is, in fact, similar to having a small computer that operates inside of the main computer. The microprogram approach breaks down every basic operation into a microinstruction. This includes commands that we may loosely describe by examples such as

get [RX]   # Read data from RX

input [RX] ALU   # Move [RX] to ALU input

except that this should properly be called **microcode instructions** and is one level deeper (more primitive) than assembly language. In fact, an assembly-level computer instruction such as the load word (lw) is created by using a sequence of micro-operations as shown in Figure 11.35. Every instruction is defined in this manner, but it will take a different number of microinstructions to create each assembler instruction.

Microprogramming allows the designer to create an instruction by combining the needed operations at the microcode level. This allows one to build a large instruction set since the microinstructions provide all of the basic movements in the system. Each microcode sequence is stored in a **microcode ROM** array. When the computer is given an instruction such as load word (lw), the microcode sequence is accessed. The overall result is a control unit that operates in the manner shown in Figure 11.36. An instruction in the IR is sent to the **sequencer**, which determines the location of the microcode corresponding to the specified operation. The sequencer reads and executes the first microinstruction, then the second, and so on, until the instruction sequence is completed. Every instruction undergoes this sequencing.

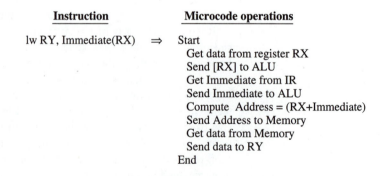

Instruction		Microcode operations
lw RY, Immediate(RX)	⇒	Start
		Get data from register RX
		Send [RX] to ALU
		Get Immediate from IR
		Send Immediate to ALU
		Compute Address = (RX+Immediate)
		Send Address to Memory
		Get data from Memory
		Send data to RY
		End

Figure 11.35 One computer instruction consists of several microinstructions

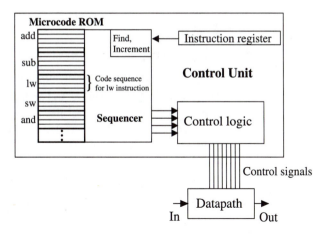

Figure 11.36 Microcode usage in the control unit

This is a powerful and flexible approach for increasing the instruction set of a computer. Adding a new instruction simply requires that additional code be written and stored in the ROM, so that modifications to the instruction set do not require extensive hardware changes. The main drawbacks are

- The internal sequencer circuit must be added to the unit.

- Each microinstruction requires a time t_{micro} to complete, so that computer-level instructions such as add and lw may have different execution times.

Microcoding is straightforward to do once the machine is defined. Perhaps the most difficult consideration in the CISC approach is the fact that the hardware tends to grow in complexity with the richness of the microcode instruction set. From a historical perspective, the increase in hardware complexity has been matched by improved technology. This has resulted in machines that were both more powerful and faster with each successive generation.

11.6.2 RISC Machines

Now let us look at the concepts behind a RISC computer. An interesting study of programs running on CISC computers led to what is now called the **80/20 rule**. To understand the rule, suppose that a CISC machine has a large number of instructions available. Then, according to the 80/20 rule,

> "80% of the program only uses 20% of the available instructions."

For example, if a computer has 500 possible instructions, then 80% of a typical program will only use about 100 of them. This, of course, is not an accurate statement of numbers, but it does illustrate an important point: most programs can run using only a small number of instructions. This led to the concept of the **reduced-instruction set computer** (**RISC**).

The philosophy of a RISC computer can be summarized by the following statements:

- Only include the most useful instructions in the datapath, and
- Insure that the datapath yields fast execution of every instruction.

This yields simpler hardware and architectural features, but the compilers must be written to produce efficient machine code. While these statements convey the important ideas, the difference between the CISC and RISC architectures is much deeper than might be imagined; the two approaches can produce radically different machines. It is worthwhile to examine the differences at the operational level, as this will make it easier to appreciate modern design.

Figure 11.37 shows the basic structure of the control unit in a RISC machine; the alert reader will recognize this as having the same features as the one we studied in our general discussion. The instruction register is connected directly to the control logic block, which in turns transmits control signals to the datapath. However, note that we have indicated that the RISC approach only permits a **single-pass** datapath structure. This means that each unit in the datapath (the ALU, memory, and the register file) can only be accessed **once** during an instruction. Restricting the datapath to be single-pass has three major consequences.

- The datapath can be optimized for the fastest throughput.
- Every instruction takes the same amount of time.
- Only instructions that can be completed in a single pass through the datapath are allowed.

Practical implementation of these concepts leads to many constraints on the architecture. For example, to add a new instruction, the datapath itself must be altered to allow the necessary connections. Also, the instruction word and the data word are chosen to be the same size to allow for easier transfer between the CPU and the memory unit.

The differences between the microprogram approach used in a CISC machine and the single-pass datapath RISC design can be understood using the block diagrams in Figure 11.38. Recall that every instruction starts with

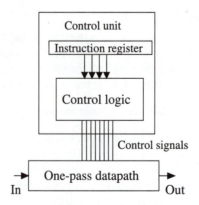

Figure 11.37 A RISC design uses simple control and a one-pass datapath

- Instruction fetch (IF), in which the instruction is brought to the IR from memory

followed by

- Instruction decode (ID), where the control signals are generated and sent to the datapath.

These are common to both designs. The distinction arises in the manner in which the computations are performed. Figure 11.38(a) represents the CISC case where microcode controls the data flow among the ALU, the memory unit, and the register file. Once the instruction is decoded in the ID stage, control signals are sequenced to provide the necessary operations. For example, this allows us to execute a command of the form

$$\text{add M[AD], M[A1], M[A2] \# M[A1]+M[A2]} \rightarrow \text{M[AD]}$$

which takes the contents of two memory locations A1 and A2, adds them, and stores the result in memory at address AD. Similarly, altering the microcode slightly would produce an instruction of the form

$$\text{add RD,M[A1],M[A2]} \qquad \text{\# M[A1]+M[A2]} \rightarrow \text{RD}$$

which places the content into register RD in the register file.

Now consider the basic RISC single-pass datapath illustrated in Figure 11.38(b). In this design, the ALU accepts inputs from the register file and can store the results either in the register file or in the memory. Since the datapath is restricted to a single pass through each unit per instruction, it does not allow

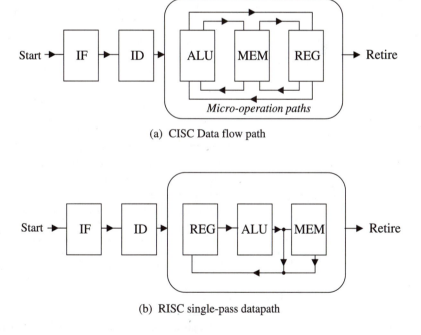

(a) CISC Data flow path

(b) RISC single-pass datapath

Figure 11.38 Comparison of data flow in CISC and RISC architectures

either of the operations discussed above for the CISC design. This is due to the fact that both use memory accesses for the ALU inputs, but the RISC datapath only allows inputs from the register file. This simple example illustrates that the RISC design is much more restrictive in the types of instructions that can be included. If the datapath is expanded to allow for more variations, then the increased complexity may slow down the system.

11.6.3 Modern Computers

Modern processor design has evolved in a manner such that it usually embraces as many ideas from the RISC philosophy as possible. This is especially true for VLSI microprocessors created in silicon. Some processors, such as the MIPS and Power PC® chips, are true RISC machines, while others, such as the Pentium and the AMD K6 processors, have a RISC core network that is embedded within a larger control structure to provide CISC-type operation. This type of design is particularly important when a new generation of a microprocessor family is desired to be backward-compatible with an early CISC product.

11.7 Floating-Point Operations

Arithmetic operations such as addition and multiplication often require that we use fractional values that may range over a wide span of magnitudes. In mathematics, these are known as **real numbers**. Some common examples are the constants

$$\pi = 3.14159 \ldots$$
$$e = 2.718 \ldots$$

(11.8)

and values such as

$$4.35 \times 10^{-4}$$
$$c = 2.997 \ldots \times 10^{10}$$
$$q = 1.602 \times 10^{-19}$$

(11.9)

These numbers have different characteristics than the integers we have been studying thus far. Although we can use the fractional conversions discussed in Section 1.5, a more efficient approach called a **floating-point** representation is used in practice.

Floating-point numbers are represented in binary by using the basic expression

$$(-1)^S \times (1 + \text{significand}) \times 2^{(\text{exponent} - \text{bias})}$$

(11.10)

where the significand provides the significant bits of the number, the exponent specifies the power of 2, and the bias is a value that is used to control the range of the number. Also, S is the sign bit defined as $S = 0$ for a positive number and $S = 1$ for a negative number. The actual implementation of the floating-point representation depends on how many bits are allocated to represent the number and, hence on the accuracy. We will study the IEEE 754 floating-point standard (FPS) that uses a

32-bit word that provides a **single precision** representation.[6] The standard also provides for a more accurate 64-bit representation called **double precision**.

Before progressing to the details of the FPS, let us examine the conversion of a decimal number to a binary floating-point representation. Suppose that we wish to represent the base-10 number $X = 4.625$ in a binary format. Using the fact that $4_{10} = 100_2$ and $0.625_{10} = 0.101_2$, we can write

$$X = 100.101_2 \tag{11.11}$$

This can be expressed in scientific notation as followes:

$$100.101 = 1.00\,101 \times 2^2 \tag{11.12}$$

by shifting the decimal point two positions to the left and multiplying by 2^2. This shows that binary floating-point representations are the same as standard base-10 ideas.

In the FPS, single-precision floating-point numbers are based on a 32-bit word that is divided into three distinct fields as shown in Figure 11.39. The exponent is represented by an 8-bit field, while the significand is allocated a total of 23 bits; the sign bit S completes the talley. The single-precision format is defined with a bias of 127. Since this admittedly looks a bit cryptic, let us examine how it works using an example.

Figure 11.39 Single-precision floating-point format

Example 11-1

Consider the base-10 number $a = 12.15_{10}$. Using the standard conversion techniques, this has a binary representation of

$$\begin{aligned} a &= 1100.00\overline{1001} \\ &= 1.100001\overline{1001} \times 2^3 \end{aligned} \tag{11.13}$$

In writing this, we have noted that 0.15_{10} has a binary representation that contains the repeating value 1001

$$0.15_{10} = 00\,1001\,1001\,1001\,\ldots$$

which we have denoted by $\overline{1001}$ in the equation. The significand to 23 bits is thus given by

$$S = 11000\,1001\,1001\,1001\,1001\,10 \tag{11.14}$$

---

[6] In case you missed this acronym in Chapter 2, **IEEE** stands for the Institute of Electrical and Electronic Engineers, a professional organization.

by direct comparison. To determine the exponent, we note that the bias is 127 so that

$$127 + 3 = 130$$
$$= 10000010_2 \tag{11.15}$$

which is the value of the exponent. Since a is a positive number, the FPS representation is

$$0 \mid 1000\ 0010 \mid 11000\ 1001\ 1001\ 1001\ 1001\ 10$$

as can be verified by reversing the process.

Double-precision floating-point representations are based on a 64-bit word as shown in Figure 11.40. The exponent is increased to 11 bits with a bias of 256, while the significand is taken to be 52 bits in length. Although the size of the double-precision FPS is twice that of the single-precision representation, the concepts remain the same.

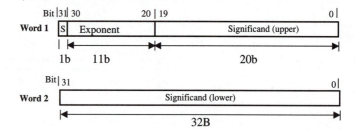

Figure 11.40 Double-precision format for floating-point numbers

11.7.1 Arithmetic Operations

Binary floating-point arithmetic is very similar to the procedure used for base-10 numbers that are in scientific notation. Consider the two decimal numbers

$$X = 4.24 \times 10^3$$
$$Y = 6.78 \times 10^2$$

To obtain the sum $(X + Y)$, we first find a common power of 10 by shifting the decimal point in one (or both) of the numbers This is called **alignment** of the exponents:

$$X = 42.4 \times 10^2$$
$$Y = 6.78 \times 10^2$$

Adding then gives

$$X + Y = 49.18 \times 10^2$$

To obtain the product

$$X \times Y = (4.24 \times 10^3) \times (6.78 \times 10^2)$$

we just add the exponents to obtain

$$X \times Y = 28.7472 \times 10^5$$

so that no alignment is necessary.

Binary floating-point operations proceed in the same manner. To add two floating-point numbers, we first align the exponents. This means that we find a common value of m in the factor 2^m. After this is accomplished, we add the significands to obtain the sum. Multiplication is accomplished by multiplying the significands and adding the exponents. Conceptually, these operations are straightforward. However, the details are actually quite involved and result in a complicated logic network. For the floating-point adder, a lot of the complexity is due to the algorithm needed to perform alignment of two arbitrary floating-point numbers. The multiplier circuits are complicated by the size of the significands and the need to perform fast calculations. Owing to these observations, we will forgo the details here and maintain our discussion at the higher, system level.

11.7.2 Application to Computers

General-purpose computers are designed to handle both integer and floating-point numbers. Since the logic networks are distinct, it is common to provide both integer and floating-point ALUs. One approach is to use the floating-point unit as a **co-processor** as shown in Figure 11.41. A co-processor is used as needed along with the main (integer) processor. In the basic system portrayed, the instruction is stored in a common IR. Decoding the instruction determines whether it is to be sent to the integer unit or the floating-point processor (**FPU**). This is achieved by using different instruction codes for integer and floating-point operations. The data bus controller insures that the inputs and outputs are sent to the proper units. In the early days of microprocessors, the co-processor was sold as a separate "add-on" chip. However, advances in VLSI have made it an integral part of the processor core circuits.

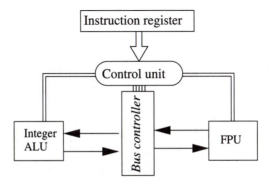

Figure 11.41 Use of a floating-point unit as a co-processor

11.8 VLSI Aspects of Computer Design

Modern computers are designed in silicon using high-density integration techniques. This applies to large systems all the way down to the microprocessor, which is an entire CPU plus L1 cache on a single chip. Although the details of designing a VLSI chip of this complexity are beyond the scope of this book, the concepts are straightforward and are based on the material presented in Chapter 7.

Two keys to completing a VLSI design of this complexity are

- A strong, cohesive group of engineers, and
- A good CAD toolset and cell library.

Both are critically important, since modern microprocessor designs continuously push the technology limits to achieve their ambitious goals. The starting point in the design process is to specify the system characteristics. This includes the details of the instruction set, hardware and clocking considerations, and some of the unit specifications. The architecture of the system is directly related to the instruction set, so that architectural features need to be designed along with all necessary logic. In VLSI, the circuit design and layout are critical aspects that affect the performance of the system, so that the technology must be fully understood and characterized by the design engineers.

Let us examine the creation of a simple microprocessor from the layout viewpoint. Using the library, we may choose a memory cell and use it to create a register. The register may then be instanced to produce the register file, as shown in Figure 11.42. VLSI design relies very heavily on the ability to design a cell and then use it many times in the chip layout, as this saves much time and energy.

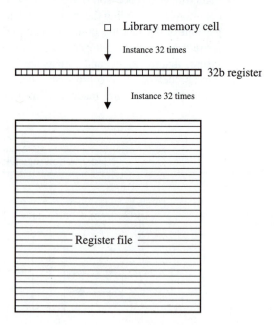

Figure 11.42 Using a cell library to create the register file

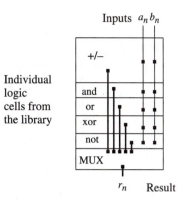

Individual
logic
cells from
the library

Figure 11.43
Creation of an ALU bit slice

Next, consider the ALU. This may be designed using the bit-slice approach discussed earlier in the chapter. A single slice might be constructed as shown in Figure 11.43, where cells have be instanced from the library for addition/subtraction and basic logic operations. For a given slice, the inputs a_n and b_n are routed into each cell, with each cell producing a particular output result. Using a MUX as the final cell in the vertical stack allows us to select the desired function result r_n. The completed ALU is then obtained by using the bit slice as a basic unit, and instancing it as many times as needed. This results in the large cell illustrated in Figure 11.44.

The cell library and instancing are used to create all other sections of the microprocessor chip. Every unit is characterized by height and width dimensions that are estimated during the floorplanning stages of the chip specification. Once the units are completed, they are pieced together and wired to give the final layout as portrayed by the drawing in Figure 11.45. This shows the use of "input/output pads," which are rectangular metal regions that can be wired to the package leads.

This overview provides a glimpse into what the field of VLSI microprocessor design is like. In the real world, however, engineering a chip with state-of-the-art

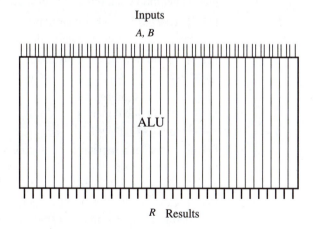

Figure 11.44 ALU formed by instancing bit-slice units

Input/output pads

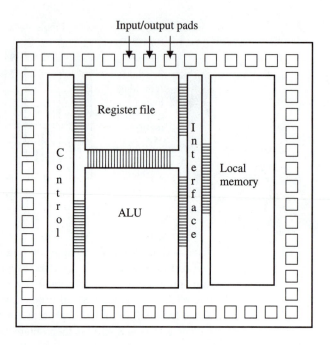

Figure 11.45 Final layout of the microprocessor chip example

features can take a team of 200+ engineers 2 or 3 years (or more!) to complete to the manufacturing stage. Challenges such as these are one of the aspects that make the field quite interesting and popular among engineering students.

11.9 Problems

[11.1] Can you list two examples of common output devices on a personal computer other than the monitor and disk drives?

[11.2] Discuss how a basic word processing program can be created using only the two basic computer operations of data movement and binary operations.

[11.3] Several sizes for register files are listed below. For each case, specify the word size for the select words used to choose a specific register.

  (a) sixteen 16-bit registers
  (b) sixteen 32b registers
  (c) eight 32b registers

[11.4] Consider a register file that consists of eight 8b registers. Draw a block diagram for the register file using the same detail shown in Figure 11.13. Be sure to specify the select words, MUX sizes, etc. in your drawing.

[11.5] Consider the simple ALU shown in Figure 11.19. Suppose that the input s is specified to be $a_n = 1$ and $b_n = 0$ for a particular bit slice. Find the output result r_n for each function select word listed.

(a) $f_2 f_1 f_0 = 010$
(b) $f_2 f_1 f_0 = 101$
(c) $f_2 f_1 f_0 = 110$
(d) $f_2 f_1 f_0 = 011$

[11.6] Consider the simple ALU shown in Figure 11.19. Suppose that the input s is specified to be $a_n = 1$ and $b_n = 1$ for a particular bit slice. Find the output result r_n for each function select word listed.

(a) $f_2 f_1 f_0 = 111$
(b) $f_2 f_1 f_0 = 000$
(c) $f_2 f_1 f_0 = 100$
(d) $f_2 f_1 f_0 = 001$

[11.7] Suppose that we want to construct a basic 2-input ALU network that only provides four operations: ADD, NOT, AND, OR. Specification of an operation is accomplished using the 2-bit word $f_1 f_0$.

(a) Design the logic for a bit slice of the ALU using the drawing in Figure 11.19 as a guide.

(b) After you have finished the design, you decide that you want to provide the SUB operation in the ALU; SUB is to be selected by another control bit f_s. Use the circuit in part (a) and provide additional circuitry for this modification.

[11.8] Construct a block diagram for a datapath that is 4b wide and allows register-to-register operations using an 8-register file. Be sure to specify the size of the register selection words.

[11.9] Construct a block diagram for a datapath that is 16b wide and allows register-to-register operations using a 16 register file. Be sure to specify the size of the register selection words.

[11.10] Let $X = 1001$ and $Y = 1011$. Compute the results of the following operations.

(a) $X \cdot Y$
(b) $\overline{X \oplus Y}$
(c) $X \cdot 1$
(d) $1 + Y$

[11.11] Let us define two 8b words with base-10 values of $A = 156_{10}$ and $B = 73_{10}$. Find the base-10 result after the following binary operations are performed.

(a) $A \cdot B$
(b) $A + B$
(c) $\overline{A}$
(d) $A \oplus B$
(e) $A + \overline{B}$

[11.12] Consider the basic datapath in Figure P11.1. The initial contents of the registers are shown with the base-10 equivalent values. Assume these values as the starting point for each part below.

(a) Calculate the contents of R2 after the operation
      add R2, R4, R5

(b) Find the contents of R6 after the following sequence of instructions is completed.

> add R5, R6, R1
>
> add R6, R2, R7

(c) Find the contents of R4 after the

> and R4, R2, R6

Give your answer in both binary and base-10 form.

Register file

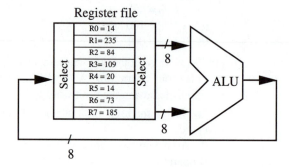

Figure P11.1

[11.13] Consider the basic datapath drawn in Figure P11.1. The values shown in the registers represent the initial contents when the following instruction sequence is run:

> sub R1, R3, R6
>
> add R2, R6, R7
>
> or R3, R5, R6
>
> add R4, R3, R1

List the contents of the register file after this sequence is completed. Give your results in base-10 values.

[11.14] Consider the basic datapath draw in Figure P11.1. What are the contents of R4 and R5 after the following instruction sequence has been executed? Give your answer in both binary and decimal values.

> add R3, R4, R6
>
> shl R4, R3, 3
>
> shr R5, R3, 3

[11.15] Consider the datapath drawn in Figure P11.1 with the ALU specified by the bit slice portrayed in Figure 11.19. For each instruction below, list the binary values of the ALU function select $f_2 f_1 f_0$ and the register select words $d_2 d_1 d_0$, $s_2 s_1 s_0$, $t_2 t_1 t_0$.

(a) and R3, R5, R7

(b) sub R6, R1, R4

(c) xnor R4, R2, R0

[11.16] Consider the following instructions. Which of these cannot be implemented on the RISC datapath drawn in Figure 11.26? For each case, provide the sequence of RISC instructions that will accomplish the same result.

add R3, R28, R32
sw R2, 4(R27)
add R4, M[R6], 0x 34
lw R17, M[R21]

[11.17] List what you think are the two most important characteristics for the CISC architecture. Then do the same for the RISC architecture. Finally, discuss the possibility of combing all four features into a single computer design. Do you think it would result in a better architecture?

[11.18] Convert the decimal number 23.625 to a binary floating-point number.

[11.19] Convert 16.875 to a binary floating-point number.

[11.20] A binary number has the floating-point representation of

$$1.001011 \times 2^3$$

Find the equivalent base-10 (decimal) value.

[11.21] Find the decimal equivalent of the single-precision floating-point number

0 | 1000 0011 | 10000 1001 1001 1001 1001 10

using the IEEE FPS.

[11.22] Suppose that we have a 32-bit integer CPU and interface it with a double-precision FPU. Why does it make sense to use separate register files for the two sections? Be specific in your answer.

[11.23] As computer architectures get more and more powerful, additional circuitry is needed to accommodate the new features. In terms of integrated circuits and VLSI, this could lead to very large chips. Discuss the problem of yield (Chapter 7) and how this can be overcome by improved silicon processing.

[11.24] Suppose that we have a technology where a 1-bit CMOS VLSI memory cell occupies a silicon area of 1.8 $\mu m \times 1.2 \mu m$.

(a) Estimate the size of a 32-bit register.
(b) Estimate the area of a register file made up of $32 \times 32b$ registers.
(c) Suppose that each circuit dimension is scaled by a factor of $\alpha = 1.9$ as defined in Chapter 7. Find the size of the resulting $32 \times 32b$ register file.

[11.25] Use the World Wide Web to look up information on specific computer chips available from different companies. For the URLs, either run a search or try the form http://www.*company_name*.com where *company_name* denotes the name of a manufacturer (e.g., *company_name* = intel will put you into the Intel home page).

CHAPTER 12

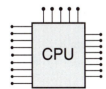

Advanced Computer Concepts

The study of computer performance deals with the techniques used to make computers faster. At the qualitative level, this means that we want to reduce the time needed to complete a specified set of calculations, or to run a particular program. In order to increase the speed of a computer, we must either modify the existing circuits or try an entirely new architectural idea for the data flow path. In this chapter, we will examine some of the most important ideas that are used to enhance the performance of modern computers. The discussion will serve to illustrate the fact that the processing speed depends upon the details of every level in the design hierarchy, from basic circuits up through advanced architectures.

12.1 Computing Speed

Suppose that we are going to purchase a desktop computer to replace our old system that we bought a couple of years ago. We want to maximize our return by selecting a new system that "runs better" than the present one, and that exhibits a noticeable improvement when running our most-used programs. Now comes the critical question that all good engineers and scientists must ask: What does it mean to "run better"? What quantitative criteria exist and can be applied to help us wade through the massive amounts of advertising associated with the various brands and models? Some things are fairly obvious: size of the main memory, disk storage space, monitor size and resolution. But how do we judge the performance of the CPU?

To compare the two computers, we can choose a program of reasonable length and complexity, run it on both systems, and record the total execution time on each system. This would result in a total time T_x for system X, and a time T_y for system Y. If $T_x > T_y$, then system Y would be judged to be faster since it takes less time to execute the same program. These times can be used to define the **performance rating** P for the two by

$$P_x = \frac{1}{T_x}, \qquad P_y = \frac{1}{T_y} \qquad \qquad \textbf{(12.1)}$$

such that

$$P_y > P_x \qquad \qquad \textbf{(12.2)}$$

tells us that system Y has a higher performance rating than system X.

This is a reasonable way to compare two similar computing systems, but it does not provide an *absolute* measure of the speed since it does not account for the different types of instructions that are used in the program. A more meaningful approach from the practical viewpoint might be to choose a set of programs, run them on both systems, and attempt to determine the differences in CPU execution times.

Regardless of the somewhat vague interpretation, a concept that does have some meaning is that of **instruction throughput**. Quantitatively, we can define this as

Instruction throughput = # instructions processed per second

This can be used to compare the two systems so long as we have defined the type of instruction(s) that are actually running. As we saw in the previous chapter, the datapath limits the instruction throughput due to the logic delays intrinsic to the hardware. We usually assume that we have used the fastest logic circuits available, so we may not be able to expect any improvement on that front. Instead, let us examine techniques for improving the instruction throughput by means of changes in the overall architecture of the computer. If this can be accomplished, then we do not need to wait for an advancement in technology to occur.

12.2 Pipelining

Pipelining is a technique for reducing the total time needed to execute a program. It is based on the idea of controlling the flow of instructions through a datapath in an effort to maximize the throughput, i.e., the number of instructions that can be processed per second. This is achieved by devising a scheme that tries to use every circuit in the data flow path during each clock period to its maximum potential.

Let us construct the CPU using the block diagram shown in Figure 12.1(a). In this drawing, each group of circuits represents the logic network needed to complete the desired task. The datapath has been divided into the standard grouping

- **IF**: Instruction fetch
- **ID**: Instruction decode
- **EXE**: Execution of the instruction
- **MEM**: Memory access (if needed)
- **WR**: Write back to register file (if needed)

which follows the sequence established in the previous chapter. The motivation for pipelining can be illustrated by examining the data flow through the network. In a basic clocking scheme, it takes one clock period to complete all five operations. If we trace the data flow through the CPU chain, we encounter the normal delays associated with each unit. The time line in Figure 12.1(b) shows the location of the data as it moves through the network. The result is available at the time t_{inst} where

$$t_{inst} = t_{IF} + t_{ID} + t_{EXE} + t_{MEM} + t_{WB} \tag{12.3}$$

is the worst-case time needed to access every unit, as obtained by summing the individual times. Using this arrangement, new data is admitted every clock cycle. The minimum clock period T_1 is thus given as

$$T_1 \geq t_{inst} \tag{12.4}$$

since this always allows the data to complete the cycle. Executing n instructions requires a total of nT_1 seconds. The key observation here is that once the data passes through a particular unit in the datapath, it is not used again until the next wave of data. Since the unit has completed its assigned task, it is idle for the remaining portions of the clock cycle. Pipelining attempts to eliminate this idle time for every unit by using a different approach to data flow control.

Figure 12.2(a) shows how the data flow path has been modified to allow for pipelining. The logic units are still broken up into the sections labeled IF, ID, EXE, MEM, and WB, except that now each unit has its own set of input registers that control the entrance of data into the unit. Since the registers are all controlled by the

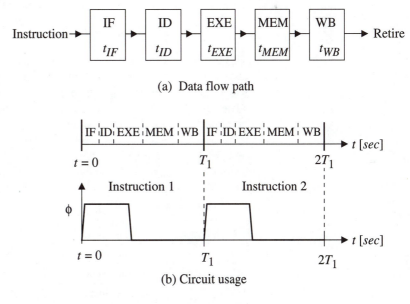

(a) Data flow path

(b) Circuit usage

Figure 12.1 Timing and circuit usage in a basic computer

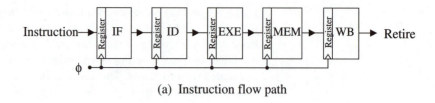

(a) Instruction flow path

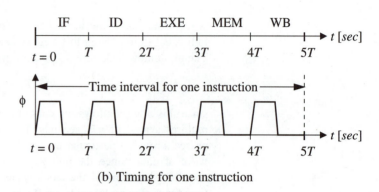

(b) Timing for one instruction

Figure 12.2 Modification to a pipelined structure

same clock signal ϕ, every unit receives new data at the same time. The characteristics of the data flow are illustrated by the timing diagrams in Figure 12.2(b). For this system, we have chosen a new clock period $T < T_1$ such that

$$T = \max\{t_{IF}, t_{ID}, t_{EXE}, t_{MEM}, t_{WB}\} \tag{12.5}$$

This insures that the slowest unit is allocated enough time to respond, and it is the fastest clock that can be used. Completing a single instruction requires a time of $5T$, which may in fact be *larger* than T_1 in the simpler clocking scheme! The speed advantage of pipelining is achieved when a long sequence of instructions (i.e., a program) is run through the system.

A simplified view of the pipeline is shown in Figure 12.3. As the instruction is processed by the computer, it "occupies" one group of circuits at a time. Tracing the instruction from start to finish through this pipeline allows us to visualize the movement. To accomplish this task, let us label the clock cycles by CC1, CC2, etc.; then, the circuit usage is as follows:

Clock Cycle	Circuit Usage
CC1	Instruction fetch (IF)
CC2	Instruction decode (ID)
CC3	Execute instruction (EXE)
CC4	Memory access (MEM)
CC5	Write back (WB)

Figure 12.3 Instruction movement in a pipeline

This shows that the instruction only uses one of the five circuit blocks at a time. Pipelining makes use of this fact by allowing a new instruction to enter the logic flow at the beginning of every clock cycle. The flow is shown in Figure 12.4 for the case where we want to perform the addition operation as denoted by the op code ADD. This clearly shows the progression of the instruction through the processor.

It is now a straightforward task to add pipelining to this system. Suppose that we want to implement a sequence of instructions *Inst1, Inst2, Inst3*, and so on. Each instruction is applied in order to the pipeline, with a new instruction entering the IF unit with each rising clock edge. This leads to the instruction flow illustrated in Figure 12.5. During CC1, *Inst1* enters the IF stage. During the next clock cycle CC2, *Inst1* is moved to the ID unit, while *Inst2* enters the IF stage. The next clock cycle CC3 moves *Inst1* to EXE, moves *Inst2* to ID, and allows *Inst3* to enter the IF unit. This sequential loading of the pipeline continues so that it is filled by CC5 when the result of *Inst1* is available. The next clock cycle CC6 retires *Inst1* and allows *Inst6* to enter the pipeline.

The name "pipeline" is used to describe this type of flow because it is analogous to filling up a pipe with balls, one at a time, until the pipe is full. After that, each additional ball forced into one side results in a ball being forced out of the other side. The alert reader will also notice that the operation is basically the same as that of a shift register discussed in Section 8.5 of Chapter 8, except that the pipeline stages consist of complex logic networks instead of single memory cells.

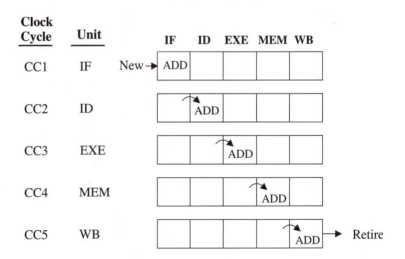

Figure 12.4 Instruction movement and timing in the pipeline

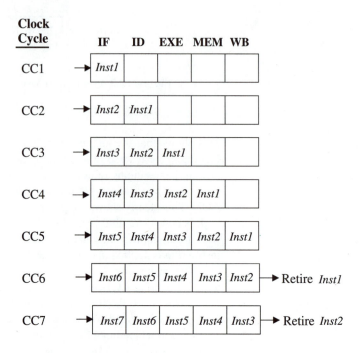

Figure 12.5 Instruction flow through a pipelined network

Now that we have seen the pipeline in action, let us analyze the improvement in the computing speed of the computer. Since it takes one clock period T to move the instruction from one unit to the next, the result of the first instruction *Inst1* is available at a time $5T$ after it is applied to the pipeline; this is verified by the simple example shown in Figure 12.5. However, since the pipeline is filled, *Inst2* is completed by a time $6T$, *Inst3* is completed by $7T$, and so on. In other words, a filled pipeline completes an instruction every clock period. This should be compared to the situation where we only have a single instruction in the processor; in this case, we must wait $5T$ seconds for every instruction.

It is possible to calculate the increase in speed that is achieved by filling a pipeline over the case where a single instruction is processed at a time. Suppose that we want to run a program that consists of N instructions. If only one instruction is processed at a time, then it takes a total of

$$t_1 = N \times (5T) = 5NT \tag{12.6}$$

seconds to complete. A filled pipeline, on the other hand, requires a total time of

$$t_{pipe} = 5T + (N - 1)T$$
$$= (N + 4)T \tag{12.7}$$

where the term $5T$ assumes that the pipeline is initially empty and accounts for the time needed for the first instruction to reach the end, while the second term $(N - 1)T$

is the time needed for the remaining $(N - 1)$ instructions to complete the cycle. For example, if $N = 1000$, then

$$t_1 = 5000T$$

$$t_{pipe} = 1004T$$

(12.8)

so that the pipeline increases the effective speed of the processor by a factor of about 5. Although these are only ideal estimates and do not account for any of the problems encountered in a realistic application, it is clear why pipelining is a popular technique for increasing the processor speed.

The number of stages used in a pipeline usually depends upon the switching speed of the slowest unit. This is because the period T must be long enough to allow every unit sufficient time to perform the desired logic. As technology evolves, it may become possible to increase the speed of the slow circuit or unit, allowing an increase in the frequency of the pipeline clock. Alternatively, it may be possible to increase the number of pipeline stages while decreasing the clock period. This creates what is called a **deep pipeline**, where the concept of "depth" is used to describe the number of stages in the cascade.

An example is shown in Figure 12.6(a). In this system, the execute unit has been split into two distinct units labeled EXE1 and EXE2. Similarly, memory access is now accomplished in the two distinct pipeline stages MEM1 and MEM2. The five-stage pipeline has thus been expanded into a seven-stage network that is clocked with a new period T_D. Due to the simpler design of the most complex units, $T_D < T$. As shown in Figure 12.6(b), it takes seven clock periods T_D to complete one instruction. If

$$7T_D < 5T$$

(12.9)

then the deep pipeline will process an instruction in a shorter time, thus increasing the throughput. Although the concept is straightforward, it must be remembered that the limiting factor on the architectural design is the speed of the individual circuits. In other words, the system performance depends on every other level in the design hierarchy. An interesting point to note here is that if $T = T_D$, then the instruction throughput is the same for both systems once the pipelines are filled.

12.2.1 Data Hazards

The idealized description of pipelining assumes that every instruction is independent of every other instruction. This means, for example, that *Inst2* does not depend upon the result of *Inst1*. If a dependence exists between closely spaced instructions, then a problem occurs in the manner in which the instructions process the data. This is due to the fact that the result of an instruction is not finalized until data is written to the memory (during the MEM cycle), or to a register (during the WB cycle). In general, we refer to this situation as a **data hazard**.

Consider the following sequence of instructions:

$$\text{add R1, R8, R9} \qquad \text{\# R8 + R9} \rightarrow \text{R1}$$
$$\text{sub R2, R1, R6} \qquad \text{\# R1} - \text{R6} \rightarrow \text{R2}$$

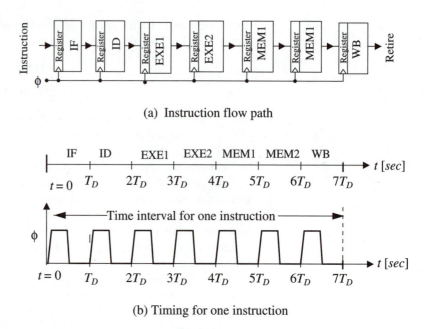

(a) Instruction flow path

(b) Timing for one instruction

Figure 12.6 Increasing the depth of the pipeline

The first instruction modifies the contents of register R1 with the sum of the contents of registers R8 and R9. The second instruction takes the difference of the numbers stored in R1 and R6 and places the result in register R2. If the SUB instruction is in the pipeline directly after the ADD instruction, then we have the problem illustrated in Figure 12.7. Figure 12.7(a) shows that the ADD instruction calculates the sum when it is in the EXE stage. On the next clock cycle (Figure 12.7b), SUB uses the value in R1 to find the value of (R1 − R6). However, the correct result from the ADD instruction is not placed into R1 until it reaches the WB stage, as indicated in Figure 12.7(c). The SUB instruction is thus based on a previously stored value in R1, not the one implied by the program. Figure 12.8 shows the problem with numerical values placed in the register. In this example, the original contents shown in Figure 12.8(a) are used for both the ADD and SUB operations. The SUB instruction stores the value of R1 − R6 = 12 − 4 = 8 in register R2. However, the value of R1 after the first ADD WB is 30, which should have been used in the SUB calculation.

This type of conflict arises because of the sequential nature of the pipeline. One straightforward approach to solving this type of problem is to **stall** the SUB instruction in the pipeline so that it does not access the contents of register R1 until it receives the correct result from the ADD operation. This is accomplished by inserting **bubbles** into the pipeline as shown in Figure 12.9. In this context, a bubble in a stage means an instruction that has no effect but is included to induce the stall. This is called a **NOP** (no operation) instruction.[1] The drawing shows that inserting three

---

[1] Nop is usually read as no-op, short for "no operation."

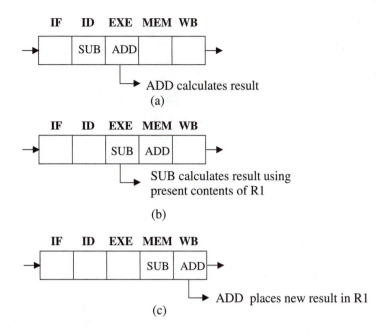

Figure 12.7 Occurrence of a data hazard in the pipeline

R1	12		R1	12		R1	**30**
R2	1		R2	**8**		R2	**8**
R3	64		R3	64		R3	64
R4	9		R4	9		R4	9
R5	27		R5	27		R5	27
R6	4		R6	4		R6	4
R7	52		R7	52		R7	52
R8	14		R8	14		R8	14
R9	16		R9	16		R9	16

(a) Original contents   (b) Contents after SUB   (c) After ADD WB

Figure 12.8 Example of register file contents with a data hazard

bubbles (indicated by shaded ovals) into the pipeline after the ADD instruction has the effect of stalling the SUB by three stages. The ADD updates the registers in the WB stage (when SUB is undergoing the decode operation). Thus, when the SUB instruction reaches the EXE stage, it uses the proper values from the previous instruction.

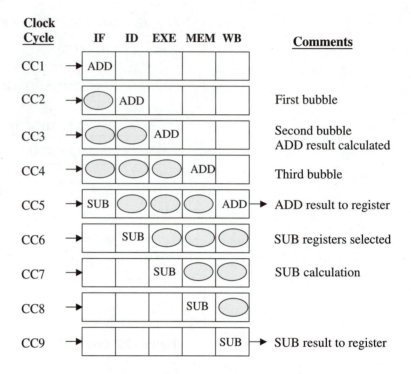

Figure 12.9 Stalling the pipeline using bubbles

12.2.2 Resolving Hazards

The drawback in adding bubbles to the pipeline is twofold. First, we must include logic and circuits at the input that can detect a hazard and induce the necessary stalls in the pipeline. Second, since a NOP bubble does not perform any useful operation, adding bubbles reduces the throughput of the system and lowers the performance. Advanced techniques have been devised to avoid this slowdown. One approach, called **data forwarding,** is illustrated in Figure 12.10. This adds logic to the system to detect the presence of a data hazard in the IF and ID stages. When the instruction reaches the EXE stage, the result is steered around the pipeline and forwarded to the MEM or WB blocks. Although this is conceptually straightforward, it introduces many other problems. For example, the MEM and WB units are already processing an earlier instruction, so the data forwarding must be timed so as not to interfere with the normal operation. One important aspect of this problem is that may data hazards can be detected by the compiler. Of course, it is problems like these that make system design interesting at all levels: hardware, software, and architecture!

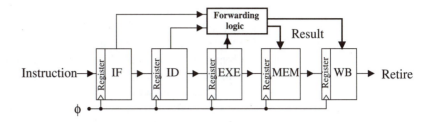

Figure 12.10 Data forwarding to avoid hazards in a pipeline

12.3 Cache Memory

The main memory of a computer is used to store the program listing, the operating system, and data. It is an integral portion of the datapath and is critical to the architecture of the processor. Large system memory arrays allow for more powerful programs and operating systems, so that even desktop PCs have 64MB or more of memory.

Memory size is usually dictated by total cost. The user generally wants a large memory, but budgetary considerations limit the size that is practical. The key to overcoming this problem is to use memory that is inexpensive. The price can be determined if we know the cost per byte C such that the total cost for M bytes is given by

$$Price = M \times C \tag{12.10}$$

An obvious conclusion is that the consumer wants C to be as low as possible (as always!). The memory chip manufacturer, on the other hand, is faced with the problem of providing the lowest cost per byte while still insuring that the product line remains profitable. This is achieved by

- manufacturing memory chips that have the highest cell density possible

and

- keeping the price per chip as low as possible.

At the performance level, however, the **access time** t_{acc} is also important. The access time is the time needed to "access" data from the memory. Small access times are desirable so that the memory unit does not slow down the instruction flow.

When we examine the options available for semiconductor random access (or read/write) memory (RAM), two types of cells emerge. Static RAM, or SRAM, is very fast, but the cell circuit is intrinsically large. In terms of electronics considerations, an SRAM cell requires 4 to 6 transistors, and the arrays have complex interconnect wiring. The alternative is to use dynamic RAM (DRAM) memories. The DRAM cell is much simpler than that used in the SRAM, requiring only one transistor and a capacitor. In addition, DRAM memories are easier to wire into a large array. However, they are slower than SRAM in that they have longer access times t_{acc} associated with them.

Which is used in a real computer? A quick examination of a commercial computer motherboard shows that DRAM is the clear winner. In desktop systems, DRAM can be purchased in **memory modules** that are designed for insertion into the computer boards. These are called SIMMs (single in-line memory modules) or DIMMs (dual in-line memory modules), depending upon their design. Memory modules allow the user to create large memory units easily while maintaining a reasonable cost budget. Figure 12.11 provides an overview of how the memory modules are located relative to the central processing unit. Although the cash outlay is reduced, there is a price to be paid in system performance, since DRAM is relatively slow. In addition, since the memory is physically separate from the CPU chip, the interconnect wiring of the printed circuit board can induce additional delays. These considerations lead us to the conclusion that

- The main memory unit is the slowest part of the data path.

Since the system cannot be clocked any faster than the slowest unit permits, a slow memory can be the limiting factor in designing a high-speed computer. Note that this affects memory access instructions such as load word and store word, in addition to the instruction fetch unit that loads instruction words.

The solution to this problem is to introduce small local memory arrays called **cache** memories that use the faster SRAM. To keep the system cost within reason, the cache size is kept relatively small, usually measured in kilobytes (KB) instead of the megabyte parcels found in DRAM. Cache memories provide for temporary local storage that can be used by the CPU as though it were system memory. This is achieved by inserting the cache between the CPU and the system memory, as illustrated in Figure 12.12. Both address and data lines are needed, and the cache itself contains internal circuits to control the operation.

The block diagram in Figure 12.13 shows the architecture and provides details of the operation. As the CPU executes an instruction such as lw or sw, it sends an address to the memory unit to specify the location. When the system is first turned on, the cache will be empty. When the address is received, the cache cannot respond with any data, so the address is forwarded to the system memory.

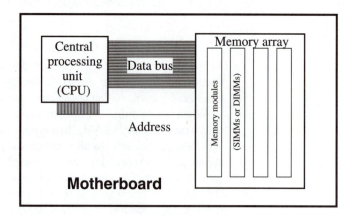

Figure 12.11 Placement of system memory on a motherboard

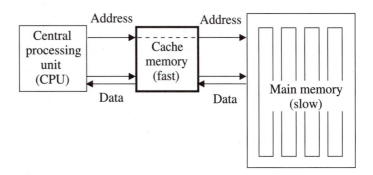

Figure 12.12 Location of cache memory in the datapath

This situation is called a **miss**; a miss occurs whenever the cache does not contain the requested data. The next step in the process is where the system memory responds with the requested data, which is fed through the cache to the CPU. However, instead of providing the data at only one location, the units interact to transfer a *block of data* around the specified address from the system memory to the cache.

Figure 12.13(b) shows the situation where the cache already contains the requested data, which was moved there after a miss. This defines an event called a

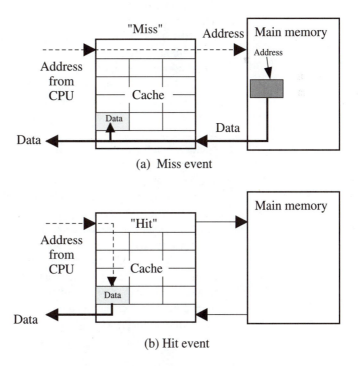

Figure 12.13 Hit and miss events in the cache memory

hit, where the address refers to data that is already present in the cache. In this case, the cache can respond very quickly with the necessary data. System speed is enhanced by having a large **hit ratio**, i.e.,

$$\frac{\text{Number of hits}}{\text{Total number of access events}} \times 100\% \qquad \textbf{(12.11)}$$

should be large. This can be accomplished by designing two features into the system. First, programs usually store groups of data in close, if not contiguous, locations in memory. Once an address is specified, there is a high probability that the next memory access will be close to the first-accessed word. Loading the cache with the data around the original address thus increases the hit rate. The second feature is that the cache is divided into several smaller segments, each of which can accommodate a block of data that has been transferred from the memory. When these segments are loaded, the hit rate increases because there are more locations that can be accessed.

Now let us consider the situation where the CPU produces a result that is directed to the memory unit. Since the cache is in between the CPU and the main memory unit, a write policy must be established to address the problem of consistency between the contents of the cache and that of the main memory. This is called data **coherency**. Figure 12.14(a) shows the situation where data from the CPU is written only to the cache. In this case, the data stored in the main memory is "old" or "stale" and does not reflect any recent updates. This does not necessarily introduce a system problem so long as the next data access is obtained from the cache, and not the main system memory. A **write-through** cache is shown in Figure 12.14(b). This type of network updates both the cache and the main system memory. Although this may seem worthwhile, it forces the system to communicate with the slower system memory on every write operation. This requires that both the data and address lines be used, which may slow down other operations. An intermediate scheme is called a **write-back** design. This approach only updates the main memory when it is necessary and uses special circuits that track the usage of the data and address buses. Write-back is more complicated to design, since it must decide when the data updates are necessary and can be made without slowing down the rest of the system.

Cache memory is often provided in different "levels" depending upon its usage. These are illustrated schematically in Figure 12.15. A Level 1 (**L1**) cache is the closest to the central processor. It is relatively small and is used to store the most frequently accessed data. In the case of a microprocessor, the L1 cache is provided directly on the chip itself and is an integral part of the design. A Level 2 (**L2**) cache is usually larger than an L1 cache and is located between the L1 and main memory. The L2 cache is used to hold the same data as the L1 but has additional storage capacity for other data blocks. The L2 cache improves the performance of the L1 cache because it acts as an intermediate high-speed interface to the main memory.

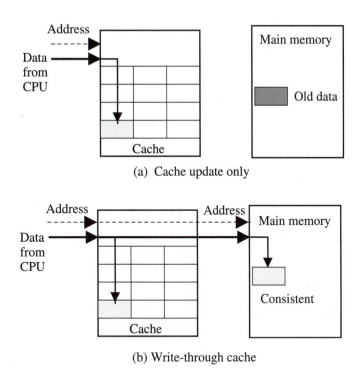

(a) Cache update only

(b) Write-through cache

Figure 12.14 Cache updating and coherency

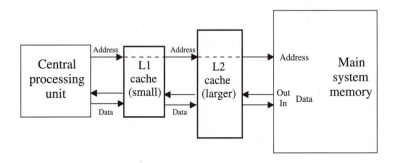

Figure 12.15 Level 1 and Level 2 cache memory

12.3.1 VLSI Aspects of Cache Memory

All high-performance microprocessor chips have an on-board (L1) cache built into their design. A Level 1 cache can be used to provide faster memory access than would be possible if only the main system memory were used. At the architectural level, a large cache is desirable; in fact, the larger the cache, the more efficient the overall operating speed. Silicon VLSI processor designs recognize the need for

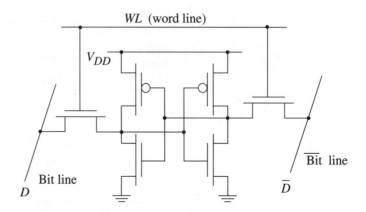

Figure 12.16 The 6-transistor CMOS SRAM cell

cache by incorporating static RAM (SRAM) arrays on the same chip as the CPU circuits. However, the size of the cache is limited by physical considerations.

Let us recall that the CMOS SRAM cell shown in Figure 12.16 requires 6 MOS-FETs. Storing an 8-bit byte thus requires $6 \times 8 = 48$ transistors. If the data bus is 32 bits wide, then the storage circuits for a single word use 4 bytes, which is $4 \times 48 = 192$ MOSFETs. For a cache array capable of storing 1024 32b words (1K words), we have to use $1024 \times 192 = 196,608$ transistors just for the storage cells. Added to this number are the FETs needed for the read/write operations and the address decoders. These simple calculations illustrate that the transistor count, and hence the complexity of the chip design, increase greatly with the size of the cache. Even if we could accommodate the large number of transistors at the circuit design level, the overall dimensions of the chip would increase, reducing the yield of the chip-making process. Smaller yields in manufacturing mean that we must charge a higher price for individual units. This may result in reducing the customer base and the future profitability of the product, as a board designer under strict cost limitations may choose a cheaper chip.

To overcome these problems, the size of the cache must be determined using metrics at both extremes of the design hierarchy. System-level architectural considerations set the desired capacity of the L1 array, which must be balanced with the VLSI concerns of die size, complexity, and fabrication yield. Since the objective is to create a high-performance design that is manufacturable with good yield, a compromise must be reached to satisfy both ends. This is another example of how the design of a complex digital system requires an understanding of many seemingly unrelated subjects. This type of interaction increases with advances in both theory and technology and is one of the keys to becoming a successful designer.

The photographs in Figure 12.17 portray the Pentium Pro® CPU manufactured by the Intel Corporation. This is an interesting example as the PGA package contains two separate silicon chips. The larger chip is the CPU itself and has L1 cache built into it. The smaller chip in the photo is an adjacent L2 cache that is mounted in this "dual-chip module" and directly wired to the main processor.

Figure 12.17 The Intel Pentium Pro microprocessor

12.4 Superscalar Architectures

A superscalar architecture uses multiple pipelines to process instructions simultaneously. This provides for increased instruction throughput at the expense of a more complicated system.

The block diagram for a basic **dual-issue** superscalar is shown in Figure 12.18. This uses two independent pipelines denoted as Pipeline 1 and Pipeline 2 that can process instructions at the same time. Note that the two pipelines share a common instruction cache, and both terminate in a common data cache that provides memory access. The ordering of the pipeline stages has been modified to provide for the common data storage unit by interchanging the write back (WB) and memory (MEM) units in our earlier model. This does not affect the overall characteristics of the pipeline itself.

The operation of the dual-issue superscalar architecture is straightforward at this level. The instruction pre-fetch cache unit is responsible for retrieving instructions from the program stored in memory and analyzing the data flow required. Figure 12.19 shows the basic subunits that handle each task. The instruction fetch network uses the contents of the program counter (PC) to track the instruction flow. Instructions are loaded into the instruction cache whenever the system data bus is free. Once the instructions are contained within the **code cache**,[2] they are analyzed by the subblock labeled as "Instruction issue logic." This has the responsibility of determining *blocks* of *independent code* that can be processed on the two pipelines

---

[2] The term **code** usually implies the lines of a program, either a portion or in its entirety.

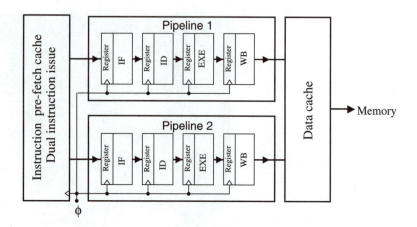

Figure 12.18 Dual-issue superscalar architecture

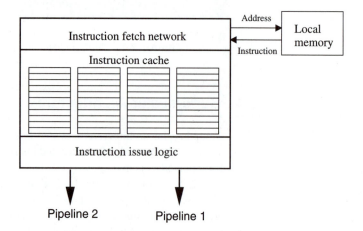

Figure 12.19 Details of the instruction cache

without reference to each other. Increasing the cache size allows for a higher probability of finding independent groups of instructions for which this is possible. Once these are found, the blocks are divided and issued to their respective pipelines. The adjective "dual-issue" simply refers to the fact that two pipelines' code sequences can be relayed simultaneously. A common data cache is provided at the end of the pipeline. As shown in Figure 12.20, this unit receives data from both pipelines. The data cache must then sort out the results and send them to their proper locations as specified by each instruction.

Figure 12.21 illustrates the superscalar architecture used in the Intel Pentium microprocessor as a real example. The Pentium uses two pipelines that are denoted "U" and "V" as shown. The V-pipeline has five main stages and provides integer operations. The U-pipeline, on the other hand, has both integer and floating-point capabilities. The fifth stage of this network is denoted by WB/FP1 and is designed

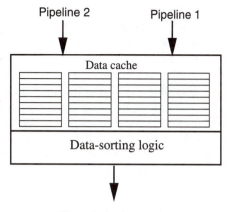

Figure 12.20 The data cache stores results for sorting

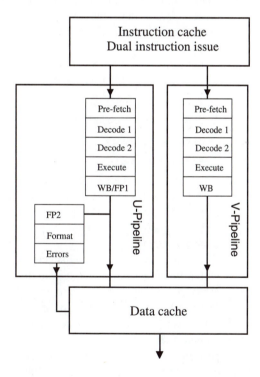

Figure 12.21 Architecture of the Intel Pentium microprocessor

to perform a write back if an integer operation is being processed. In the event that the instruction is a floating-point command, this stage prepares the data to proper format and then passes it to the floating-point stages. The FP2 stage contains the execute circuits for the floating-point operations, and the format stage does round-off calculations. Error reporting is the responsibility of the final stage in the pipeline circuits. As an important observation, a Pentium running at a clock frequency of 60 *MHz* outperforms an earlier-generation 486 processor running at 66 *MHz*, even if the same CMOS technology is used to make both chips. This illustrates the advantage that changing the architecture can introduce.

This simple representation illustrates the usefulness of the superscalar architecture. Of course, analyzing the operation at the system level masks all of the details of the logic networks and timing. Designing these machines is a very challenging task that is accomplished by large engineering teams. These teams consist of specialists in logic synthesis, system architecture, electronic circuits, and VLSI integrated circuits.

12.5   Basic Concepts of Parallel Computing

Parallel computing deals with the concept of using several processors simultaneously to solve a problem. A similar idea is contained in **distributed** computing, where the sections of the overall task are delegated to various processors that all process data at the same time. For reasons that are discussed below, the fields of parallel and distributed computing are often viewed as key concepts for the 21st century.

Let us recall the process of running a program in a classical von Neumann computer. For each instruction, the computer performs the four main operations of

1. Instruction fetch (IF)
2. Instruction decode (ID)
3. Execute (EXE)
4. Storage (STO)

where "Storage" means either a write-back operation (with the results placed in a register) or storage in the main memory. Since a computer program is a sequence of many individual instructions, the speed with which a computer can execute the entire program depends upon the processor throughput, i.e., the number of instructions per second.

Let us examine the consequences of this structure by comparing two processors with different speeds. Let T_a be the time needed to complete an instruction for Processor A, and let T_b be the time needed to complete an instruction for Processor B. Figure 12.22 compares the two for the case where $T_b < T_a$. Processor B has a higher performance rating than Processor A. Since both processors go through the same sequence (IF → ID → EXE → STO), it is obvious that Processor B has a higher performance rating than Processor A. This in turn implies that the individual circuits in Processor B are faster than the circuits in Processor A. These arguments are of value even if the systems are pipelined, since a pipeline does not decrease the time needed to implement a single instruction.

This simple comparison illustrates a crucial point about the von Neumann architecture. In order to increase the computing speed of the system, we need to decrease the delay through the individual units. This is achieved by increasing the clock speed while simultaneously improving the circuits. This characteristic is called the

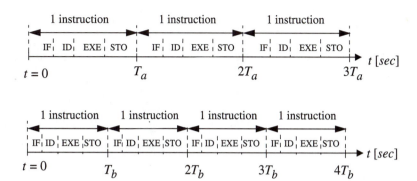

Figure 12.22 Increasing the speed of a von Neumann processor

von Neumann bottleneck. It shows the fact that the limit to the computing speed is due to the architecture.

A parallel computer overcomes the von Neumann bottleneck by providing several processors that act simultaneously to execute the program. While this sounds similar to introducing a pipeline, the general concepts are much more powerful. In particular, the speed of a parallel processing network is not limited by the clock speed of an individual processor. Instead, the speed depends upon the number of processors used in the system and how they are connected together.

Parallel computers consist of basic entities called processor elements (**PE**s) that are connected together to share data. In general, a PE has inputs, outputs, and control signals and performs some type of operation; the symbol that we will use to represent a PE is shown in Figure 12.23. A **processor element** can be very simple or very complex, depending upon the machine. The actual functions are chosen according to the type of calculations that are to be performed. Figure 12.24 provides a few basic examples. A PE might consist of a single gate, while more powerful designs may define a PE as an adder or even a general-purpose microprocessor. With our current knowledge of parallel computing, we usually do not choose the most complex PE possible, as this tends to increase the system complexity. However, as the field evolves, more powerful PEs are destined to become integral components in powerful systems.

Figure 12.23
Symbol for a general
processor element

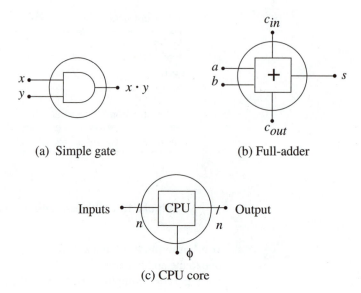

(a) Simple gate   (b) Full-adder

(c) CPU core

Figure 12.24 Examples of processor elements

12.5.1 Classifications of Parallel Machines

Parallel computers can be grouped according to two primary quantities:

- The number of instructions N that can be carried out at one time, and

- The number of data segments S that are being processed at one time.

A classical von Neumann computer issues one instruction that operates on a single data segment, so that both $N = 1$ and $S = 1$. This type of computer is referred to as having **SISD** (**single-instruction, single-data-segment**) architecture.[3] In the present discussion, a general SISD machine will be represented by the simple diagram shown in Figure 12.25. The processor element (PE) accepts the input D and produces a result $R_X(D)$ according to the instruction Inst X. The simplicity of this drawing is indicative of the high level it occupies in the system hierarchy.

More complex parallel architectures can be described by introducing three more groupings. These are

- **SIMD**: Single-instruction, multiple-data-segment ($N = 1$, $S > 1$);

- **MISD**: Multiple-instruction, single-data-segment ($N > 1$, $S = 1$);

- **MIMD**: Multiple-instruction, multiple-data-segment ($N > 1$, $S > 1$).

Each type of architecture has it own characteristics that make it appealing for a particular application.

---

[3] To pronounce these classifications, read the first three letters as a word, and then say the letter "D" afterwards. For example, SISD is read *sis-dee*, SIMD is read *sim-dee*, and so on.

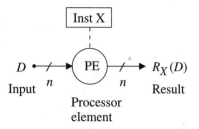

Figure 12.25 A single-instruction, single-data-segment (SISD) machine

SIMD

A SIMD architecture is depicted in Figure 12.26 for the case of four individual processors labeled PE0, PE1, PE2, and PE3. In this architecture, every processor performs the same operation, but each acts on the different data segments. With inputs of $D0$, $D1$, $D2$, and $D3$, the results are given by

$$R_X(D0), \qquad R_X(D1), \qquad R_X(D2), \qquad R_X(D3) \qquad \qquad \textbf{(12.12)}$$

from processor elements PE0, PE1, PE2, and PE3, respectively. For example, if the instruction is ADD, then each processor performs the addition, but the inputs (and therefore the results) are different for each processor.

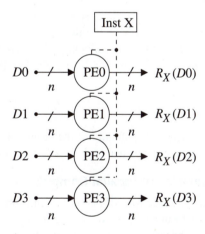

Figure 12.26 A single-instruction, multiple-data-segment (SIMD) machine

MISD

In a MISD architecture, each processor receives a distinct instruction, but every processor uses the same input data. An example of a MISD machine is illustrated in Figure 12.27. In this case, the input data D is subjected to four different operations, yielding four different results. With the notation shown in the drawing, the four outputs are

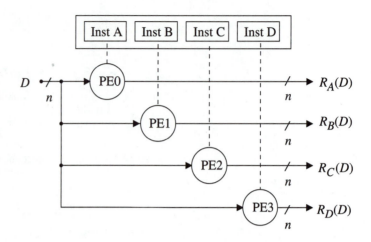

Figure 12.27 A multiple-instruction, single-data-segment (MISD) machine

$$R_A(D), \qquad R_B(D), \qquad R_C(D), \qquad R_D(D) \qquad \text{(12.13)}$$

providing the operations for the instructions labeled as Inst A, Inst B, Inst C, and Inst D, respectively.

MIMD

This is the most general type of parallel machine. As shown in the example of Figure 12.28, a MIMD architecture allows for several operations to be carried out using different data segments. In the drawing, the inputs into the parallel network are $D0$, $D1$, $D2$, and $D3$. Each processor performs a different instruction, so that the results are

$$R_A(D0), \qquad R_B(D1), \qquad R_C(D2), \qquad R_D(D3) \qquad \text{(12.14)}$$

This type of system allows the programmer the maximum level of flexibility but is the most complicated to design, build, and use.

12.5.2 Examples of Parallel Computations

The classifications of parallel architectures introduced above are fairly straightforward to understand. It is less obvious where these types of computing systems might be useful. Let us examine a few examples to provide you with a better understanding of what can actually be done in a parallel computer.

Vectors and Matrices

Parallel machines are particularly useful for certain vector and matrix calculations, such as those that arise in **digital signal processing (DSP)**. In the most general case, a vector is a one-dimensional list of numbers. For example, a 4-element **column vector** $[A]$ can be represented by

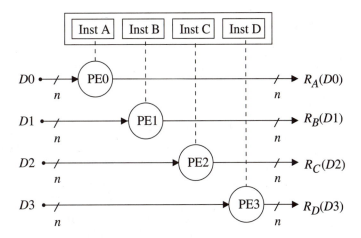

Figure 12.28 A multiple-instruction, multiple-data-segment (MIMD) machine

$$[A] = \begin{bmatrix} A_1 \\ A_2 \\ A_3 \\ A_4 \end{bmatrix} \qquad \textbf{(12.15)}$$

where A_1, A_2, A_3, and A_4 are the elements. A 4-element **row vector** $[B]$ is written as

$$[B] = \begin{bmatrix} B_1 & B_2 & B_3 & B_4 \end{bmatrix} \qquad \textbf{(12.16)}$$

with the elements denoted by B_1, B_2, B_3, and B_4. A 4×4 **square matrix** $[C]$ is portrayed as

$$[C] = \begin{bmatrix} C_{11} & C_{12} & C_{13} & C_{14} \\ C_{21} & C_{22} & C_{23} & C_{24} \\ C_{31} & C_{32} & C_{33} & C_{34} \\ C_{41} & C_{42} & C_{43} & C_{44} \end{bmatrix} \qquad \textbf{(12.17)}$$

The elements are denoted by $C_{\alpha\beta}$ where α is the row number (1 through 4) and β is the column number (1 through 4).

Now let us review some simple matrix algebra. Consider the case where we multiply a column vector $[A]$ by 2. Since this means that every element in the vector is multiplied by 2, it results in

$$2 \times [A] = \begin{bmatrix} 2A_1 \\ 2A_2 \\ 2A_3 \\ 2A_4 \end{bmatrix} \qquad \textbf{(12.18)}$$

In a classical SISD computer, the sequence for calculating $2 \times [A]$ would be

Calculate $2 \times A_1$, then
calculate $2 \times A_2$, then
calculate $2 \times A_3$, and then
calculate $2 \times A_4$.

This would take four instruction cycles to complete. However, in a SIMD (single-instruction, **multiple-data-segment**) architecture, this would be a single operation since we can identify the single instruction as "multiply by 2" and the multiple data inputs as A_1, A_2, A_3, A_4. The computing array in Figure 12.29 would thus complete the matrix calculation in one cycle. If we make the simple assumption that each processor takes the same amount of time as the PE in the SISD arrangement, the SIMD array would complete the entire calculation in the time a SISD computer needs to compute one entry. The speed advantage of a parallel architecture is clear in this case.

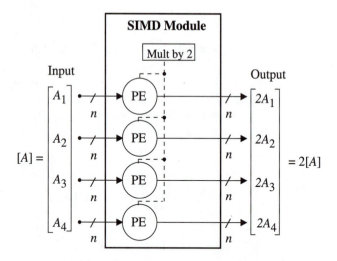

Figure 12.29 SIMD network for multiplying a column vector

Next, consider the multiplication of a row vector by a column vector that produces a scalar P through the expression

$$[B][A] = \begin{bmatrix} B_1 & B_2 & B_3 & B_4 \end{bmatrix} \begin{bmatrix} A_1 \\ A_2 \\ A_3 \\ A_4 \end{bmatrix}$$ **(12.19)**

$$= A_1 B_1 + A_2 B_2 + A_3 B_3 + A_4 B_4$$

$$= P$$

A SISD machine would compute P by first computing each term $A_k B_k$ for $k = 1, 2, 3, 4$ and then summing:

$$P = \sum_{k=1}^{4} A_k B_k$$ **(12.20)**

This would require several instruction cycles, one for each multiplication, and then the subsequent addition of terms.

A SIMD alternative is shown in Figure 12.30. In this network, we use the matrix elements A_k and B_k as inputs to the module and issue the multiply command to every PE. It should be noted that every element is really an n-bit binary word. Each PE provides one of the product terms, and the scalar product P is obtained from the adder network (denoted by Σ in the drawing) as a binary word that is $2n$ bits wide. An alternative approach would be to incorporate adders directly into each PE, thus eliminating the external circuit.

As a final example of a matrix operation, consider the vector product of a column vector times a row vector as computed from

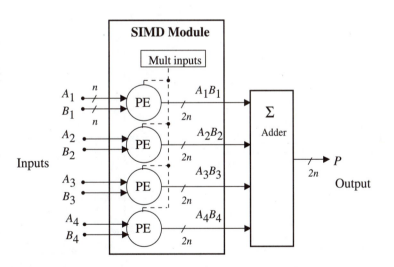

Figure 12.30 SIMD network for calculating the scalar product

$$[A][B] = \begin{bmatrix} A_1 \\ A_2 \\ A_3 \\ A_4 \end{bmatrix} \begin{bmatrix} B_1 & B_2 & B_3 & B_4 \end{bmatrix}$$

$$\quad (12.21)$$

$$= \begin{bmatrix} C_{11} & C_{12} & C_{13} & C_{14} \\ C_{21} & C_{22} & C_{23} & C_{24} \\ C_{31} & C_{32} & C_{33} & C_{34} \\ C_{41} & C_{42} & C_{43} & C_{44} \end{bmatrix}$$

The elements of the square matrix $[C]$ are given by the products

$$C_{ij} = A_i B_j \quad (12.22)$$

for $i, j = 1, 2, 3, 4$. Using a standard SISD computer would thus require 16 individual instructions cycles, one for each of the 16 entries. However, if we instead create a SIMD processor using 16 processor elements, as in Figure 12.31, then the entire matrix can be computed in one cycle since each processor can calculate one of the entries.

SIMD Module

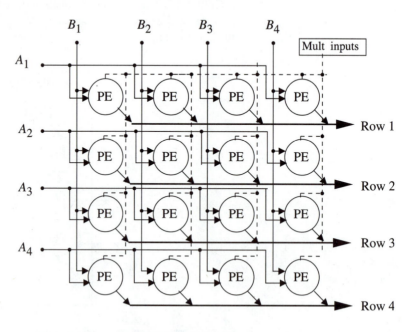

Figure 12.31 SIMD network for obtaining the matrix product of two vectors

The examples based on matrix calculations illustrate the main idea behind parallel architectures. In each case, the task is broken into distinct parts and assigned to individual processors that work in unison. The results are then combined to produce the outputs. A little thought will verify that this approach can be expanded to more general systems design by using different wiring schemes and more powerful processor elements.

12.5.3 General Architectures

A parallel computer consists of several processor elements that are interconnected in a manner that allows for easy communication and data flow. At the conceptual level, there are a huge number of variations that can be visualized, depending upon considerations such as

- The number of processors,
- The manner in which the processors communicate with each other, and
- Control of the data flow.

Parallel computers can be analyzed according to common features in their architectures. To understand the concept, let us briefly examine each consideration in more detail so that we can see how the ideas are related.

Number of Processors

The number of processors chosen for a particular design depends upon the specifics of the application. Suppose that we want to design a very powerful system. At one extreme, we can delegate the processing power to the PEs. This allows us to use a few complex processing elements to accomplish the task. At the other end, we may choose to use many simple PEs and place the computational burdens more on the system architectural features. These two cases are shown in Figure 12.32. The tradeoff is the complexity of the individual units versus the interconnect wiring needed to link many simpler units.

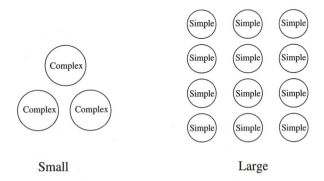

Figure 12.32 Number of processors versus PE complexity

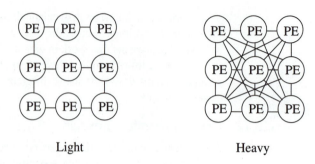

Light   Heavy

Figure 12.33 Interconnect levels

Level of Connectivity

A related aspect is the choice of an approach to interconnect the PEs together. The extremes are shown in Figure 12.33. In a lightly connected topology, we allow the processors to interact in a very limited manner. This means that a given PE can only "talk" to a small subset of the total number of processors in the system. System design is complicated by the fact that the order in which the PEs are connected becomes a factor in how tasks may be divided and allocated to the individual elements.

A heavily connected design is also shown in Figure 12.33. This shows the case where most PEs can communicate with most other PEs in the system. At the extreme end of this design aspect, every PE is connected to every other PE in the system. While a heavily connected system simplifies the allocation of individual tasks, it increases the physical design problems due to the large number of wires that are needed for the communication paths.

Control

The third aspect of interest to us is the level of control provided at the system level versus the amount of control built into the individual processor elements. This is portrayed in Figure 12.34. A system with "loose" control is defined to have most of the control logic contained within the processor elements. This allows the individual PEs to specify their own operations without regard to what other PEs are doing, but it requires more complex circuits. Data flow among the processors is controlled at the system level, which acts to coordinate the inputs and outputs.

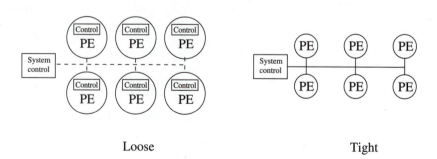

Loose   Tight

Figure 12.34 Control of the parallel machine

A tightly coupled design employs the system control logic to provide individual control signals for the PEs while also coordinating the data flow in the network. Since the PEs do not control themselves, simpler circuits can be used than in a network that uses loose control. On the other hand, a tightly coupled design increases the wiring complexity of the network.

12.5.4 General Design Variables

To simplify the concepts involved in parallel architectures, let us consider three system parameters discussed above: the number of processor elements N, the level of connectivity I, and the degree of control C. Any processor may be represented by a point in a three-dimensional coordinate system with these three variables used to define the axes. This is portrayed in Figure 12.35. Designing a parallel computing arrangement requires that the variables be considered individually and as a group. As with any complex system, changing viewpoints to all levels in the design hierarchy illustrates the tradeoffs and helps point the designer to a reasonable solution.

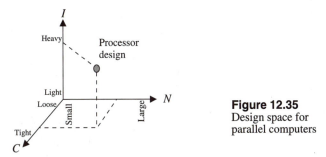

Figure 12.35
Design space for parallel computers

12.5.5 Interconnection Networks

One of the key problems in creating a parallel computer is the approach to interconnecting the processor elements. As discussed above, increasing the communication paths among the processor elements allows more general processing, but the interconnect wiring gets very complicated. In addition, interconnect delay along different lines is more difficult to deal with.

Several general interconnection techniques have been studied, resulting in some standard topologies that are worth mentioning here. Perhaps the simplest approaches are those portrayed in Figure 12.36. The **ring** network (Figure 12.36a) is structured in a linear closed loop with every processor connected to its two neighbors. Data flow is thus restricted along these paths. A **star** (Figure 12.36b) allows every processor element to communicate with every other PE by increasing the interconnect.

Another interesting topology is the **n-cube** structure. An n-cube is defined to have 2^n processor elements, with each processor connected to n other PEs. Figure 12.37(a) shows a 2-cube that is just a square with a PE at each corner. To construct a 3-cube, we start with a three-dimensional cube and add a PE to each corner, resulting in the topology shown in Figure 12.37(b). If $n > 3$, the structures become

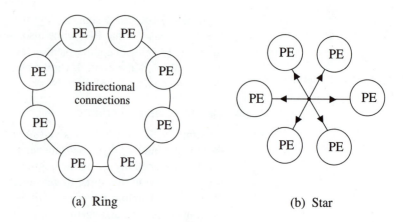

(a) Ring   (b) Star

Figure 12.36 Basic interconnect configurations

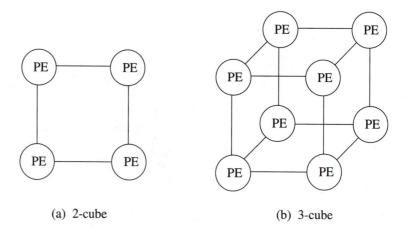

(a) 2-cube   (b) 3-cube

Figure 12.37 $n = 2$ and $n = 3$ architectures

more difficult to draw and are called **hypercubes**. A 4-cube is shown in Figure 12.38. This consists of $2^4 = 16$ separate processor elements, and each PE is connected to 4 other PEs. This can be visualized as a 3-cube embedded within another 3-cube, with communications provided between the corners as shown.

In a hypercube architecture, each processor node is labeled with a binary address; an n-cube requires a 2^n-bit address. A node can only communicate with nodes whose address differs by a single bit. For example, an $n = 3$ cube address has the form $(n_2n_1n_0)$. A node with the address (000) communicates only with node addresses (001), (010), and (100). To get to any other node, the signal must pass through other processors. This type of limitation dictates the type of programs and the nature of the programming structure that the architecture will be efficient at.

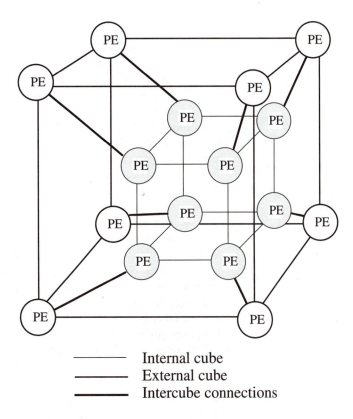

— Internal cube
— External cube
— Intercube connections

Figure 12.38 An $n = 4$ hypercube

12.5.6 The Challenge of Parallel Computing

This short introduction to the field of parallel computing was written to illustrate some of the basic ideas involved in the next generation of computers. As with any field of this complexity, progress can be slow or it can move at the "speed of light." It all depends upon the intelligence and ingenuity of the engineers and scientists working on the problems.[4]

There are many hurdles that must be surmounted before the true power of parallel computers can be harnessed. Included in the basic problem set is the choice of architecture, detail design of the processor elements, interconnect strategies, and efficient techniques for parallel programming. Why are we interested in attacking such a large problem? It is because the computing power that appears to be within our grasp is virtually unlimited! An intriguing prize, wouldn't you agree?

---

[4] Luck also tends to play a major role in the success of many research endeavors!

12.5.7 Optical Interconnects

As our final topic, let us examine a research area that has direct consequences in parallel computing: using laser beams as data transmission lines. We define an **optical interconnect** as a data path between two circuits that is created by a beam of light. As the complexity of computing systems has increased, so has the problem of complicated wiring and crosstalk. In fact, the slang term for the massive bundles of wires found in large computing systems is the **rat's nest**! Optical interconnects have been proposed as an alternative to conventional wiring schemes because light beams have some particularly nice characteristics. Although optical interconnects have not yet appeared in commercial systems, they are worth a brief examination here as either a view to the future or a brief look into the advanced research.

Data transmission is accomplished by providing a light emitter such as an LED or a laser that can be turned ON and OFF to represent logic 1 and 0 levels, respectively. The beam is directed to another point in the system, where it is detected using an optoelectronic device called a photodiode. When light shines on the photodiode, it forces electric current through it. Optical interconnects are identical in concept to the fiber-optic systems introduced in Chapter 6 except that we do not use a guiding structure such as the optical fiber. Instead, we guide the beam using mirrors and other techniques. These are called **free-space** interconnects to distinguish them from guided-type lines.

Free-space optical interconnects are of great interest due to a few basic characteristics of light. These are summarized in Figure 12.39. First, optical beams can cross each other without any interactions. As illustrated in Figure 12.39(a), this means that there is no crosstalk, which makes the interconnect problem much easier. Another useful characteristic is that a single beam may be split into several individual beams, each pointed in a distinct direction, using an optical element called a **hologram**. The drawing in Figure 12.39(b) indicates that we can broadcast a beam of modulated light to several receivers at the same time.

In a practical implementation, we must provide optoelectronic devices to the standard logic circuit (which can be either a printed circuit board or an integrated circuit). A simple configuration is shown in Figure 12.40. A laser array can be used to produce the output light beams, while a photodetector array provides the circuits necessary to detect incoming light beams and change them into electronic impulses. Adding the laser and photodetector array increases the complexity of the network.

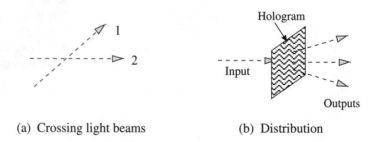

(a) Crossing light beams   (b) Distribution

Figure 12.39 Properties of optical interconnects

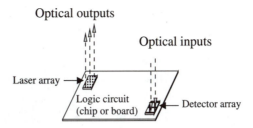

Figure 12.40 Optoelectronic light-emitter and light-detector arrays

An example of the use of optical interconnects is to provide clock distribution as shown in Figure 12.41. In this network, the system clock ϕ is fed to a laser, and the laser beam turns on and off at the same rate. The beam is then pointed to a hologram, which breaks it up into several different beams. Detector circuits placed on the logic circuit then detect the signal and change it back to electrical pulses that can be used by the network. This type of system avoids the wiring problem and also insures that all clock pulses are synchronized to each other (since there is no interconnect delay).

Optical interconnects have shown great potential in parallel computers as they provide the third dimension of wiring that is not normally possible with pure electronic techniques. The drawing in Figure 12.42 illustrates the general concept of allowing several logic boards to communicate using optics. Each circuit board in the stack has light-emitter and light-detector arrays that act as output and input channels, respectively. Holograms are inserted between adjacent boards to steer the light beams to their proper locations. The light communication paths, shown as dashed lines in the drawing, are free to transmit data between two boards without the problem of crosstalk or electromagnetic interference. The strongest point of this entire setup is that the number of interconnect beams is limited only by the ability to align the detectors and emitters properly. From the optical standpoint, a hologram can easily distribute a single input beam into a million or more outputs. This technology may then solve the problem of creating a heavily interconnected parallel processor.

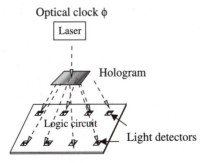

Figure 12.41
Optical clock
distribution system

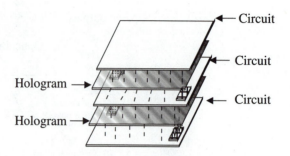

Figure 12.42 System-level optical interconnects

Work on optical interconnects is only one of many research areas related to the field of parallel processing. All have the same goal in mind: to address the problems that must be solved to turn a dream into a reality . . .

12.6 Problems

[12.1]   A performance test is run to compare two computers. Computer A runs an accounting program that has 110,000 instructions in a time T_A. Computer B, on the other hand, performs quantum mechanical calculations on molecular energy levels using a program that has 84,000 instructions and takes a time T_B to complete. What conclusions can you make about the relative performance of the two computers?

[12.2]   A computer is built using the single-instruction-per-clock-cycle timing detailed in Figure 12.1. Calculate the time needed to process a sequence of 100,000 instructions with the following clock frequencies.

(a) $f = 66\ MHz$
(b) $f = 100\ MHz$
(c) $f = 300\ MHz$
(d) $f = 333\ MHz$
(e) $f = 450\ MHz$

[12.3]   Study the prices of personal computers in a recent newspaper or magazine and prepare a plot of Price versus Processor Clock Speed for similarly equipped systems based on the ads you find. Can you make any statements and/or projections?

[12.4]   A computer is built using the 5-stage architecture illustrated in Figure 12.1. The delay times through the individual units are given as $t_{IF} = 1.6\ ns$, $t_{ID} = 1.4\ ns$, $t_{EXE} = 3.2\ ns$, $t_{MEM} = 3.7\ ns$, and $t_{WB} = 1.8\ ns$.

(a) What is the clock frequency f_1 needed for the single-instruction-per-clock cycle timing approach?

(b) Suppose instead that the clock drives each stage simultaneously as in Figure 12.2. What would be the clock frequency f_2 for this case?

(c) Compare the times t_{inst} for the two approaches. Which one is shorter in a straightforward comparison?

[12.5] Consider the 5-stage pipelined architecture in Figure 12.4. Suppose that this structure is used to build a computer with a clock frequency of 300 *MHz*. A program with 80,000 instructions is run on the system.

(a) Calculate the time needed to execute one instruction.
(b) Calculate the execution time for the entire program.
(c) What is the percentage increase in speed over the case where a 5-stage non-pipelined architecture is used?

[12.6] Let us think about the concept of a pipeline by making an analogy with cars passing through a tunnel.

(a) Describe how the cars would pass through the tunnel if they moved in a manner analogous to the data flow path described in Figure 12.1?
(b) What would the flow of cars be like if we used the pipeline concept instead?

[12.7] A 5-stage pipeline operates with a maximum clock frequency of 200 *MHz*. The system is redesigned to a 7-stage pipeline in an effort to increase the system throughput.

(a) Find the frequency of the 7-stage clock needed to obtain a 5% increase in instruction throughput.
(b) What is the percentage difference between the time period allocated per stage in the 5-stage design and that in the 7-stage design?

[12.8] Study the computer advertisements in a newspaper or magazine and see how many references to cache memory you can find. What are the typical sizes and applications? Can you find any direct references to L1 or L2 cache?

[12.9] A computer with local cache memory is running a large program. The system is exhibiting a large hit ratio when it suddenly drops to almost zero.

(a) What would cause this situation to occur?
(b) Suppose that you want to track the hit ratio as the program continues. What would you expect to happen? Why?

[12.10] Consider a dual-issue superscalar machine as shown in Figure 12.18. When only pipeline 1 is active, it takes 88 *ns* for an instruction to move from the IF to the data cache unit. Pipeline 2 has the same characteristics.

(a) Calculate the time t_{pipe} needed to run a program that consists of 225,000 instructions through one of the pipelines.
(b) Suppose that the system acts as a true superscalar with instructions being issued to both pipelines 1 and 2. What statement can be made about the time needed to run the same program in this case?

[12.11] Consider the two matrices

$$[a] = \begin{bmatrix} a_{11} & a_{12} \\ a_{21} & a_{22} \end{bmatrix} \text{ and } [x] = \begin{bmatrix} x_1 \\ x_2 \end{bmatrix}$$

Design a SIMD network that calculates [a] · [x].

[12.12] Consider the two vectors

$$[A] = \begin{bmatrix} A_1 \\ A_2 \\ A_3 \end{bmatrix}, \quad \text{and} \quad [B] = \begin{bmatrix} B_1 & B_2 & B_3 \end{bmatrix}$$

which are inputs to a parallel computer.

(a) Design a subsystem that will produce the matrix

$$[X] = \begin{bmatrix} A_1 & 2A_1 & 4A_1 \\ A_2 & 2A_2 & 4A_2 \\ A_3 & 2A_3 & 4A_3 \end{bmatrix}$$

(b) Now design a subsystem that will produce the matrix

$$[Y] = \begin{bmatrix} B_1 & B_2 & B_3 \\ 3B_1 & 3B_2 & 3B_3 \\ 5B_1 & 5B_2 & 5B_3 \end{bmatrix}$$

(c) Finally, combine the subsystems to create a system that has an output of $[Z] = [X] + [Y]$.

[12.13] Consider the two matrices

$$[a] = \begin{bmatrix} a_{11} & a_{12} & a_{13} \\ a_{21} & a_{22} & a_{23} \\ a_{31} & a_{32} & a_{33} \end{bmatrix} \quad \text{and} \quad [b] = \begin{bmatrix} b_{11} & b_{12} & b_{13} \\ b_{21} & b_{22} & b_{23} \\ b_{31} & b_{32} & b_{33} \end{bmatrix}$$

(a) Design a SIMD module that calculates the matrix sum $[c] = [a] + [b]$.

(b) Cascade your circuit in part (a) into another SIMD module that multiplies $[c]$ by a constant K.

(c) Can you design a single SIMD module that produces $K[c]$?

[12.14] Consider the row matrix

$$[U] = \begin{bmatrix} u_1 & u_2 & u_3 \end{bmatrix}$$

Suppose that we want to design a system that uses $[u]$ as the input and simultaneously produces $2[u]$, $3[u]$, $10[u]$, and $[u/2]$ as outputs.

(a) What type of parallel architecture would this be classified as?

(b) Design the system described.

[12.15] Consider the concept of programming a MIMD machine that uses general-purpose PEs. Programming is usually done in a sequential manner, where we per-

form tasks by listing instructions in a particular order. What types of problems can you envision if this type of programming style is applied to a MIMD computer? [Don't despair. This is a difficult thinking problem with no single answer!]

[12.16] Draw and label the nodes for an $n = 3$ hypercube design. Use the addresses (0xx) for the interior cube and (1xx) for the outer cube. Discuss the longest communication path between two nodes in this architecture.

[12.17] Consider an 486$n = 3$ hypercube parallel computer where the nodes are labeled by $(n_2 n_1 n_0)$.

(a) List the nodes that the processor at node (101) can communicate with.

(b) Suppose that node (100) needs to communicate with node (111). What are the most efficient paths between the two?

[12.18] Construct a design for a 2-cube network that uses optical interconnects and vertically stacked computer boards. Use mirrors as needed to steer the laser beams to their proper planes.

[12.19] A computer that uses optical interconnects may provide inputs and outputs as either electrical signals or light beams. What would be the advantage of having data in the form of a pulsating laser beam? [Hint: Recall the discussion at the end of Chapter 6.]

12.7 References

This chapter has introduced many concepts in the theory of advanced computing. The list below provides a few textbooks that go much deeper into the subject matter. To aid the reader in accessing information, each title has been assigned a letter to indicate the level of the presentation. The notation (I) means an intermediate-level treatment, while (A) implies an advanced discussion. In terms of university class levels, (I) should be interpreted as implying an undergraduate junior/senior-level book; (A) means a graduate-level text.

[1]. Don Anderson and Tom Shanley. *Pentium™ Processor System Architecture,* 2nd ed. Reading, MA: Addison-Wesley, 1995. (I). A well-written discussion of the Intel Pentium microprocessor and applications.

[2]. James M. Feldman and Charles T. Retter. *Computer Architecture.* New York: McGraw-Hill, 1994. (A). This is an advanced textbook that emphasizes high-level design.

[3]. David A. Patterson and John L. Hennessy. *Computer Organization & Design,* 2nd ed. San Francisco: Morgan-Kauffman Publishers, 1997. (I). An outstanding textbook written by two pioneers in the development of RISC computers.

[4]. Tom Shanley and Don Anderson. *ISA System Architecture,* 3rd ed. Reading, MA: Addison-Wesley, 1995. (I). A very readable treatment of PC board-level architectures and concepts.

Epilog

And so, dear reader, we have come to the end of our journey. In the past few hundred pages, you have learned about the world of digital systems design, from bits to computers, from electrons to microprocessor chips. I hope that you have profited from this trek and have come to see the field as it really is: an interdisciplinary activity with something for almost everyone. If this book has sparked your interest in pursuing more studies in any of the areas we have examined, then it has done its job. If, on the other hand, you have had your fill of digital systems design, please remember that digital systems are now a permanent fixture in our modern world, and the lessons you have learned here will serve you well.

Index